ABHANDLUNGEN DER DEUTSCHEN AKADEMIE
DER WISSENSCHAFTEN ZU BERLIN

Klasse für Philosophie, Geschichte,
Staats-, Rechts- und Wirtschaftswissenschaften
Jahrgang 1955 Nr. 2

RAMKRISHNA MUKHERJEE

THE PROBLEM OF UGANDA

A Study in Acculturation

1956

AKADEMIE-VERLAG · BERLIN

ABHANDLUNGEN DER DEUTSCHEN AKADEMIE
DER WISSENSCHAFTEN ZU BERLIN

Klasse für Philosophie, Geschichte,
Staats-, Rechts- und Wirtschaftswissenschaften
Jahrgang 1955 Nr. 2

RAMKRISHNA MUKHERJEE

THE PROBLEM OF UGANDA

A Study in Acculturation

1956

AKADEMIE-VERLAG · BERLIN

Vorgelegt von Hrn. Meusel in der Sitzung vom 5. Mai 1955
Zum Druck genehmigt am gleichen Tage, ausgegeben am 15. August 1956

Erschienen im Akademie-Verlag GmbH, Berlin W 8, Mohrenstraße 39
Lizenz-Nr. 202 · 100/216/56
Kartengenehmigung (K. 11) MdI. Nr. 2761
Gesamtherstellung: Druckerei „Thomas Müntzer“ Langensalza
Bestell- und Verlagsnummer: 2001/55/V/2
Preis: DM 29.—
Printed in Germany

CONTENTS

LIST OF TABLES

PREFACE

I began to write this book in 1950 when I was in Uganda, and completed it in 1955 in Germany. During these years many changes have taken place in Uganda. They have not all been included in this study. A description of Uganda in minute details is not, however, the subject-matter of this book. Since it deals with an analysis of *the problem* of Uganda, its usefulness should be examined from that standpoint only. Many more new facts will emerge in relation to Uganda in the coming years. Will they vitiate the analysis presented in this study or will they substantiate it? Upon this standard this book should be judged.

For any study of a living society there is the problem that spectacular events crop up, wonderful assurances for future development are given by politicians, and complacence or pessimism prevails according to the appearance of stray events. If this book can give some idea as to how one should judge the relevance and importance of such events in relation to Uganda, it will maintain its usefulness. In a sense, however, as in almost all sociological analysis, only history can prove whether the deductions made in this study are valid or not.

An apology I should make at the outset. While reading this book the reader may feel that some of my comments are too sharp. This could have been partly avoided with a better command over a foreign language, which, I regret, I do not possess. Also partly this is due to the fact that while being brought up in a colonial country I was perhaps in a better position than many to understand *the problem* of Uganda (as I could visualise to some extent what the earlier generations in my country had undergone and link it up with my own experience), this has also contributed to the intensity of my feelings which will not escape any one when reading this book. It is true that scientists are no diplomats; yet even when presenting the stark reality in course of a scientific analysis one need not express also his own feelings. For this I hope to be excused. I should state, however, that this personal factor has not affected the objective analysis of *the problem* of Uganda. The facts will now speak for themselves.

While writing this book I received help from many individuals. To all of them I convey my thanks. I also gratefully acknowledge the kindness of the Editors of *Man in India* (India), *Interagra* (Czechoslovakia), *Wirtschaftswissenschaft* and *Mitteilungen des Instituts für Orientforschung* (Germany) for publishing previously several articles on Uganda, the contents of which are now incorporated in this study.

Berlin, 31st January, 1956

Ramkrishna Mukherjee

ERRATA

Page	Line	In place of	Please read
2	40	poeple . . .	people . . .
5	24	central east Africa . . .	central *and* east Africa . . .
5	44	in this respect is a . . .	in this respect *there* is a . . .
6	18	to destitute of . . .	*too* destitute of . . .
7	(Footnote, line 3)	fo . . .	of . . .
8	30	deserted him on this way . . .	deserted him on *his* way . . .
14	45	ans why have they . . .	*and* why have they . . .
15	31	Her water power recources . . .	Her water-power resources . . .
19	(Note 52 a)	Vol. I, pp. 34—38, . . .	Vol. I, pp. **19**, 34—38, . . .
22	33	Achoil, . . .	Acholi, . . .
25	19	cash crops like cotton and coffee . . .	cash crops like *African-grown* cotton and coffee . . .
26	24	for thesecondment . . .	for the secondment . . .
156	12	discussed on earlier . . .	discussed earlier . . .
178	47	The Table VI. 6 . . .	The Table *IV*. 6 . . .
182	7	as always under colonial rule. Uganda . . .	as always under colonial rule, Uganda . . .
183	1	by the Foreign interests . . .	by the foreign interests . . .
183	(Footnote, Table 4)	Percentage of the value in 1950 to that in 1951 . . .	Percentage of the value in **1951** to that in 1950 . . .
213	(Note 359)	For details why ,,rosts" . . .	For details why ,,*riots*" . . .
219	(Footnote, line 21)	As how the United States . . .	*And* how the United States . . .
221	47	Uganda began . . .	Uganda, began . . .
253	(Footnote, lines 1—2)	is not the same given by . . .	is not the same *as* given by . . .
270	39	as it became quite evident the case of . . .	as it became quite evident *in* the case of . . .

CHAPTER I.

GENERAL BACKGROUND

The one who roars in the distance,
The one who crushes the young men and smashes them,
The one who debauches our wives.
They desert us, they go to the town to live bad lives.
The ravisher! And we are left alone.

[Soliloquy of an African servant in a White Man's camp in Zululand as quoted by H. A. Junod in *Life of a South African Tribe* (1927, vol. ii, pp. 196—7).]

1. Africa and Uganda

To many people Uganda was merely a small patch on the map of Africa until recent events in the post-war world brought it to their attention. The uprisings of the Uganda Africans in 1947 and 1949 and the deportation of the Kabaka (the African King of Buganda) in 1953 have made the world public conscious that even in this little territory in the heart of Africa the people are not sleeping in the "bush".

Nor is this stirring restricted only to Uganda. The whole of the great African continent is in ferment. This is one of the most distinctive features of the post-war world. Everywhere in this vast continent it is the same story. The Africans are resolutely standing up against unjust laws and actions of the foreign rulers or their local representatives. They are striving for their legitimate claim to run their own country, shape their own destiny, and march on towards freedom and peace, progress and prosperity.

Starting from the southern tip of the Continent, an observer will find that in the Union of South Africa, where the White Government has openly come out with racial, *apartheid*, laws, the Resistance Movement sponsored by the solidly united Africans, Indians and Coloured peoples has shown the world that they will not tolerate racialism in any form.

Moving a little to the north of the Union, he will witness the strong opposition of the people of the Bechuanaland Protectorate (a British colony) to the unjust moves of the British Government. Here, in 1948, the British Labour Government banished the chief of the African people, Seretse Khama, for the crime of having married an English girl. To one man the African population, representing 98 per cent of the total in the Protectorate, resented this decision of the colonial government. So strong has been this opposition that the Labour Government, then in office, and lately the Conservative Government have consistently failed in their attempt to instal anyone in place of Seretse Khama as a Chief in the Bechuanaland Protectorate.

Entering central Africa — in Southern Rhodesia, Northern Rhodesia and Nyassaland, the observer will come across the fight of the people to secure their land from the White land-grabbers. Here he will witness the united opposition of all sections of the African population (including the "tribal chiefs") to the formation of the Central Africa Federation, a federation which has the status of a Dominion, and a Dominion "of the type of South

Africa, in which a White minority holds power over an African majority deprived of democratic rights"[1].

The same sort of resistance would also be noticed in Tanganyika, the Kenya Colony and the Uganda Protectorate where the scheme is to establish an East Africa Federation, mainly with the same purpose in view. The story is the same in West Africa with regard to the proposed West Africa Federation*.

Moving to the north east of central Africa, that is, in the Kenya Colony of British Empire, the movement of the Africans will be seen to be high-lighted by the universal desire of the Kenya Africans to regain their land which was taken away by the White Settlers. Kenya, the ancient land of the Africans, has been turned into a "White Man's Country" in the twentieth century. In their own land the Africans can now remain only as serfs of the Europeansettlers (overwhelmingly British) who, in the words of a reputed journalist, have created "the last real stronghold in the world today of ancient English feudalism"[2]. As a matted of fact, the situation for the Africans is even worse. On them, severity of capitalist oppression has been added to the barbarities of feudal despots. And the Africans have the alternative either of meeting a slow starvation death on the impoverished land in the much too congested "tribal reserves" (with a population density of 1,000 per square mile) or to face the brutalities of the White lords[3]. So their demand is to get back their land.

In the Uganda Protectorate, where as described above the situation is stirring up all the while, the people will be found to demand that (a) the British Government should not push Uganda into the East Africa Federation, (b) Uganda being a Protectorate, and not a Colony, should be transferred to the Foreign Office instead of being under the Colonial Office of the British Government, (c) a time schedule should be announced for the independence of Buganda, the largest and the most developed province of Uganda both economically and politically, which was previously a separate kingdom but became a part of the Uganda Protectorate under the Uganda Agreement of 1900.

Moving a little northwards into what previously was the Egyptian Sudan, an observer cannot fail to notice that in the post-war years the demand for the unconditional withdrawal of the British from the territory became so powerful that even the pro-British Umma Party had to subscribe to that demand of the people, and that recently the Sudanese people have attained their full independence in the Republic of Sudan.

Further westwards it will be seen that such has been the force generated by the movements all over Africa that even in the densest forests of Equatorial Africa and in the northern deserts (that is, in the Belgian Congo, French Equatorial Africa and Sudan, etc.) the rumblings of discontent are heard. Even here the popular awakening is gaining strength.

Coming to the west, the observer will realise the fact that, on the Gold Coast, the Convention People's Party won the postwar election because of its demand for *Self-Government Now*. Here, on account of the powerful post-war upsurge of the Gold Coast Africans, the foreign rulers were forced to agree in 1951 to the formation of an African Government; and now the Government and the poeple of Gold Coast are expressing their desire to form a republic. This development has greatly contributed to stirring up national ideas in other parts of Africa and has provided a positive proof that the Africans can do many things if they fight unitedly for their birth-right of freedom, progress and prosperity.

Visiting Nigeria next, he will learn that although as a sop to the people's movement the foreign rulers here agreed to the participation of Africans in the Nigerian Government, several outbreaks were reported from this territory after 1951. The growth of a militant

* It is also said that this scheme of forming the three Federations which can be integrated far more easily in case of an emergency than the number of States existing in British Africa today, has been designed to facilitate turning Africa into a strategic base in the event of a Third World War.

Mukherjee, The Problem of Uganda

Akademie-Verlag, Berlin · Kartengenehmigung (K. 11) MdI. Nr. 2761

trade union movement and the expression of national revolt through the National Council of Nigeria and the Cameroons, established in 1944, are two outstanding features of these people on their march towards freedom, prosperity and progress.

In West Africa, as in other parts of the continent, the movement is not restricted only to the British colonies. It has also penetrated deeply into the French territories. In the post-war period the African Democratic Rally was formed by gathering together the democratic organisations in all thirteen French African territories south of the Sahara. This organisation, with a membership of a million, soon became such a formidable force against French colonialism in West Africa that the French Government fell upon the people and their organisation with military and police campaigns of unexampled brutality. Yet, total strikes of industries and commercial enterprises took place in Dakar and throughout Senegal on October 13 and November 3, 1953, in order to force the colonial administration to implement the Labour Code for Overseas Territories (which the French National Assembly had passed on November 23, 1952). This strike movement signalised a great victory of the African workers in explicitly banning forced labour, recognising trade union rights including the right to strike, the principle of the 40-hour week in industry and in the administrative services, the right to holidays with pay, and the principle of the right to family allowances and benefits to expectant mothers.

And when an observer reaches the Mediterranean Coast, he does not need to look for "incidents" in connection with the vigorous struggles of the peoples of Algeria, Morocco and Tunisia for democracy and human rights. The powerful movements of these peoples are frequently featured in newspaper headlines.

Such is the situation in Great Africa today. It is obvious that there is „trouble" for the foreign rulers everywhere in this rich continent. From the Cape in the south to the Mediterranean coast in the north, from the Horn of Africa in the east to the Atlantic coast in the west, Africa is shaking her vast angry body to release herself from the skackles of degradation, poverty and bondage. The spread of popular movements all over the continent in the form of peasants' and workers' resistance and risings and general strikes (both for economic and political ends); the fight of all non-White peoples against repressive laws and the colourbar (which has taken both parliamentary and sometimes more militant forms); and the striving of the Africans for their birth-right of freedom; all these features leave no shadow of doubt that the native population of Africa (as well as the immigrant Asians, provided they ally themselves with the Africans in their just cause, and the so-called "Coloured" people born out of marriages between black and white) will not bear the yoke of subjugation for long.

So Africa is no more merely something on the world map. It is now very much in the mind of world public.

2. Africa and World Public

But what do the common people outside Africa think about the life in this continent? Is it not a fact that their views are coloured by century-old distortions of facts made by the Colonial Powers which went to Africa as harbingers of so-called "Western Civilisation" in order to uplift the benighted Africans? Africa, the land of ancient civilisation, has since then been described by them as the "Dark Continent" and her people as "Black Savages". The picture of Africa which was conjured up by these aggressive European Powers in the nineteenth century was as untrue as was their characterisation of all colonial countries.

To take an example, in the sixteenth century when the first English "merchant traveller" Ralph Fitch visited India and her neighbouring territories in the east, he was amazed by the highly developed civilisation of these countries. Fatehpur Sikri, "the wonderful red sandstone city which Akbar had just built", struck him with awe; he found the much less important town Pegu in Burma as "so much bigger than London"; and such were his reactions to other places he visited[4]. Later also, until the British ruling class became the masters of India, their representatives who were expanding their interest in India by means of intrigues and guile[5] described it as a great and highly civilised country and expressed the opinion that the Indian people were intelligent, industrious, hospitable, and of a strong moral character[6]. But, when British power had consolidated itself in India after last decades of the eighteenth century, the spokesmen of the foreign rulers began to describe her as the immemorial "village continent"[7], even though city-life existed in India in as remote a past as five thousand years ago[8]. They also discovered that the Indians were indolent, extravagant, superstitious, lacking in initiative or intelligence and moral character[9]. In short, the people were characterised by them as primitive and backward, and uncivilised. Not only that. To avoid responsibility for having caused serious agrarian crisis and grinding poverty in India by ruthless loot and plunder of her wealth, destruction of her industries and trade, and extraction of colossal tribute from the land while utterly neglecting the state of productive forces[10], they propagated the view that such was the condition of India at all times. For, according to the expert opinion of the *Royal Commission on Agriculture* which visited India in 1926: "The desire to accumulate money is not characteristic of rural society"[11]. On the basis of such facts and arguments, now they found India to be peculiarly obscurantist and mystic, and thus they "orientalised" the Orient in order to maintain their rule. No wonder that from their revised opinion of India there emanated the decision that such a country could not easily be made amenable to such "modern" concepts as democracy and self-government. The *Indian Statutory Commission* of 1930 declared[12]:

> "Any quickening of general political judgement, any widening of rural horizons beyond the traditional and engrossing interest of weather and water and crops and cattle, with the round of festivals and fairs and family ceremonies, and the dread of famine or flood — any such change from these immemorial preoccupations of the average Indian villager is bound to come *very slowly indeed.*" [Emphasis added]

Such is the view propagated by the foreign rulers with regard to the lands they have turned into colonies. Hence, just as it is the solemn duty of the Orientals to "de-orientalise" the mind of the world public who have for a long time been fed with lies manufactured by the aggressive European Powers regarding the "oriental" countries of China and India, and all countries in the "Far East" (the compass obviously being located where the Colonial Powers have traditionally lived, namely, Western Europe, and therefore geographical directions being Middle, Near, or Far East), so should it be the duty of social scientists to describe objectively what exactly were the stages of civilisation in Africa before the "Whites" came. Were all Africans hunting cannibals? Were they really "diabolical" creatures, bereft of all sense of human values? Were they always as poor as they are now, and did they undergo chronic starvation as they do now all over Africa[13]? In short, how far are the widespread stories of savage life of the Africans true? And, even if there is some truth in the primitive ways of living and thinking of the Africans today, how much of it is an after-growth during the "civilised" rule of the whites? Is it not rather due to a blockade of the progressive development of the African societies since they came under Colonial Powers?

A study of these questions is not the main theme of this book. But they have a serious bearing on any study of African people, for in the back of many people's minds, especially in Europe and America, beats the wild drum of the African savage in the "bush". Thanks to the barrage of propaganda from the exporters of "western civilisation", the mental vision of the world public is often of the blood-thirsty Africans killing one another until the "European" invaders brought "civilisation" to them.

3. Myth and Facts about Africa

How many of the general public know that long before the European „mercantile" powers set their foot in this continent there were many rich and prosperous feudatory states in Africa? Of the great civilisation which flourished in Egypt five thousand years ago a good deal is known. Some also know of the Arab civilisation spreading over Northern Africa from about the seventh century of the present era. But how many of us know of the civilisation of the territories, now known as Abyssinia, between the first and the fourth centuries, or of the Sudan since the tenth century of the present era? How many of us know of the flourishing states in West Africa after the ninth century of our era? Do we all know of the Ghana State, or that of Mosi, Mali, etc., etc.? Do we know that before the Portuguese (as the first European Power) visited central Africa in 1492, there were well-developed kingdoms in that region? How many people know of the State of Angola, of the kingdoms of Luba and Lunda in Congo, of Monomotapa in Rhodesia, and such other feudatory and slaveholding states in that and other regions[14]?

Even though very little information is available about ancient Africa, it is known that since about the eleventh century of the present era commercial relations existed between East Africa and Asia, transactions being made by Arab, Persian, and Indian traders. South of the Sahara, with the discovery of iron by the local population, handicraft and agricultural production, which were also flourishing previously, now reached a higher stage of development from the fifteenth century of the present era. Textile and metal work spread in extent, and money currency and credit systems were developed. By the sixteenth and seventeenth centuries there were established in central east Africa kingdoms like that of Buganda, Bunyoro, Ruanda, Urundi. The kingdom of Buganda had an impressive army and a strong navy[15]. The Englishman Speke, who was the first European explorer to visit the kingdom of Buganda in 1861, was distinctly impressed by the grandeur of the "Palace of Uganda" and the beautiful huts of the African people. Narrating his first visit to the king's palace, Speke recorded[16]:

> "The palace or entrance quite surprised me by it extraordinary dimensions, and the neatness with which it was kept. The whole brow and sides of the hill on which we stood were covered with gigantic grass huts, thatched as neatly as so many heads dressed by a London barber, and fenced all round with the tall yellow reeds of the common Uganda tiger-grass; whilst within the enclosure, the lines of huts were joined together, or partitioned off into courts, with walls of the same grass. . . . The first court passed, I was even more surprised to find the unusual ceremonies that awaited me. Three courtiers of high dignity stepped forward to greet me, dressed in the most scrupulously neat fashions. . . . This, then, was the ante-reception court; . . . ."

Later Speke visited other chambers of the Royal Palace — the Grand Court Room, the King's Toilet Chamber, Bed Room, and so on as he described these rooms in his narrative. All these rooms were richly decorated with skins, blankets, cowrie-shells, bark-cloth, etc. He had also the occasion to visit the equally impressive palace of the king's mother.

And how did the African peoples live in those days? Even taking as recent a period as the last century, it is found that in this respect is a good deal of difference between the views expressed by the sympathetic visitors who first visited Africa in the nineteenth century and the views commonly propagated today. It is true that those who took a favourable view of African life also described the Africans as savages whom they had come to civilise and turn into Christians, but they did not always put forward the view so common these days that before the Europeans came the Africans had no moral character, no stable home life, and so on; in short, that they did not live as human beings should.

Early traces of the present view-point could be found in the writings of those Europeans who had links with the colonial system. Thus Speke, who gave vent to a good deal of spleen in his description of the life of the African people, was a true stalwart of the British colonial system. "He had entered the army in 1844, and completed ten years of service in India, serving through the Punjab Campaign"[17]. Besides exploring the "source of the Nile", he was entrusted with opening "trade" between Britain and the Kingdoms of Buganda, Bunyoro, etc., for the purpose of which he wrote in his book that he carried a message from the Queen Victoria[18]. His picture of Mutesa (the King of Buganda) as a tyrant, and of Kamrasi (the King of Bunyoro) and of some African chiefs as cruel, avaricious and treacherous, not only confused public-mind by presenting these as authentic illustrations of all Africans, but he also categorically stated in the introduction to his book that the Africans "think only of self-preservation in this world"[19]. They are therefore "too avaricious", "to destitute of fellow-feeling", possess a great "proficiency for telling lies most appropriately in preference to truth", etc., etc.

While Speke thus painted a nasty picture of African life, having failed to understand the Africans primarily because of his haughtiness as representing a "higher civilisation" (evidence of which abounds in his narratives), the description obtained from sympathetic visitors to Africa at about the same time is totally different. Thus, Reverend Henry Rowley, who as a member of the Universities' Mission to Central Africa spent long years in this continent, wrote in 1876[20]:

> "I have no wish to appear in the light of a special pleader of the brighter side of the African's life, but it is right that it should be known. The revolting features of their existence have received a wide publicity, but they no more indicate the whole than the disgusting portraits of individuals represent the ordinary physical characteristics of the entire race.
>
> "Speke has given us a sketch of Mtesa, the King of Uganda, and the brutality of that one man is accepted by many as indicating the disposition of his tribe. It shows what individual Africans whose power is unrestrained may become, but such men are not numerous. When excited by tribal animosities, or by superstitous fears, or when under the influence of superstitious feeling of any kind, the Africans will all do horrible things, but such causeless, wanton cruelty as Mtesa's you do not often meet with. The same may be said with regard to the man whom Sir S. Baker describes as wearing a spiked bracelet, with which it was his habit to claw his wives. This diabolical disposition may have been true of the individual, but is no more true of the whole of the tribe than that the wife-beaters and kickers are the representatives of English husbands generally."

Rowley was not alone in his estimate of "African Capacity and Character". Glowing tributes were paid by those who wanted to understand the Africans, such as Pieter Nielson, Sir Thomas Graham, Dr. Donald Fraser, Sir Sidney Oliver, etc.[21].

Referring to the position of women in African society, Rowley wrote[22]:

> "There is a great deal of error in the ideas which obtain amongst ourselves with respect to the position held by the African women. Their occupations are to my mind less degrading, and the labours imposed upon them less severe than those to which many women in England, France and Germany are condemned. Field work in Africa is a light occu-

pation for a brief period of the year. This work entails far less drudgery than that to which many women in Europe are exposed. The manufacture of clothing does not as with us employ hundreds of thousands of women and girls from morning to night all through the year. The African ideal of womanhood may be revolting to our sensibilities, yet it is in harmony with the actual state of things, and though it exhibits itself in different ways, the woman's influence is as great in Africa as in more civilised lands."

As regards the "conjugal affection of Africans", in contrast to Speke's and similar other descriptions of the Africans treating their wives as slaves whom they may even "cut up into pieces", Rowley noted[23]:

> "Their affection for their wives is perhaps unparalleled among people of so low a grade. A husband will spare no sacrifice to redeem an imprisoned wife; of which characteristic the ivory traders (mostly Arabs — RKM) make a good use, for whoever possesses a female hostage can obtain almost any amount of compensation."

And about "filial affection of Africans"? Rowley recorded[24]:

> "Filial affection is manifested in a remarkable degree by these tribes. It ends only with life. 'Ah! my mother!' is the common expression of sorrow, and men will make any personal sacrifice for their mothers. I believe this to be true of almost all the African races."

"Cannibalism in Africa"? Rowley came to the conclusion from his extensive experience in Africa[25]:

> "Generally, there can be no doubt cannibalism is abhorred by the Africans. I saw the greater part of a tribe die of hunger, without a single instance of cannibalism occuring"*.

Not that starvation was a regular feature of life in Africa in those days as an interested mind may like to picture from the above extract. All Europeans who visited this continent before the European invasion, whatever might have been their attitude to the standards of African civilisation, spoke of the prosperity of the African people. There were differences between rich and poor, between the nobility and commoners in the societies, and also the territories and regions were at different stages of development, but the people in general certainly did not starve in those days, as they are found to do now. Referring to the Baganda in the then Kingdom of Buganda, Roscoe has noted[26]:

> "The plantains grow so freely that a woman can supply the need of her family with a minimum labour, and with the barkcloth trees a man can supply their clothing. The country had all its needs supplied by its own products for many years, and the people were happy and healthy before the introduction of Western civilisation."

And this was in the country where the "despot" Mutesa reigned!

Speke and some other European visitors who had similar ideas had announced to the world that the Africans have no "fellow feeling". The chiefs were not hospitable to the guests to their country, and entertained them only to extort presents from them. Moreover, such visitors had asserted that without arms and without reliable escorts one could not move about in Africa, for the African peoples are untrustworthy and they may rob and kill with impunity for the smallest gain. But Rowley noted[27]:

* It may be of interest to mention here that, as Speke wrote in his narrative, Kamrasi, the King of Bunyoro (that is, of the Kingdom of Kitara) was at first unwilling to allow him and his men to visit the kingdom for he was told that the White Men "were not satisfied until they got a dish of the 'tender parts' fo human beings three times a day" (cf. J. H. Speke's *Journal of the Discovery of the Sources of the Nile*, Everyman's Library, London, 1922, p. 397). Here the laugh was on the other side. The "Black" man thought the Whites to be cannibals!

"They are generous in the distribution of food; even when starving I have known them share the last morsel with their friends. They are hospitable to strangers. When once you are known as a friend, every village you enter is made your home. When travelling I did so without fear of being plundered, though I sometimes had property with me that in the estimation of the natives was great wealth. ... Travellers sometimes have their tempers vexed by chiefs who will not speed the departing guest, but keep him as long as they can, that they may gain more and more of the good things which he possesses. I can not, however, think this a crime which distinguishes the Africans above all people. It is but making use of their opportunities, a characteristic which we have elevated into a virtue amongst ourselves. I am very sure, however, that the desire for property is not always the motive which leads a chief to detain his guest; for the desire to avoid the responsibility of any mischance that may be associated with his sojourn amongst the people to whom he is going, is a frequent cause of such delay. While the traveller is fretting and fuming, the chief is sending and receiving messages to prepare the way for his departure, and is immersed in all the arts of African diplomacy to avert trouble from himself. For if misfortune should befall a tribe while a traveller is with it, those who sent him onwards are held to be responsible, rather than he."

And thus Rowley differentiated between these visitors who appeared in this continent with an aptitude for understanding the people and those who were interested in imposing their own attitude on the hosts[28]:

"You must live some time amongst the Africans to know what they really are, and what are the motives which influence their actions. Travellers pass through the land as strangers to the people, while the people are strange to them, and thus easy occasion is given for mistrust and misunderstanding. But when you really know them, are sympathetic with them, and they know and like you, your experience is far more pleasant."

It may be worth stating here that Speke himself noted, although he had very little to say in favour of the Africans, that the kings of Buganda and Bunyoro made careful preparations to conduct his party safely from their respective kingdoms to the next destination, lest any misfortune befell the travellers. He also noted that when some members of his party who were given to him by the king of Buganda deserted him on this way to Bunyoro from Buganda and with guns returned to the monarch, the king (the same diabolical creature Mutesa) told them[29]:

"Did I not command you to take Bana (Speke—RKM) to Gani at all risks? If there was no road by land, you were to go by water; or, if that failed, to go by underground, or in the air above; and if he died, you were to die with him: what, then, do you mean by deserting him and flying here? You shall not move a yard from this until I receive a messenger from him to hear what he has got to say on the matter."

Mutesa was not only so concerned about his previous guest, but, although Speke pictured him as an avaricious man, he refused to take the guns from the deserters even though in those days guns were the most highly prized articles in Africa. He kept them for his *friend* Speke to fetch them later.

Similarly, although Speke characterised Kamrasi (the King of Bunyoro) as a "cunning beggar" who was interested only in fleecing him, he noted that once on hearing the rumour that Mutesa's men were on the way to snatch the *White* men away, "the king asked his men if they would ever permit it; and, handling his spear as if for battle, said at the same time he would lose his own head before they should touch his guests"[30]. Such was the sense of hospitality of the "barbarian" king who, furthermore, to seal his friendship permanently with Speke proposed to exchange blood so that "we might be brothers till death"[31]. Speke, of course, commented: "As to exchanging my blood with a black man's, it was a thing quite beyond my comprehension; . . ."[32]. And that about summs up the attitude of those from whose lips the world public have heard such terrible stories of Africans before they were "civilised" by the whites.

4. Africa and the Dawn of Capitalist Era in Europe

Undoubtedly, there is no need unduly to glorify the standards of civilisation of Africa in those days, which were less developed than those of Europe. But it must be noted that they were painted in much darker colours and were also grossly distorted by the proponents of "white civilisation" who indeed have consistently played a much worse role in this "Black Continent" than the local despots. To begin with the first phase of large-scale contact between Europe and Africa, from the sixteenth century Africa became a prey to the European traders who began to consider her as the "warren for the commercial hunting of black-skins". Those were the days of primary accumulation of capital, of which Marx wrote[33]:

> "The discovery of gold and silver in America, the extirpation, enslavement and entombment in mines of the aboriginal population, the beginning of the conquest and looting of the East Indies, the turning of Africa into a warren for the commercial hunting of black-skins, signalised the rosy dawn of the era of capitalist production. These idyllic proceedings are the chief momenta of primitive accumulation."

But here again justification was put forward to make the slave trade palatable to the sensitive taste of "civilised" folks. It waw declared by several European Powers, who were engaged in the profitable business of catching African slaves and exporting them to the New Hemisphere, that slavery was not an unknown institution among the Africans. On the contrary, it may be of interest to note what Rowley wrote in this connection[34]:

> "Slavery amongst the Africans who are removed from foreign influence, is in many things very unlike slavery amongst civilised nations. It is far less harmful in its immediate results. It presents various aspects ,according to the different characteristics of the tribes.
>
> With the agricultural tribes it assumes a purely patriarchal aspect. Those who are in bondage are not called slaves, but children; those to whom they are in bondage are called not masters, but fathers. The slaves have rarely been acquired by conquest, but mostly by inheritance, or in some less obnoxious form than war, and bad blood, therefore, is rarely found to exist between the 'father' and his 'children'.
>
> The chiefs and men in authority sometimes exercise an authority over their children that might be considered severely patriarchal, but generally within the boundaries of their master's territory the slaves have as much liberty of action as the free men. They are not compelled to labour for their chiefs' profit; but when he goes abroad on a visit to a brother chief, they are required to accompany him, or if he goes to war they are bound to fight for him. If inclined to stray, the chief can punish them according to his discretion. If such vagabonds get into trouble, their chiefs have to pay the penalties imposed; and if any slave-dealer is near at hand, they sometimes sell such rebellious children, in order to reimburse themselves for the expense which they have incurred. If no such means exist for disposing of the incorrigibly disobedient, to avoid future trouble, they will probably put them to death. . . .
>
> Under certain conditions a man might sell his wife, his children, and even himself. And this is the case with the Manganja, and also, I believe, with the agricultural tribes generally. But the difference of condition between the bond and the free was not usually accompanied by any marks of degradation. All lived alike, all followed the same occupations. A stranger passing through their land would not know, from anything he saw to remind him of it, that there existed amongst them such a distinction as bond and free.

With the exception of the power which it gave to the chiefs and others to sell their slaves, or to dispose of them in any other way, slavery as it exists amongst the agricultural tribes of East Central Africa, when left untouched by foreign influences, is productive of no great harm. It does not degrade morally, as it scarcely seems to inflict a stigma socially; and what is true of these tribes, holds good, I believe, of all others under similar circumstances."

Similar views have been expressed by others as well, such as Roscoe, Harris, etc.[35].

Without in any way justifying slavery in any form, one can see what havoc was wrought in African life when foreigners began to indulge in the lucrative trade of slave-hunting in Africa. To quote Rowley again[36]:

"... wherever the slave-trader penetrates, there ensues a different state of things. The cupidity of the chiefs is excited; and the desire to get guns and gunpowder, beads, brass wire, and Americano (unbleached calico), tempts them not only to make war upon one another, to steal men, women, and children from one another, but to sell their own people also. The old peaceful order of things is broken up. The strong destroy the weak. War desolates the country; villages are burnt; the fields are left uncultivated; a fruitful land becomes a desert; and men, women, and children are everywhere killed or enslaved. It is not possible to exaggarate the evil effects of the slave trade in Africa itself. The reality surpasses the wildest flight of the imagination."

This phase of disintegration of African life, begun by the Arabs, reached its height with the keen competition between the European Powers in trafficking in African slaves. In that period of primary accumulation of capital, the "mercantile" powers of Europe swooped on the free Africans, and shipped them from their soil to America and the West Indies to slave in plantations and gold and diamond mines where the labour force of the indigenous population was found unable to cope with the white looters' drive to plunder the wealth and resources of these countries[37]. In the later half of the sixteenth century, Sir John Hawkins, who was the first from Britain to join the slave trade, captured 400 Africans from the Guinea Coast and sold them in the West Indies for £ 25 a head. The great powers of Europe, avowedly Christians, had then "no more scruple [in indulging in slave trade of the African 'savages'] than they had about enslaving horses"[38]. Rowley noted in this connection[39]:

"The African slave trade and slavery, with which we are more immediately concerned, is the work of comparatively modern times. The discovery of America opened out a new career for slavery. Before there was any thought of enslaving the Africans on a large scale, the aborigines of the West Indies were brought into bondage by their European discoverers. Ill-treated and over-worked, they soon passed away. Africa was then resorted to for slaves to supply the place of the Caribs, who had been killed in war, wasted in the mines, or worked to death in the plantations. And thus it came to pass that the Africans have so largely partaken of the miseries of slavery.

The Portuguese were the first of all nations of Europe to engage in the slave trade, and they have been the last to abandon it. But it was not long before other European nations entered upon this inhuman traffic. Spain and France, Holland and England, competed with Portugal, until, with characteristic energy, England became first and foremost in this iniquitous traffic. She not only supplied her own colonies with slaves, but, by a treaty with Spain, she obtained the monopoly of supplying the Spanish colonies also, both on the continent of America and on the islands.

The history of the European African slave trade in Western Africa needs but little consideration here, it is well known. The extent to which it reached may be estimated by the fact that in the year 1840, fifty years after Great Britain had commenced an offensive warfare against it, 150,000 slaves were said to have been imported into the Brazil, Cuba, and Porto Rico. This implies a loss to Africa of at least 200,000 souls, for it may safely be assumed that 50,000 died before they reached their destination, or were slain and harassed to death in the wars that led to their capture. Furthermore, it is estimated that there are 14,000,000 of persons of African descent living on the mainland and islands of America. These people are slaves or the descendants of slaves. The majority of those who are free have but lately acquired their liberty. They represent in their numbers probably not more than a third of what were taken from Africa. For though the Africans have an irrepressible vitality, and multiply under circumstances that would be death to a feebler race, yet the wear and tear of the slave trade and slavery, as carried on by Europeans, was most exhausting to life."

Available figures indicate that in spite of their tremendous vitality half the number of Africans captured as slaves perished before they were put to "back-breaking toil and inhuman treatment in the plantations". In the British-run Jamaica trade, slave losses were "12½ per cent in harbour, 4½ per cent before sale, and 33 per cent in 'seasoning'". Yet the trade was very lucrative indeed, as the following extract suggests[40]:

> "By 1680 wealthy and respectable merchants of Bristol, Liverpool and London were exporting 15,000 Africans yearly. Later, the total increased. Britain alone seized, transported and sold over two million slaves between 1680 and 1786. At the height of the trade there were 192 British ships engaged, carrying 47,000 Africans between them on each trip. By 1791 there were forty slaving stations, euphemistically called 'factories', on the West Coast alone."

The civilised governments gave support to this inhuman business in opposition to all efforts of humanitarians to have the slave trade abolished. In 1775, Lord Dartmouth, the then Colonial Secretary of the British Government, declared[41]:

> "We cannot allow the Colonies to check or to discourage in any degree a traffic so beneficial to the nation."

And the Christian Church also made money out of this trading in heathens. Kartun noted[42]:

> "The Portuguese made a good thing out of the trade — their Bishops standing by to baptize each man, woman and child to be driven in chains to the ships so that their souls might find salvation in the very probable event of their death on the high seas. The Church did well out of the transaction, charging 300-rei baptism tax per head — suckling infants excepted."

There was indeed a significant difference between the "higher" civilisation of Europe and the less developed one in Africa.

5. Mirage of Civilising Mission

It is true that direct slavery is now condemned by everyone. But not all recognise the game of indirect slavery in the colonies. Therefore, while Malan's panacea of "White Civilisation" may leave them cold or even rouse their just sentiment against the revival of Nazi doctrine in another form, and the declared evils of mixed marriages between Black and White may draw a sceptical grin or irritate their liberal outlook, such is the extent of distortion of the mutual role of the Africans and Europeans in Africa that they are often confused when they hear that the vast majority of Africans do not believe in the good intentions of their European masters.

To start with, did not the European Christian missionaries make good men out of the Black people? — some may be inclined to ask. Would they believe what the Kenya Africans say about the missionaries today? They say, "When they came, they had the Bible, we had the land; and now we have the Bible, they have the land". So they demand, "Give us back our land, and we shall give you back 'your Bible'"[43].

Such demands may be regarded by some as unbecoming of a people who must have benefited greatly by their association with the European missionaries. Do they know that these missionaries as a body (individually some of them may be genuine friends of the Africans) were not infrequently one of the spearheads of aggression and ultimate colonisation? History abounds in such examples from many colonies, and especially in Africa. Some instances will be given in course of the present study.

Next, take for example, the European settlers. One often hears the argument in the West, "Did they not bring civilisation to the Africans living in the 'bush'?" Yes, they did in some parts of Africa, as for instance, in the Union of South Africa, Rhodesia, in the Kenya Colony, etc., where the climate was even better then in many parts of Europe. But the "civilisation" they brought with them, was it for the benighted Africans or for themselves flourishing in a better situation than they had in Europe?

Indeed, the first effect of European settlement in Africa was to narrow down the living space of the Africans. They who once possessed all the land are now encircled in small reserve areas where the "traditional" life is "preserved". As early as in 1916, the Native Land Commission noted in its report that in the Union of South Africa, "Native Reserves" accounted for only 7.13 per cent of the total area[44]. The situation has no doubt been worse since then with the normal increase of population and continual encroachment of the Europeans. The same is the story in other parts of Africa wherever Europeans have settled in the bracing climate and on rich lands. Thus, the Kenya Highland, one of the most beautiful and fertile parts of Africa, is completely in the possession of the British settlers, while the Africans have to eke out their living in congested "tribal" reserves[45].

And so the Africans, driven away from their ancient land, have to live in narrow reserves, while the European settlers have gobbled up such vast areas that individual farmers cannot possibly utilise their estates for proper cultivation[46]. In Kenya, while the Africans are dying from slow starvation in the over-crowded "tribal reserves", vast stretches of land are kept reserved by the foreign rulers for "future settlers"[47]! It has been estimated that some five and half million of Africans who once owned the whole of what is today the Kenya Colony are now living in 52,000 square miles of inferior land, while some 30,000 white settlers hold one-half of the cultivable land of the colony but actually farm only one-tenth of this area[48].

Furthermore, being ground down by poverty and having little assistance from the governments to maintain themselves in the reserves, large numbers of Africans are forced to leave their "bush" and come to the "civilised" parts of their countries to work for the European masters for a mere pittance. On the basis of government reports, Aaronovitch has shown that while thousands of pounds sterling are spent as Government subsidies to the European farmers under the Increased Production of Crop Ordinance in Kenya, such subsidies "are not available to African cultivators"[49]. The situation is similar in other parts of Africa where the Europeans have settled[50]. For instance, referring to the "peasant farmers on the rich wheat-growing land of Herschel" in the Union of South Africa, it has been stated that the cultivable area available to the Africans is so inadequate that for an ordinary peasant farmer "unless he has a grown-up *son or daughter* able to go out to work and supplement his income by *their* wages (sic) he must leave his croft and go out himself to work"[51].

Indeed the European masters have not remained content only by the arrangement for the supply of cheap labour-power from the "tribal reserves". By inducing the governments, run by them and for them, to impose enormous taxes on the Africans, they have so adjusted the situation that the Africans must acquiesce to "voluntary" or forced labour on the

European estates and mines at very low wage-rates[52]. With characteristic frankness, a stalwart of this policy suggested in 1849, in the earlier phase of European settlement in South Africa, how to guarantee this supply of labour[53]:

> "Permanent locations should be established within the Colony; and in selecting the sites of these locations, sufficient intervals should be left between each of them for the spread of white settlements; each European immigrant would thus have it in his power to draw supplies of labour from the location in his more immediate proximity."

Lord Delamere, the well-known spokesman of the British settlers in the Kenya Colony (who started with a share of 100,000 acres of Kenya land as the leader of the first settlers), declared before the East African Native Labour Commission, of which the report was issued in 1912—13[54]:

> "If the policy was to be continued, that every native was to be a landholder of a sufficient area on which to establish himself, then the question of obtaining a sufficient labour-supply would never be settled."

Other witnesses to this Commission were even more vocal. Mr. Fletcher of Kyambu said[55]:

> "If the Reserves were cut down sufficiently, it would undoubtedly have the effect of turning off a large number of natives, who would be made to work for their living."

Another witness suggested "making the native wear European clothes, in order to buy which he will be compelled to seek work from the whites"[56]. Quite truly Lord Oliver declared with regard to the white settlers[57]:

> "The attitude of the Kenya Die-hards towards the natives and the Indian population is a direct lineal derivative of the attitude of Afrikander theory in South Africa; and its active emergence in Kenya is the remote but recognisable outcome of the slave-policy forced by the Dutch East India Company on the first Cape Colonists."

And what Lord Oliver noted in another place fairly summs up the role of the whites in Africa, although his attitude to the question is not quite identical with that of the author[58]:

> "To a student of the psychology of Imperial expansionism Kenya Colony offers a clinical subject of peculiar interest. There are two motives which bring the white man into contact with coloured races. White men go to uncivilised lands to make money or a livelihood, desiring to benefit themselves: and they go also as missionaries, desiring to benefit the natives. From the time of the Spanish colonisation (which purported to combine the two purposes — the benefit to the white being economic, to the native, religious) until the period of the Partition of Africa very few people imagined or pretended that the two purposes were harmonious: in fact both parties — colonisers and the missionaries — generally thought the contrary and were continually at issue with each other. Then emerged the brilliant idea that colonisation was The White Man's Burden, and that those who extended the Empire were serving both Mammon and God. The British Public took kindly to this belief: its sentiments towards coloured and native peoples being fundamentally decent. Lord Balfour and other high-minded Englishmen, during the War and during the Peace negotiations, made great running with the doctrine that English Colonial Government was a Providence to the natives and that German colonies had better be handed over to her to govern. President Wilson and the allies jumped at the notion and invented the dogma of trusteeship. But this was really a little hard on the men of the old-fashioned school of colonists who were simply out to make their own living as settlers or make money as company-floaters out of other white men who might attempt to do so. East Africa and Uganda were not, in fact, annexed for the sake of the natives: but to get cotton and trade.

Kenya Colony was merely a by-product of the Uganda-Railway. Sir Charles Eliot thought it would make a good white man's country. The missionaries co-operated, not at all with the same idea as that which encouraged settlers. The Protectorate Government started off with unrestricted forced labour for "public" purposes: the leading settlers fought against a fair land policy for the natives: Sir Percy Girouard clapped on heavy taxation to make the native work. All the time these unfortunate settlers, with their simple primeval colonising mentality — identical with that of their Anglo-Saxon forefathers — feeling no question of their right to take the native's land or to exploit his labour, have had to contend with these latter-day notions which seem to them merely the old missionary nonsense in a new dress — reinforced with all the prestige of an international doctrine of Trusteeship. They cannot understand or endure it. Even about Tanganyika they write furiously to the *Times* complaining that Germans (in a mandated territory) are given equal privileges with Englishmen, and natives allowed to grow coffee."

Much worse has happened in Africa since the above lines were written. In this continent, where before the flowering of colonial rule the European "pioneers" found the people to have plenty to eat, today the same people suffer from chronic malnutrition. The Africans, who used to be praised for their wonderful physique and who had such a stamina that they could withstand centuries of devastation by slave traders, are now unable to do a full day's work. Today, although "even with her inefficient methods Africa's yearly share of the total value of agricultural products grown in the world very nearly equals her share of the world's population", hunger and chronic malnutrition have so much devitalised the people and arrested the growth of surviving children that to extract a full day's work from the African workers has become a serious problem for the colonial governments[59]. This the experts are forced to admit in their studies on African labour efficiency in different parts of British Africa. Ridden with horrible diseases, having hardly anything to eat, living like beasts in the overcrowded ,,reserves" or in the hellish shanty towns, acquiring hardly any education and culture, the healthy and prosperous peoples of pre-colonial Africa now lead a sub-human existence.

Such are the benefits conferred upon the Africans by the importers of "western civilisation" to their country.

6. Latest Propaganda

True, the arguments about the civilising mission of the European settlers, the missionaries, and the commercial enterprisers are becoming stale in many quarters where the common people are beginning to see, at least partially, through the game of the Colonial Powers to defend their rule in Africa and in other colonies. So the Colonial Governments and their intellectual allies are now coming out with a slightly different argument. It may be, they will admit, that "mistakes" were made during the first phase of ruling the colonies, but now they are fully conscious of the needs of the Africans and so are steadfastly discharging their responsibility to them.

This game of playing the Big Brother to the Africans has been very well executed in the post-war years. Books have been written, periodicals have been flooded with such stories and reportages, and even full-length films have been made in Britain, France, and Belgium, — the home-lands of the principal Colonial Powers in Africa today. All these help to confuse the mind of the common people outside Africa. They feel inclined to argue that after all the governments, however slowly, are improving the condition of the Africans. And, if that be true, what then is the need for the Africans to be so impatient, ans why have they no faith in their present-day rulers?

Playing exactly on this naive attitude of many people, the "political" traveller Elspeth Huxley has come out with her widely-publicised book entitled *Sorcerer's Apprentice*[60]. Like many such writings in recent times, which are published in order to uphold the colonial rule in Africa, this book is intended to explain the futility of the African attitude in East Africa and the patient "welfare" work done for them by the British Colonial Administration. Reminding the readers of the Egyptian story that the sorcerer's apprentice brought disaster to himself by trying to be independent of his master, this able apologist for British Rule attempts to prove that the Africans are not yet fit for self-government, and therefore, for the present at least, they must remain under the protective wings of Britain. According to this story-teller of witchcraft, Britain is no doubt doing her best to "educate" her apprentices in the arts of life in the strenuous conditions of the modern world, while assuring in the meantime a very happy future for them.

7. Why This Study?

Are such contentions true? Do the Colonial Governments really look after the interests of the subject peoples? This is the central point which this book endeavours to answer. Its main purpose is to show that there is no fundamental difference between the earlier and the recent phases of colonial rule. There is an unbroken continuity throughout the system, the sole aim of which is to fatten the foreign ruling class at the expense of the native population, while offering a few crumbs to the local allies in order to keep them contented with the colonial rule. Only the system of exploitation takes different forms, direct or concealed, — depending upon the overall circumstances. For this reason, the demand of the Africans expressed by their slogan — *Africa for Africans* — is a genuine and legitimate demand.

Africa is a very rich continent. Her reserves of iron ore are the largest of any continent on the earth. Only one of her many coal fields, namely the Wankie in Southern Rhodesia (under the British Commonwealth), is regarded to be the richest single coal deposit in the world. She possesses 98 per cent of world's industrial diamonds, 80 per cent of the cobalt, more than 50 per cent of the gold, nearly 25 per cent of the manganese, 20 per cent of the copper and tin, and substantial deposits of other minerals — some of which are of great strategic importance, such as uranium. To add to this, are the agricultural products — coffee, cocoa and cotton, palm and other oils, tobacco and tea, etc.; and timbers, ivory, and many such important commodities. Her water power recources are also three times those of Europe, although so far only one-three hundredth part of them have been developed. Naturally, therefore, giant monopolies, supported by the Colonial Powers, have caught her in an octopus-like grip and are further spreading their tentacles. In the last sixty years only enslaved Africa has yielded to these monopolies in London, New York, Paris, Brussels, etc., antimony, asbestos, coal, cobalt, copper, chrome, diamonds, gold, iron, lead, manganese, platinium, tin, uranium, vanadium, and zinc to the total value of £ 4,300 millions. In addition, there are coffee, cocoa, palmoil, tobacco, sisal, ect., etc.[61]. And in return her people are inhabiting filthy places, suffering from disease and hunger, and are living without any culture or education. And if they protest at such a fate they receive beating and torture and death at the hands of white masters. This has now become the regular occurence in the Union of South Africa, in Kenya, Uganda, and several other parts of Africa. No wonder then that the Africans are now seen to forsake their faith in the civilising mission of the Europeans and are coming out ever-more resolutely with the slogan of *Africa for Africans*.

Needless to say, this desire of the African peoples to become free and lead a prosperous life must come true, for before the mighty wave of the people's movement for freedom and democracy, for progress and prosperity, all obstacles will be swept away. And this success will no doubt come to them without any help or advice from other quarters. Even so it is necessary to analyse the problems of Africa either as a whole or of her separate territories, for colonialism is a canker in world society and so the world public must be interested in the happenings in the colonial countries. There is also a special reason for undertaking such an analysis at this stage of world history, for, as the Prime Minister of the Republic of India, Pandit Nehru, stated in New Delhi on the eve of the Bandung Conference[62]:

"A change has come all over Asia and in large parts of Africa, and it is clear that the static period is over. Something is happening; all kinds of forces are at work in the minds of people, millions of people. We want to build up our countries as rapidly as possible and to bring peace and happiness to our peoples. And yet while we seek to build, other forces come in our way. While the world becomes more and more unified, in a sense disruptive forces also work. The world becomes one world and yet that one world lives in the shadow of possible disaster."

Nehru's voice was echoed and his viewpoint was further elaborated by the President of the Indonesian Republic, Dr. Sukarno, when in his opening address to the Afro-Asian Conference at Bandung he stated on the 18th of April 1955:

"I say to you that colonialism is not yet dead. Vast areas of Asia and Africa are still not free. Colonialism is a skilful and determined enemy and appears in many guises. .... I beg of you, do not think of colonialism only in the classic form which we of Indonesia and our brothers in different parts of Asia and Africa knew. Colonialism has also its modern dress in the form of economic control, intellectual control, and actual physical control by a small but alien community within a nation. It is a skilful and determined enemy and it appears in many guises. It does not give up its loot easily. Wherever, whenever and however it appears, colonialism is an evil thing and one which must be eradicated from the earth."

No doubt, the world public will be in accord with the above views of Nehru and Sukarno, and it thus becomes a particular responsibility of the social scientists to study the problems of Africa. For, since the last war, Africa has become the main seat of the Colonial Powers. Other territories of the colonial world, notably in Asia, have forced themselves beyond the sphere of direct control by the Colonial Powers, and therefore these powers as well as the monopolist concerns flourishing under their protection are striving to clamp their grip even more firmly on Africa, and bleed it dry. Also Africa has become of special importance to them because of her rich possession of military-strategic materials, such as uranium in the Congo and others — some noted before and others to be noted later. The Colonial Powers, therefore, want to hold tenaciously to Africa as their last stronghold. Furthermore, there is the possibility that in the event of a Third World War the Colonial Powers may use Africa as their most important rear base. Hence, their vigorous attempts to form federations of colonies, whereby African territories would be organised into "large, stable units capable of supporting war industries, military establishments and strategic communication networks"[63]. Hence, their plans to connect different parts of Africa by a network of railways (which will also serve more efficient economic exploitation); to turn East Africa into the "main atomic-age training ground of the British Army and a main support base in the new Empire defence system"[64]; and to make Kenya "the new centre of Commonwealth defence, and South Africa its arsenal"[65].

Taking the situation as a whole, nobody therefore can remain passive or neutral today. One of the major battles of world humanity — freedom or bondage, exploitation or prosperity, degeneration or progress, war or peace — is being fought in this continent. Africa, a bastion of the Colonial Powers, is a threat to world peace, for it breeds discontent and disorder through internal oppression and it facilitates preparations for a war of world destruction. On the other hand, Africa — a land of free, progressive and prosperous people, is a bulwark for world peace and security, for the future progress of humanity. Hence, it is the duty of social scientists to unravel her problems and acquaint the world public with the results. That is the basis for this study.

8. Scope of Study

This book, however, does not deal with the whole of Africa. It tells the story of the Uganda Protectorate, a small part of the vast continent, which is under British Colonial Administration. The question may therefore be asked: why Uganda and not any other territory in Africa? The obvious answer is that while the Uganda Africans are feeling restless and striving to change the present state of affairs, not a single serious study has so far been made of the specific problems of this territory and its people. For the Union of South Africa, or the Kenya Colony, or for the Gold Coast, Nigeria, and other important parts of Africa there are some studies; but for Uganda there is none.

There is another reason for telling this story of Uganda. This is the personal contact of the author with the people of this Protectorate. In recent times he had the opportunity to visit Uganda, and see for himself how the people live there. He could meet the peoble from various walks of life, — the African peasants, workers, intelligentsia and native officials, as well as Asians and Europeans. He lived in the "bush" and also in towns. The barrier of suspicion and distrust of foreigners, inevitable today in the polluted atmosphere of colonial rule, was diligently overcome. Sympathetic understanding opened the gate to the mind of the African people. And then came out in a flood torrent the pathos of their present life, their aspirations for a future of prosperity and progress. The throbbing of Africa in the heart of every African, who only wants to live in his own country as his own master, shouted louder and louder.

This eternal desire of every people in all parts of the globe, the author intends to convey to the world public from this small part of the world. As one who has lived under colonial bondage, he wants to bring this cry for freedom to the ears of other freedom-loving peoples. However limited may be the scope of this study, however weak and incompetant may be this elucidation of the *problem of Uganda*, the author would feel justified in his attempt if this story of a people in bondage echoes their freedom song in the hearts of the peoples of the world.

REFERENCES

1 Dutt, R. P. — "The Crisis of Britain and the British Empire", Lawrence & Wishart, London, 1953, p. 47

2 Reid, Alec — "Awakening in Africa", India Information Services, New Delhi, 1953, p. 42

3 Aaronovitch, S. and K. — "Crisis in Kenya", Lawrence & Wishart, London, 1947, pp. 40—131

4 Collis, Maurice — "British Merchant Adventurers", William Collins of London, 1942, pp. 8—15

5 Mukherjee, Ramkrishna — "The Rise and Fall of the East India Company", Verlag der Wissenschaften, Berlin, 1955, Chapter 4

6 See, for instance, Lajpat Rai's "Unhappy India", Banna Publishing Company, Calcutte, 1928, Chapter XXIII, giving extracts from studies on India by V. A. Smith, Rhys Davids, Mountstuart Elphinstone, Holwell, Stewart, Malcolm, Bishop Heber, India Reform Society, etc.

7 Dutt, R. P. — "India Today", People's Publishing House, Bombay, 1949, p. 189, etc.

8 For instance, the findings at Mohen-jo-daro and Harappa, dated from the third millenium before the Christian era.

9 See, for instance, the writings of Thomas Babington Macaulay on India. Perhaps the most virulent book on the subject, published in the present century, is Katherine Mayo's "Mother India" (Jonathan Cape, London, first published in July 1927). This book was much appreciated by Mr. (now Sir) Winston Churchill (cf. *loc. cit.* 6, p. 255)

10 *loc. cit.* 5, Chapters 5 and 6; *loc. cit.* 7, pp. 94—120, 187—261; etc.

11 The Report of the Royal Commission on Agriculture in India, Government Central Press, Bombay, 1928, p. 6

12 The Report of the Indian Statutory Commission, Government of India Publication, 1930 p. 15

13 See, for instance,
(a) Buell, Raymond Leslie — "The Native Problem in Africa", The Macmillan Company, New York, 1928;
(b) *loc. cit.* 1, pp. 62—64, 222—224;
(c) *loc. cit.* 3, pp. 41—51;
etc., etc.

14 See, for instance,
(a) Westermann, D. — "Geschichte Afrikas", Greven-Verlag, Köln, 1952;
(b) Leo Frobenius — "Das Unbekannte Afrika", C. H. Becksche Verlagsbuchhandlung, München, 1923;
(c) Leo Frobenius — "Civilisation Africaine";
etc., etc.

15 (a) Speke, J. H. — "Journal of the Discovery of the Sources of the Nile", Everyman's Library, J. M. Sent & Sons Ltd., London, 1922;
(b) Roscoe, J. — "Twenty-Five Years in East Africa", Cambridge at the University Press, 1921;
etc., etc.

16 *loc. cit.* 15a, p. 234

17 *ibid.*, p. vii

18 *ibid.*, p. 379

19 *ibid.*, p. 1

20 Rowley, Henry — "Africa Unveiled", London: Society for Promoting Christian Knowledge, New York: Pott, Young & Co., 1876, pp. 49—50

21 See, for instance,
(a) Oliver, Lord — "The Anatomy of African Misery", Leonard & Virginia Woolf, London, 1927, pp. 233—234;
(b) Harris, J. H. — "Africa: Slave or Free?", Student Christian Movement, London, 1919, p. xi;
etc., etc.

22 *loc. cit.* 20, p. 46

23 *ibid.*, p. 40

24 *ibid.*, p. 46

25 *ibid.*, p. 40

26 Roscoe, J. — "The Baganda: An Account of Their Native Customs and Beliefs" Macmillan and Co., Ltd., London, 1911, p. 5

27 *loc. cit.* 20, pp. 47—48

28 *ibid.*, p. 48

29 *loc. cit.* 15a, p. 415

30 *ibid.*, p. 412

31 *ibid.*, p. 420

32 *ibid.*, p. 421

33 Marx, Karl — "Capital: A Critical Analysis of Capitalist Production", George Allen & Unwin Ltd., London, 1949, p. 775

34 *loc. cit.* 20, pp. 174—176

35 (a) *loc. cit.* 26, pp. 14—15, etc.;
(b) Roscoe, J. — "The Bakitara or Banyoro", Cambridge at the University Press, 1923, pp. 11, 303, etc.;
(c) Roscoe, J. — "The Banyankole", Cambridge at the University Press, 1923, pp. 151, 153 etc
(d) *loc. cit.* 21b, pp. 59—71;
(e) Weidner, Fritz — "Die Haussklaverei in Ostafrika", Verlag von Gustav Fischer, Jena, 1915, pp. 3, 16—17, 54—55, etc.;
etc., etc.

36 *loc. cit.* 20, p. 176

37 (a) *ibid.*, pp. 171—200; etc.
(b) Zechlin, Egmont — "Heinrich der Seefahrer und die Entdeckung von Negerafrika", *Afrikanistische Studien*, Herausgegeben von I. Lukas, Akademie-Verlag, Berlin, 1955, pp. 18—29;
etc., etc.

38 (a) Senior, Nassau — "Slavery in the U.S.", quoted by Maurice Dobb in "Studies in the Development of Capitalism", George Routledge & Sons Ltd., London, p. 114;
(b) *loc. cit.* 21b, pp. 89—93;
(c) *loc. cit.* 37b, pp. 24—25;
etc., etc.

39 *loc. cit.* 20, pp. 172—173

40 Kartun, Derek — "Africa, Africa!", Lawrence & Wishart, London, 1954, p. 9.
See also, *loc. cit.* 35e, pp. 54—56, etc.

41 *ibid.*, p. 9

42 *ibid.*, pp. 8—9. See also 37b, pp. 24—25.

43 *loc. cit.* 2, p. 48

44 Report of the Native Land Commission, vol. I, p. 3; also quoted in *loc. cit.* 13a, vol. I, p. 74

45 *loc. cit.* 3, pp. 67—69

46 (a) *loc. cit.* 13a, vol. I, p. 85, etc.;
(b) *loc. cit.* 3, pp. 77, 81, 91—97, etc.;

(c) *loc. cit.* 2, pp. 8—9
(d) Thomsen, Hans — "Die Verteilung des landwirtschaftlichen Grundbesitzes in Südafrika", Verlag von Gustav Fischer, Jena, 1927;
etc., etc.

47 Hailey, Lord — "An African Survey", Oxford University Press, 1938, p. 751

48 *loc. cit.* 40, p. 62

49 *loc. cit.* 3, pp. 79—80

50 (a) *loc. cit.* 13a;
(b) *loc. cit.* 21a;
(c) *loc. cit.* 21b;
(d) Leubuscher, Charlotte — "Der Südafrikanische Eingeborene als Industriearbeiter und als Stadtbewohner", Verlag von Gustav Fischer, Jena, 1931;
etc., etc.

51 *loc. cit.* 13a, vol. I, p. 77

52 (a) *loc. cit.* 13a, Vol. I, pp. 34—38, 77, 213, 240, etc.;
(b) *loc. cit.* 3, p. 93ff.;
(c) Bursian, Alexander — "Die Häuser- und Hüttensteuer in Deutsch-Ostafrika", Verlag von Gustav Fischer, Jena, 1910, pp. 50—62, etc.;
etc., etc.

53 Brookes, Edgar M. — "The History of Native Policy in South Africa from 1830 to the Present Day", Cape Town, 1924, p. 327; also quoted in *loc. cit.* 13a, vol. I, p. 72

54 quoted in *loc. cit.* 21a, p. 179

55 quoted in *loc. cit.* 21b, p. 125

56 *loc. cit.* 21b, p. 125

57 *loc. cit.* 21a, p. 64

58 *ibid.*, pp. 187—189

59 *loc. cit.* 40, pp. 28—31, 92

60 Huxley, Elspeth — "The Sorcerer's Apprentice: A Journey through East Africa", Chatto and Windus, London, 1948

61 For an account of the wealth and resources in Africa, estimated as early as in 1917, see K. Dove's *Wirtschaftgeographie von Afrika*, Verlag von Gustav Fischer, Jena, 1917.

62 Extract from a speech given by Pandit Nehru in New Delhi on the 15 th. of April, 1955, at a State banquet given in honour of the Egyptian Prime Minister and the Afgan Deputy Prime Minister (cf. *India News*, Weekly Paper published by the Office of the High Commissioner for India in London, Vol. 7, No. 17).

63 *loc. cit.* 40, p. 84

64 Reported in *Daily Express*, London, 1946; and quoted in *loc. cit.* 1, p. 301

65 Reported in *Daily Mail*, London, 1946; and quoted in *loc. cit.* 1, p. 301

CHAPTER II.

INTRODUCING UGANDA

Those who despised his youthfulness
Today are paying the penalty,
Those who despised his youthfulness
Today are giving gifts.

[A song composed in the Kingdom of Buganda during the reign of Mukabya Walugemba Mutesa: 1852—1884; quoted in *The Customs of the Baganda* by Sir Apolo Kagwa, Columbia University Press, New York, 1934, p. 146.]

1. Land and People

Uganda, a British Protectorate in East Africa, with an area of ninety-four thousand square miles (of which 13,689 square miles are water), forms a part of the central African tableland. In the Protectorate the greatest distance from north to south is about 400 miles and from east to west 350 miles. The nearest seaboard is just under 500 miles distant. The greater part of the Protectorate has an altitude of between 3,500 and 4,500 feet above sea-level.

The Protectorate is like an island surrounded by colonies and states either directly under the British Crown or until recently under its direct influence. The only exceptions are the western frontier of Uganda which meets the Belgian Congo and a small portion of the southern frontier meeting Ruanda-Urundi (under Belgian Trusteeship). Otherwise, there is Tanganyika in the south, held by Germany before the first world war and thenceforth by the British; the Kenya Colony in the east; and the Republic of Sudan on the north, which until very recently was under British control.

The Protectorate itself is divided into four provinces. Buganda with an area excluding open water of 17,295 square miles and a population of 1.3 millions, is the most important of all provinces. By its land frontiers it is open to the three other provinces of the Protectorate in the east, north, and west; only the southern frontier being linked with Ruanda-Urundi. The province is enriched with very fertile lands, and here is situated the main town of the Protectorate, namely Kampala. This is the leading commercial centre of the Protectorate, and is connected with the port of Mombasa in the Kenya Colony by the Kenya-Uganda Railway. Its population, according to the latest estimate, is nearing 40,000. The capital and seat of the Protectorate Government, Entebbe, with a population of 8,000, also belongs to this province. The province is divided into three districts of Mengo, Masaka and Mubende, and its main inhabitants are called since pre-British times Baganda. Their language which belongs to the Bantu stock is known as Luganda. Socially, economically, and politically the Baganda are the most developed people in the whole of the Protectorate. Although some authorities, like Lord Hailey, are of the opinion that "no one could claim that the peoples of the Protectorate as a whole have yet developed the essential characteristics of nationhood"[1], there would be little doubt after perusing this study that the Baganda have developed into a nation.

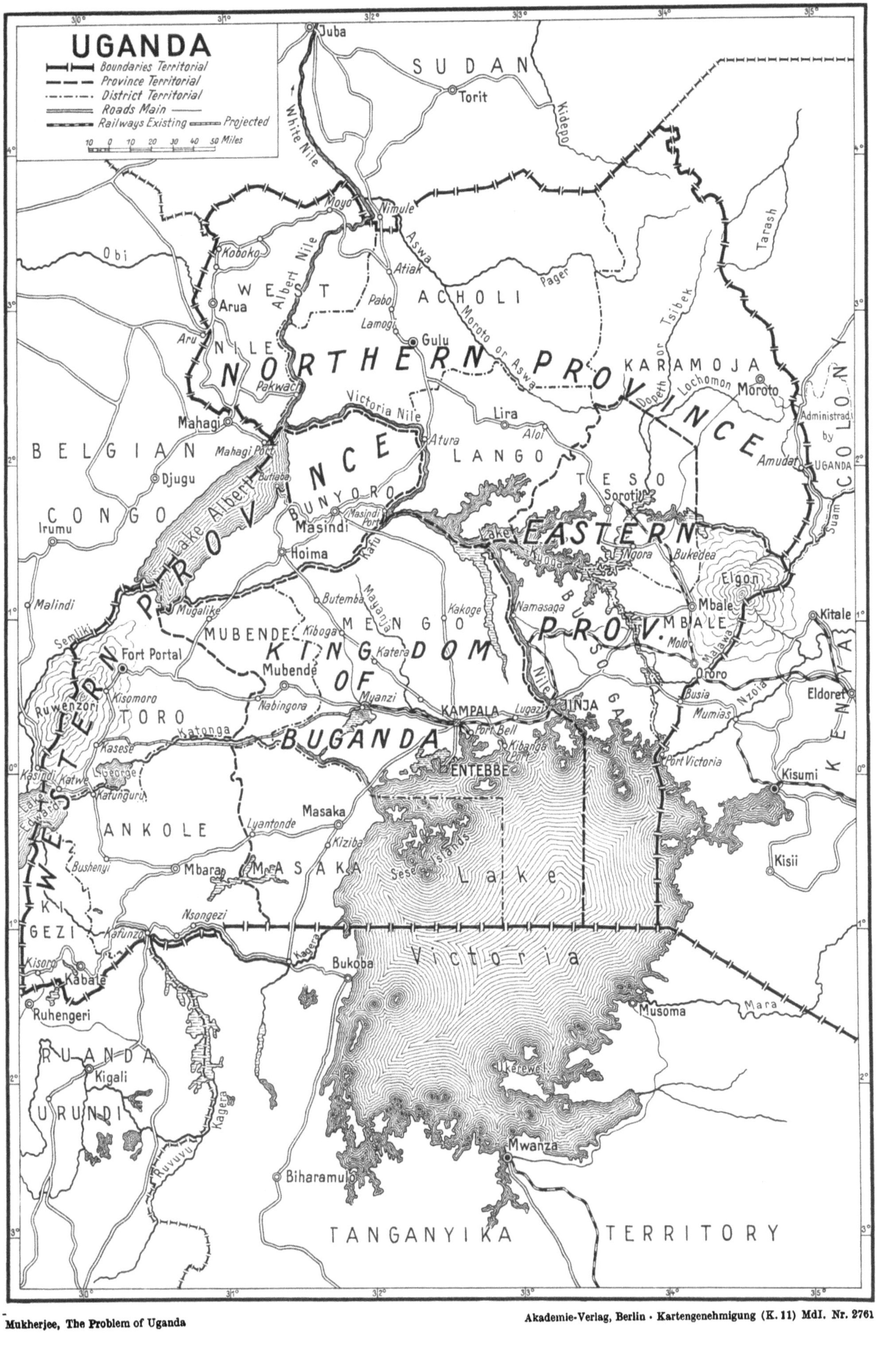

Mukherjee, The Problem of Uganda

Akademie-Verlag, Berlin · Kartengenehmigung (K. 11) MdI. Nr. 2761

Next to Buganda, the most important province in the Protectorate is known as the Eastern Province. To the east of this province is the Kenya Colony, to the north and west the Northern and the Buganda Provinces of the Protectorate, respectively; and the south opens out to Lake Victoria. This province is very rich in mineral wealth as well as in fertile lands and substantial cattle population. It covers an area of 11,698 square miles excluding open water, and has a total population of 1.5 millions. The province is divided into three districts, namely Busoga, Teso and Mbale.

The district of Busoga is chiefly inhabited by the Basoga people whose language belongs to the Bantu group. The rapidly growing town of Jinja with a population of 19,200 is situated in this district. The district itself contains a population of a little more than half a million in an area of 3,709 square miles excluding open water. Although agriculture is virtually the only source of livelihood for the Basoga (as in all other parts of the Protectorate), only about two-thirds of the land-area is available for cultivation "owing among other reasons to infestation by fly"[2].

The Teso district is inhabited mainly by the Iteso people. This district has a much lower density of population than in Busoga or Mbale, and it is only here that a large number of ploughs are used for cultivation. In other parts of the Protectorate the hoe is virtually the only means of preparing the soil, mainly due to high cattle mortality from the infestation by Tsetse fly which the British Government has not yet been able to cope with in spite of more than half a century of its rule. This district also possesses fairly large herds of cattle maintained in the savannah areas in the north of the district. The total population of the district is 405,189 within an area of 4,649 square miles excluding open water.

The Mbale district, previously known as the Central district, is the abode of many peoples, of whom the Bagishu with a population of 243,742 is the most notable today. There are also the Bakedi, Sebei, Bagwere, Banyuli, Jopadhola, Bagwe, Basamia peoples and a sprinkling of the Iteso. As in other parts of the Protectorate, agriculture is the main source of livelihood of these people. In all, the district contains a population of 603,839 within a total area of 3,340 square miles excluding open water.

The next province worthy of note is the Western Province where is situated the capital of the ancient kingdom of Kitara, which is believed to have had spread over a very large area before the advent of the British in this parts of Africa. The seat of this kingdom was in Bunyoro which is now one of the districts in the province, the other districts being the Toro, Ankole, and Kigezi. The province meets the Belgian Congo in the west, Ruanda-Urundi in the south, the Buganda Province in the east, and the Northern Province in the north. This province is also enriched with considerable mineral wealth, and at present a good deal of extraction is being undertaken at Kilembe in the district of Toro near the western border of the province. The province as a whole contains a population of about 1.2 millions in an area of 18,131 square miles excluding open water.

The district of Bunyoro with a population of 109,306 in an area of 4,847 square miles excluding open water has the lowest density of population of all the four districts. It is inhabited mainly by the Banyoro people who also used to be known as Bakitara because of their kingdom of Kitara. Their language is known as Lunyoro, and is considered to belong to the Bantu group. Next to Baganda, and perhaps on a similar level with the Basoga, the Banyoro are one of the most important peoples in Uganda today. In their district, in the county of Kibande there is a group of people the Chopi, speaking a Nilotic language of the Shilluk group. Compared to the Banyoro, however, they are in an unimportant minority.

The district of Toro has a population of 259,374 in an area of 5,143 square miles excluding open water. This district is inhabited mainly by the Batoro people whose language also belongs to the Bantu stock. According to some authorities, the Batoro are an offshoot from the Banyoro, the separation having been finally established at the time of British invasion of Uganda. In their district there are also the people known as Kabonjo, "a primitive Bantu people living on the slopes of Mount Ruwenzori", and the Baamba, "an even more primitive people living west of Ruwenzori"[3]. Both these peoples are however small in number.

The district of Ankole has a population of 401,938 in an area of 6,172 square miles excluding open water. It is inhabited mainly by the Banyankole people who had formed the kingdom of Ankole in pre-British times. These people speak a language belonging to the Bantu group, and were previously largely pastoral. The greater part of the district lying in a comparatively dry belt with extensive grass lands was particularly suited for a pastoral livelihood. Since the British occupation, however, agriculture has become the chief source of livelihood of the people. Besides the Banyankole this district contains another people known as Bakunta who are Baganda fled from Buganda in the last century but who are now much intermixed with the indigenous Banyankole.

The district of Kigezi with an area of 1,969 square miles excluding open water but with a population of 396,023 is the smallest district in the Protectorate but with a high density of population. This is partly due to the immigration of people from Ruanda-Urundi since the British occupation, which in some areas resulted in a density of population of 720 per square mile. The district has a heterogeneous population, but the majority belong to the Bakiga people who are said to have affinities with some of the "primitive tribes in the forest belt of the Belgian Congo"[4]. Besides them there are the Banyaruanda of the Bufumbiro county in the south-west corner of the district who formed originally a part of the kingdom of Ruanda and are said to have "a small element of the Hamite Batusi, but a preponderance of Bahutu, who are Bantu agriculturists"[5]. There are also a number of Bahororo of Hima origin in the Ruzumbura county. Previous to 1901 this county was included in the Ankole Confederacy. Agriculture is the chief source of livelihood of the peoples in the district, much of the cultivation being carried on on the hill slopes.

The last province in the Protectorate to be described here is known as the Northern Province. This province merges on the Republic of Sudan in the north, Kenya Colony in the east, and the Belgian Congo in the west. It has a total area of 33,168 square miles excluding open water and a population of nearly a million. This province constitutes the largest area among the provinces in the Protectorate, but its population is somewhat lower. It is divided into four districts, namely the Achoil, Lango, Karamoja, and the West Nile and Madi.

The Acholi district, inhabited predominantly by the Acholi people, has an area of 11,200 square miles excluding open water and a population of 216,133. The language of the people is also known as Acholi. Agriculture is now their main form of livelihood, though passion for hunting still plays an important part. This district came under regular administration under British rule between the years 1912 and 1914.

The Lango district is inhabited mainly by the Lango people who have cultural affinities with the Acholi but are generally regarded to be racially related to the Iteso, Karamojong and Jie rather than the Acholi. The district includes a total population of 266,657 in an area of 4,650 square miles excluding open water. There also, agriculture is now the predominant source of livelihood.

The Karamoja district has an area of 11,472 square miles excluding open water but is inhabited by only 125,620 individuals. The district came under regular civil administration under British rule in 1919, but "as it is in touch with the unsettled marches lying between Kenya, the Sudan and Ethiopia, it is still treated as a 'Closed District' under the Outlying District Ordinance of 1904"[6]. The main inhabitants of the district are known as the Karamojong who are said to be of Nilotic origin but show a strong Hamite strain, "derived from contact with the Galla of Ethiopia"[7]. Besides the Karamojong there are the Dodoth and Jie who are regarded as ethnically connected with the Karamojong but their languages are different and they do not wish to be classified as Karamojong. There are also the Tepeth (found mainly in the Moroto and Kadam Hills), the Upe, and the Labwor. Though the district is thus seen to have a heterogeneous population, their economic life is similar and is still mainly pastoral. But the progress of the economy is seriously affected by infestation by the *morsitans* fly which has a very harmful effect on the cattle population. Communications in the district are very poor, and social and cultural standards are found to be noticeably deficient compared with other areas of the province or the Protectorate as a whole.

The West Nile district is inhabited by a number of peoples who are said to be ethnically Nilotic, but they are distinguished by separate languages and habitation of different regions. The Alur, Lugbara, and the Jo-nam are probably the most notable of these peoples. Part of the district previously belonged to the Belgian Congo during the lifetime of King Leopold II, but after his death in 1909 it passed to the Sudan, and in exchange for the Nimule and Gondokoro districts it was transferred to the Uganda Protectorate in 1914. The district today has the largest population of any district of the Province, although its area is less than that of Acholi or Karamoja. Its total area is 5,846 square miles excluding open water, but has a population of 336,498. The district suffers from comparative isolation, with large areas infested by fly, and is one of the least developed districts in the Protectorate.

The Madi sub-district, a part of the West Nile district, is also known as Madi-Moyo. It is divided into East and West Madi. Because of the fly-infested uninhabited region there is little contact between the Madi-Moyo and the West Nile proper. The sub-district is inhabited predominantly by the group of people known as the Madi who are claimed to be of Sudanic origin and regarded as belonging to the same linguistic group as the Lugbara.

The Uganda Protectorate is thus seen to be inhabited by many peoples. Some of them have developed into nations, such as the Baganda, and probably also the Banyoro and Basoga, etc.; and others are in this process of development. The following table gives the numerical strength of these people according to the East African Population Census of 1948.

Table II.1

African peoples living in the Uganda Protectorate	Male	Female	Total	Percentage of grand total
(1)	(2)	(3)	(4)	(5)
Baganda	411,284	424,807	836,091	17.0
Iteso	224,692	237,972	462,664	9.4
Basoga	208,768	217,840	426,608	8.7
Banyankole	186,416	201,113	387,529	7.9
Banyaruanda	166,711	122,340	289,051	5.9
Bakiga	127,779	143,959	271,738	5.5
Lango	131,443	133,853	265,296	5.4
Bagishu	119,894	123,848	243,742	5.0
Acholi	103,237	105,924	209,161	4.2
Lugbara	89,847	93,264	183,111	3.7
Banyoro	87,977	92,633	180,610	3.7
Batoro	78,708	83,951	162,659	3.3
Karamajong	52,196	56,086	108,282	2.2
Other minor ones	469,451	421,562	891,013	18.1
Total	2,458,403	2,459,152	4,917,555	100.0

While Uganda is thus acknowledged in official literatures as a land of many "tribes" (or more strictly of many nations and nationalities, as will be explained in the following chapters), all these people can be described together as Uganda Africans. For there is the underlying economic and political unity of all these people living under the same rule, and thus their aspirations for a better future being possible only through a common undertaking.

2. Administration

How is the Protectorate run? Here one finds the ingenious policy of dual government, viz. the central administration directly under the control of the colonial government and the local administrations run by selected Uganda Africans under the guidance and control of representatives of the colonial government.

Until the second half of 1955 the central administration and the legislative organisation of the Protectorate was as follows. The central administration was run by the British Governor assisted by an Executive Council of fifteen members. In this Council there were 11 European (British), 2 African, and 2 Asian members. The non-British members were "nominated unofficial members". On the British side, the Chief Secretary, the Attorney General, the Financial Secretary, the Director of Medical Services, the Director of Agriculture, and the Resident of Buganda were *ex-officio* members; the Secretary for African Affairs, the Secretary for Social Services and Local Government, and another person (in 1953, the Hon. C. B. Spencer, C.M.G.) were official members; and then there were two European unofficial members.

Laws for the Protectorate were made by the Governor in consultation with the Uganda Legislative Council which is, however, more like a debating society than a law-making body. For the Governor retains the veto power, so that he can override any view expressed in the Legislative Council and pursue his own. In 1953, the Council consisted of 33 members, of whom 21 were British, 4 Asians (Indians), and 8 Uganda Africans. The British constituent of the Council was made up of the Governor as President, six *ex-officio*, ten nominated official members, and four unofficial members. The non-British members were all unofficial, of whom the Africans were selected as two from each of the four provinces.

Since the second half of 1955, however, some changes have been made in the administrative and legislative organisation of Uganda. For instance, the Executive Council has been replaced by a Council of Ministers and it has been declared that the Legislative Council will be *Africanised*. These very recent changes will be discussed in Chapter V, where it will also be explained that they do not envisage any fundamental departure from the previous attitude towards the administration of the Protectorate.

The function of the central administration can be best understood from the following extract from the Annual Colonial Report of the British Government on Uganda for 1953. It should however be noted again that for the future some of the official designations may be altered in the same way as the Executive Council has been made into the Council of Ministers. To quote[8]:

> "The apparatus of Government consists of the Secretariat, the Provincial Administration and a number of technical departments (Agriculture, Veterinary, Medical, Public Works and others).
>
> The Secretariat, presided over by the Chief Secretary, the head of the Protectorate Civil Service, is the machine by which the Governor, without undermining the authority vested in departmental detail, can exercise general control over the direction and conduct of public affairs. It is through the Secretariat that he transmits the principles and policies by which the Administration is to be guided and it is through the Secretariat that, from all quarters of the Protectorate, knowledge reaches him of events of every description, of changes in local conditions and of the success and failings of departments and individuals.

The Secretariat also forms a repository for records of rulings by which the administration is carried on and is thus able to lay promptly before the Governor all information bearing on any subject on which his decision is required.
The Provincial Administration consists in the provinces of the Provincial Commissioner and in the districts of the District Commissioner who works under the Provincial Commissioner's supervision and control. They are the principal executive officers of Government in their areas. Without interfering in technical detail, they are generally responsible for the efficient conduct of public affairs, and Provincial and District Teams have been set up under their chairmanship to act as informal consultative bodies for the shaping of policies and schemes of local application. The teams comprise representatives of the technical departments and, at district level, representatives of the African Local Governments.

The technical departments, for example, medical, agriculture and veterinary, usually consist of a headquarters directorate supported by a field organisation, with representatives at provincial and district level and a number of out-stations."

The principal sources of revenue for the Protectorate Government are direct taxation of the African population under the system of Poll Tax, Income Tax and other non-native taxes, and several forms of indirect taxation, such as custom and excise, export taxes on the cash crops like cotton and coffee, licences, court fees, etc.

Under the Protectorate Government are the local administrations run by the African chiefs and kings at the pleasure of the Protectorate Government. The principal sources of revenue of African Local Governments are local government taxes, fines and fees from the native courts, and local fees and licences. "In addition, subventions are received from the Protectorate Government towards the cost of primary education and the maintenance of roads; and, in all but four of the richest districts, grants towards the payment of salaries and wages."[9] The main heads of expenditure of the local governments are on the salaries of officials, the upkeep of the county and sub-county headquarters, and subsidiary roads. The main capital expenditure is on the construction of permanent buildings.

The province of Buganda as a whole is ruled by the Kabaka (the title of the traditional ruler when Buganda was an independent kingdom) with the help of three Ministers — the Katikiro (Prime Minister), the Omulamuzi (Chief Justice) and the Omuwanika (Treasurer). The rule of the traditional king and the ministers is, however, "in the manner approved by Her Majesty's Government", as it was laid down by the Uganda Agreement of 1900 during the reign of Queen Victoria in Britain*. Subject to the approval of the Governor of the Protectorate, the Kabaka and his Council, the Lukiko, have the power to make laws binding on all Baganda. The Lukiko in 1953 consisted of 89 members, of whom 40 were elected through electoral colleges at parish level, 20 by county councils from local notables, and the rest were official members. Recently some changes have been made in the position of the Kabaka as the ruler of the kingdom of Buganda and in the position of Buganda itself as a part of the Uganda Protectorate. Since his return to Buganda in October 1955 from his exile in England during 1953—55, the Kabaka has been made into a constitutional monarch and the position of the kingdom of Buganda as an inseparable unit ot the Uganda Protectorate has been further confirmed. Details of these changes will be found in Chapter V. Here it may be noted that by these changes British rule in Buganda has become even more direct than it was before.

In the other three provinces of the Protectorate, the African Local Government represents somewhat different forms. In the Western Province, the districts of Bunyoro, Ankole and Toro are ruled by kings with the title of Mukama for Bunyoro, Kabaka for Toro, and Mugabe

* Details of Buganda Native Administration under the Uganda Agreement of 1900 are given in Chapter IV.

for Ankole. They are assisted by their appointed ministers and councils, and all of them are of course subject to the approval of the Protectorate Government. The district of Kigezi in the Western Province and all districts of the Eastern and Northern Provinces are locally administered by "senior executive officers and a hierarchy of chiefs exercising executive authority, assisted by local councils at the village, parish, sub-county and county levels"[10].

In all these districts, that is, except in the province of Buganda, there are District Councils consisting of a number of chiefs, some elected members and some nominated persons of standing. Above these District Councils there are the Provincial Councils in these three provinces which include members officially nominated and members elected by the District Councils. All these councils are however only in an advisory capacity. They have no law-making powers.

The African administrations have their own treasuries and funds, but these are under the over-all supervision of the Protectorate Government. In Buganda the annual estimates are prepared by the Finance Committee of the Lukiko; in other provinces the District Councils' Standing Committees are entrusted with the task. The estimates for Buganda have to be approved directly by the Governor, those of other provinces are normally approved by the Provincial Commissioners.

A change in the system of local administration is now envisaged in order to develop Uganda "as a unitary state with a central government, and with local government bodies functioning at various levels beneath it. Responsibility for certain services is to be transferred from the Central Government to the local authorities who will be expected to increase their revenue to meet their shares of the expenditure involved. ...Arrangements will be made for thesecondment of staff from the Protectorate Government to the Buganda Government and African Local Governments to enable them to carry out efficiently the responsibilities they are assuming"[11]. These recommendations were underlined in the *Report of an Inquiry into African Local Government in the Uganda Protectorate*, which was prepared in 1953 by Mr. C.A.G.Wallis of the British Colonial Office. Implementation of this policy will eventually put an end to the dual government which, as it will be evident from chapters IV and V, is losing its importance at the present time. It will be seen while perusing this book how advisable it was to institute this policy with the introduction of British rule in Uganda, and why it is no longer necessary today.

3. Present-day Social Structure

The Africans, however, are not the only inhabitants of Uganda today. The area within the Protectorate has been attractive to foreigners in many ways. Firstly, the climate is good. The annual range of temperature is very small and over the greater part of the country the mitigating effect of the altitude prevents either excessive heat or excessive cold. In no part of Uganda does the temperature reach above 95° F. at any time of the year, while the daily variation is between 20° F. and 30° F. Rainfall varies somewhat from place to place and from year to year, and there is, on the average, rain in every month with two wetter periods and two drier periods. In inches the rainfall varies from the average of between 20 and 45 in Karamoja, the driest region in the Protectorate, to between 50 and 60 inches in Buganda and Western Uganda, with the Sesse Islands in the Lake area averaging 90 inches.

The climate is of course of much less importance than the other reason attracting foreigners to Uganda, namely that Uganda is potentially a very rich country. Being blessed over a large proportion of its area with fertile soils and a reliable rainfall, Uganda is in many ways potentially the richest agricultural producer of the East African territories. Her industrial potential is also substantial. Uganda possesses iron, copper, cobalt, tin, wolfram, gold, and there may be other precious deposits which would have been discovered if the government took seriously to geological exploration. She is also enriched with ample water resources for the generation of hydro-electric power and irrigation for agriculture. And then there are forests which can provide an immense number of hides, skins, ivory, and other articles from the beasts dwelling in them, and also various forest products. No wonder then that in 1890 Emin Pasha declared Uganda to be "the pearl among the countries all round here"; and in search for the "pearl" came "enterprising people" from some countries, and they brought along with them some other peoples. The upshot is that the present population of about four millions in Uganda is composed of three communities which are officially lebelled as African, Asian, and European.

The Africans are the autochthonous people of this territory. Their migrations and movements in ancient times need not be discussed here. Today they are generally classified on linguistic grounds, as Bantu, Nilotes and Nilo-Hamites. As stated before, notable of the Bantu-speaking peoples are the Baganda, Banyoro, Banyankole, Batoro, and Basoga; of the Nilotic-speaking peoples, the Acholi and Lango. The Lango are also sometimes classified as Nilo-Hamitic; the most important representative of the Nilo-Hamites in present-day Uganda being the Iteso, and then perhaps the Karamojong.

The bulk of the Asians are Indians, meaning previous inhabitans of the Indian subcontinent. They came to Uganda in the wake of British occupation of this territory, either as indentured labourers or as petty traders. Some of them were present even in earlier periods as assistants of Arab merchants, and the British explorer Speke, who was probably the first European to visit this region of Africa, is of the opinion that this part of Africa was known to the Indians since ancient times. To quote Speke[12]:

> "Colonel Rigby now gave me a most interesting paper, with a map attached to it, about the Nile and the Mountains of the Moon. It was written by Lieutenant Wilford, from the 'Purans' of the Ancient Hindus. As it exemplified, to a certain extent, the supposition I formerly arrived at concerning the Mountains of the Moon (Ruwenzori) being associated with the country of the Moon, I would fain draw the attention of the readers of my travels to the volume of the 'Asiatic Researches' in which it was published. It is remarkable that the Hindus have christened the source of the Nile *Amara*, which is the name of a country at the north-east corner of the Victoria N'yanza. This, I think, shows clearly, that the ancient Hindus must have had some kind of communication with both the northern and southern ends of the Victoria N'yanza."

Today, the Indian settlers in Uganda can be divided into three categories: (a) those who owe their allegiance to the Indian Union, (b) those who consider themselves as citizens of Pakistan, and (c) the remaining ones, such as the members of the Khoja community who following the advice of their spiritual leader, Aga Khan, have decided not to become subjects of either the Indian Union or Pakistan.

Along with the Indians there are a substantial number of Goans, the previous inhabitants of Goa — the Portuguese colony on the western coast of India. Here the Goans should be considered as Indians, for while living in the same subcontinent, the separate term *Goans* only reflect their servitude to a colonial power other than the British. Moreover, it is not unlikely that in the near future the Portuguese possessions in India will be merged with the Indian Union, just as has happened with the French settlements in India.

In the Asian community besides those mentioned above, there is a sprinkling of Arabs. Other Asians are hardly present at all.

The European community is formed in bulk by the British. There are some Poles who refused to be repatriated to Poland after the last war, and there is a sprinkling of Greeks, Italians, French, Germans, and other European nationals in this community. Apart from the British, the other Europeans in Uganda are however few in number. Moreover, these non-British Europeans hold very little authority. Their future in the Protectorate evidently depends upon the British colonial policy.

According to the census of population conducted in 1948 the three communities in order of their numerical importance were as follows: Africans — 4,917,555 (or 99.1 % of the total), Asians — 36,690 (or 0.8 % of the total), Europeans — 3,448 (or 0.1 % of the total), and others — 827. Since then the government policy has taken a definite direction to prohibit further immigration of Asians, but the European population has somewhat increased. By 1951, there were 6,624 "Europeans" in Uganda, a figure nearly double that of 1948.

A major part of the European population, however, does not settle down permanently in Uganda. This is particularly true for the British. Most of them come on overseas service as representatives of monopolist companies or of the British Government. After the end of their contract or service period, they therefore return to their motherland. For this reason, Uganda cannot be considered a "White Man's country" in the same sense as neighbouring Kenya Colony where the settled British landlords constitute a major economic and political force. However there cannot be any doubt that, like in Kenya, until today the British policy decides everything in Uganda. And, what is of greater importance to note for the present study, like Kenya, Uganda remains a land of Africans. The *problem of Uganda* is therefore essentially an African problem.

4. Views on Uganda

What are the views of the authorities on the life in Uganda ? For a century, Uganda, like the rest of British East Africa, has been the happy hunting ground for the white men interested in "primitive peoples". The pioneers, viz. the explorers, missionaries and administrators at the beginning of British rule in Uganda, always indulged in an ethnological survey of the country in their memoirs[13]. Some of them wrote what are still the best available ethnographic literatures on the peoples of Uganda[14]. No doubt, such studies were necessary before and during the process of subjugation, as otherwise the foreign rulers would not have known how effectively to introduce their administration*; but they also had a useful aspect. However partial and tinged with the subjective judgement of the representatives of a "superior" civilisation, the information collected by these authorities can still furnish an idea of the stage of development these people had reached in the pre-British period of Uganda's history. Also, these studies give an idea of how the Africans fared during the occupation of Uganda and consolidation of British rule in the Protectorate.

Later came the anthropologists, economists, historians, and political "travellers"; in fact, all who can broadly be labelled as social scientists today. They studied the function of some institutions among certain groups of people, or presented an economic or historical review of the happenings in these territories, or wrote travellers' tales to acquaint the world public with the present-day situation in this part of the world[15]. Many of these studies are useful because they contain valuable information. But, apart from this usefulness, almost all

* For instance, in 1689, before the English conquest of the Indian subcontinent, the Directors of the English East India Company quite categorically demanded such studies from their representatives in India and praised the Dutch East India Company for having done so already. (cf. Ramkrishna Mukherjee, *The Rise and Fall of the East India Company*, Verlag der Wissenschaften, Berlin, 1955, pp. 71—72).

these studies appear to have the same purpose in view. This is either to find out the points of discord in the colonial rule in order to implement the regime more efficiently or to justify its need to the international public, or both.

Thus, whether dealing with Uganda in particular or with East Africa in general, and whether these studies are from the "pioneers" or from the social scientists of later times, there is a continuity in the attitude of most of these writers who made Uganda known to the outside world. Either by an implicit admission or sometimes by a categorical statement, these writers have expressed their belief that the British Colonial Administration is fulfilling its duty of trusteeship for people "not yet able to stand by themselves under the strenuous conditions of the modern world"[16], and is therefore not avoiding the "White Man's Burden" to uplift the backward Africans from their primitive socio-economic organisation to the present level of world civilisation. Only an insignificant minority of these writers have so far struck a discordant note.

To mention some of the notable authorities on Uganda and East Africa, the belief of the "pioneers" in white supremacy and in their Christian duty towards the "savage" Africans is well known. The administrator, Sir Harry Johnston, noted in 1902[17]:

> "I have, of course, pointed out that we commenced these Protectorates from motives (Continental nations may laugh, but it is true) of pure philanthropy."

Whether the continental nations laughed or not, such views have become rather threadbare these days. Nevertheless, sometimes even today the same sentiment is re-echoed by some stalwarts of British rule in Africa. Thus Elspeth Huxley wrote in 1948 on the British occupation of Uganda[18]:

> "For centuries, the king and his henchmen have ruled these countries autocratically, with the help of a hierarchy of chiefs. The British came to these countries not, as a rule as conquerors but as protectors accepted by the rulers, with whom they signed treaties. . . . British authority took on itself the right to stop abuses and maintain law and order, but not to do away with the existing form of government and substitute a new one. . . . A bloodless revolution, in fact. . . . And the British have brought it about, as usual, without breathing a word of it to the outside world."

How this "bloodless revolution" was brought about in Uganda will be discussed later, and it will also be shown how these "protectors" (not "conquerors") stopped abuses and maintained law and order, professedly for the good of the Africans. But it is worthy of note here that besides proclaiming before the world the British virtue in annexing Uganda "without breathing a word of it to the outside world", Huxley added in her *Reflections on Colonial Policy*[19]:

> "Is this a place, like Palestine and India, where Britain has attempted the impossible? Can one change natures moulded over centuries by their environment in the wink of an eye? Can one race pass on to another its own experience? Can any halt be called to the disintegrating pull of racialism? In a world where the last surplus is that of good will, can suspicion be stilled or interests blended? Can you create unity out of discordance ever, save by the slow fusion of history? Even were all the answers negative, that would not damn the attempted as wasted. . . . The British effort is something the British are impelled to make by inner forces, not by entreaties from without; something undertaken, unaware no doubt, to heal a conscience roughened by memories of slavery and exploitation, not in response to any African demand."

It appears from the above that the "bloodless revolution" brought about by the British in Uganda was "not in response to any African demand". Even so, what the British were "impelled to make by inner forces", was it good for the Africans? The following study will answer this question. It is, however, worthy of note here that the back of Huxley's mind is haunted by the spectre of Palestine and India where the people had seen through even this "last surplus" of good will, knowing fully well for whose good this will is directed.

Such are the views of the direct proponents of colonial rule in Uganda.

There are, however, some social scientists today who, unlike the "pioneers" and their ideological descendants like Huxley, would admit that the British administration in the earlier period was not always so good for the natives as it might have been. But the general consensus of their opinion is also that it was not *fundamentally* bad and that the present system of Native Administration and Indirect Rule in British East Africa, particularly in Uganda, is certainly beneficial to the Africans. Only a few would disagree. Thus, Lucy Mair, an anthropologist of the Functional School of thought, concluded in her studies on the Baganda people[20]:

> "In sum, then, the history of Buganda presents a justification for the system of Indirect Rule — the preservation of native society as a basis for new developments, and its transformation only to the extent which these developments necessitate."

In the course of this study it will be shown to what extent such a conclusion is justified. Now to cite another social scientist, Malinowski, the eminent anthropologist (who was "in sympathy with the supporters of Indirect Rule" but was also convinced that "the destiny of indigenous races [of Africa] has been tragic in the process of contact with European invasion"), thus concluded his essay on *Culture Change, Its Promise and Fulfillment*[21]:

> "It is hardly necessary to affirm here explicitly that no criticism of the specifically British methods of colonial policy is implied in the statements of this book. As one acquainted in Africa with British territories only, I had naturally to deal with culture contact in that area. Colonization is never a simple and smooth process. The nation with the longest and most extensive experience in these matters has necessarily had to solve some of the most difficult problems by the method of trial and error — for its own benefit and that of others. As a Pole born and bred, I may be allowed to say here that in my opinion the British colonial system is second to none in this capacity to learn from experience, its adaptability and tolerance, and above all, in its genuine interest in the welfare of the natives."

A similar opinion was voiced by Coupland, the well-known historian. In his historical review of the situation in East Africa, he wrote in 1939[22]:

> "To decide the fate of Africa without considering the interest of the only people who had a natural right to be there, to slice it up as if it were as inanimate as the map on which the boundaries were drawn, to 'barter' millions of Africans 'about from sovereignty to sovereignty as if they were chattels and pawns in a game' — such practices, it seemed, were part of an old order which the war [1914—18] was to bring to an end; and among the post-war dreams of a new age of freedom and goodfellowship among the peoples of the world was the dream of the backward peoples being helped to make the best of their own lives in their own lands and enabled, as time went on, to take their proper place as members of international society. ... Thus the British people were invested with a trust for East Africa."

"Trustworthy" views of the same nature have also been frequently expressed in the last few years by the Colonial Office in London, and by the Government of the Uganda Protectorate[23]. Also, A. N. Poynton, the United Kingdom Representative to the U. N. O. General Assembly, said in a speech on the Non-Self-Governing Territories[24]:

"We may admit that in the past, the share of proceeds between commercial firms operating the resources and the government of the territory was not always equitable. But we are now dealing with the present . . ."

And the Governor of Uganda noted in 1946[25]:

"The established policy [in Uganda] aims at developing Uganda for the benefit, not of imported Europeans or Asians, but of its African population."

One can put forward hundreds of such statements echoeing the same tune that the interest of the Uganda Africans is in the heart of her rulers and that Uganda is a happy member of the British Commonwealth of Nations. But do the facts bear out these statements?

5. Facts about Uganda

Even to a casual observer daily life in Uganda reveals a good deal of segregation between the communities in all spheres of life, leading to the development of tension between them. For example, there is very little social intercourse between the three communities except through the missionaries and the few units of the governmental propaganda department. There is not a single African house in Kampala, the main town of the Protectorate, as the author found in 1950. Similarly, in other towns also (like Soroti, Lira or Gulu) the Africans live on the outskirts of the towns. The Asians and Europeans live within the townships, but they live in separate zones. In small towns, like Gulu, this is very marked. The Europeans occupy the best sites, the next best go to the Asians and a few African employees of the government, and then to the Indian traders. The foreigners are concentrated to a high degree in urban centres and particularly in the three principal towns of the Protectorate, viz. Kampala, Jinja and Entebbe, where practically all amenities of modern civilisation are available. According to the 1948 census, 45 per cent of the European, 37 per cent of the Asian, and 30 per cent of other non-African population, live in these three towns. As in the pre-British days, the Africans live in the "bush".

Almost every town has an European club, like the Gulu Sporting Club, which is forbidden to non-Europeans. They are admitted occasionally by special invitation. The Asians have not organised many clubs except in large townships, like Kampala or Entebbe. There are several African clubs; but, as Elspeth Huxley, who declared that "our aims are the aims of idealism" and that "we wish to change men for the Better", noted about the Acholi Association in Gulu[26]:

"In the debating club, politics are barred. At the last meeting the topic 'Where does the rainbow come from' was discussed."

Such precautions are considered necessary even for the comparatively "primitive" peoples like the Acholi, of whom she was told by "a man who had spent all his working life among Nilotic tribes" that[27]:

"We Europeans can never really get to the bottom of the Bantu mind. It twists and turns into all sorts of queer channels where we can't follow. But these Nilotic peoples — they think as we do. Our minds work in the same way."

It may not be fortuitous that the experienced administrator's inability to understand the Bantu mind has developed at a period when the Bantu people (like the Baganda) are expressing some definite political views. The "pioneers", like the administrators Lugard or Johnston or the missionary Roscoe, had no such difficulty with the Bantus. It may be of interest to note that, later, Elspeth Huxley had occasion to meet and hear from the Bantu people and their intelligentsia that the Native Administration is reactionary and bureaucratic and that the Protectorate Government does not look after the interests of the Africans. The Baganda college students at Makerere told her that they want to be lawyers, for they cannot trust the European and Indian lawyers to fight their cases.

Evidently, such social and political inequalities have contributed to the development of tension between the communities, but the economic life of Uganda appears to have given rise to much more tension than has the social or the political life. This point will be treated in detail in this study, but here it may be worthwhile to mention that it is only in the economic sphere that the three communities regularly meet with one another. Most of the Africans live by peasant-cultivation. A few earn their living as wage-labourers. They are employed in the railways, in the copper, cobalt and gold mines, and in plantations of tea, coffee, rubber and sugar. In addition, there is the seasonal cotton ginning industry which early in the year employ a labour force of about 21,000. (For the cotton industry, however, there is also a considerable number of immigrant African workers from Ruanda-Urundi, Tanganyika, and to a lesser extent from the Belgian Congo.) Besides these African workers, there are some who live on performing menial services for the European and Asian households, and are also similarly employed in the governmental departments and commercial organisations. And then there are an almost insignificant number of Africans who have taken up a professional career or a clerical job. In 1953, out of the total 4.9 millions of Africans in Ugnada today, or out of about 2.8 millions of Africans aged sixteen and upwards, only 250,000 (or 9 per cent of the adult population) were in paid employment at any one time in the year. Out of this small number of wage-earners, again, "the numbers permanently employed are very much smaller"[28]. As will be explained later, lack of industrialisation decides this peasant-existence of the Uganda Africans, just as poor educational facilities are a serious limitation even to the upper stratum of the African societies in choosing a professional career*. In this way, the Africans occupy the lowest stratum of Uganda society, — under the Europeans and Asians. Not only they perform the basic jobs of pro-

* The following data from the *Uganda Blue Book 1945* shows how little is the educational development in Uganda. The total number of college students in the only African college in Uganda, where students come also from other parts of East Africa, was 45 in 1923 and 172 in 1945. Out of the total estimated African population of 3,901,017 in 1945, the number of African teachers was 9,459, and students 268,239. The following table gives the details.

Table II.2

Institutions	Teachers	Students
(1)	(2)	(3)
Secondary School	130	4,914
Primary Teachers' Training Centres	21	254
Vernacular Teachers' Training Centres	60	1,057
Full Primary Schools	1,307	42,593
Primary Vernacular Schools (Aided)	1,262	35,636
Selected Sub-grade Schools (Aided)	1,248	38,066
Sub-grade Schools (Unaided)	5,371	144,561
Vocational Schools	60	1,158
Total	9,459	268,239
Number of school-age children not attending schools		511,964
Total number of school-age children		780,203

Later reports, such as *The Economic Survey of the Colonial Territories-1951: Volume II, The East African Territories* (Her Majesty's Stationery Office, London, 1954), do not indicate that the situation has appreciably changed since 1945.

duction (there being no Asian or European peasant or wage -labourer in the Protectorate), but even those who are employed as clerks or professionals in the educational or medical departments of the government, for instance, are generally put into lower grades. In government departments the average salary of the Africans in 1948 was £ 29 for the whole year[29].

The Asians, barring a few very wealthy Indians owning sugar factories, cotton ginneries or some form of wholesale trade, are in control of the petty retail trade (mainly grocery stores). They also work as subordinate officers and clerks in government departments and commercial establishments. With them the Africans therefore come in contact in almost all forms of daily transactions, and thereby the Asians' position is intermediate between the Europeans and Africans. In economic gain also the bulk of them is in this "middle" position. In government departments the Asians' annual salary in 1948 was £ 220 on the average, which was about eight times higher than the average salary of the Africans but one-third that of the Europeans[30].

The Europeans, of whom more than 80 per cent were British according to the 1931 census, are the high-salaried officials who run the government departments and all the important banks and commercial organisations such as the National Bank of India, the Standard Bank of South Africa, Barclays Bank, the Uganda Company, Smith Mackenzie and Company, etc. In government departments their average annual salary in 1948 was £ 660, which was 23 times that of Africans[31]. Some of the Europeans are also engaged in export or wholesale trade and a few in missionary activities. They control the entire political, and the major part of the economic power in Uganda.

Although taking all activities in the Protectorate into consideration this is an understatement, for such segregation is even more marked in the commercial and other enterprises, it may be of interest to examine how the three communities are economically segregated in the government services. The following table has therefore been prepared from the data available in the 1948 *Budget Estimate* of the Uganda Protectorate Government, which was passed by the Legislative Council on the 18th of December, 1947[32].

Table II.3

Community	Percentage of total government employees	Percentage of total personal emoluments	Percentages of the personal emoluments in each community to the provisions in the budget as:		
			Recurrent expenditure	Total expenditure	Ordinary revenue
(1)	(2)	(3)	(4)	(5)	(6)
European	5	47	15	11	12
Asian	5	16	6	4	5
African	90	37	12	9	9
Total	100	100	33	24	26

In the 1948 estimates the Revenue was derived as (a) Ordinary, (b) Colonial Development and Welfare Grants, and (c) Land Sales. Since the Ordinary Revenue, obtained from the people, is directly concerned with personal emoluments of the employees, this has been used as a basis for the percentages in the above table. Similarly the provisions under the total budgetary expenditure as "Special", "Extraordinary" and "Colonial Development and Welfare Schemes" are not so directly concerned with "Personal Emoluments" as the "Recurrent Expenditure"; the latter has therefore been used as a basis to indicate the proportion of government expenditure on salaries[33]. The table shows that out of the total government expenses on salary, a few Europeans (British) take the largest share, about the same proportion of Asians (Indians) receive much less, and the large number of Africans very little. Moreover, several allowances are due to the European officers which go under the head of "Other Charges" of each department, and so have not been considered in the above table. Otherwise, it would have shown how even in the government departments the economic differentiation of the three communities is sharper than it appears from the above calculations. Furthermore, it should be borne in mind that in the commercial establishments and other enterprises of the Protectorate the salary-scale of the Europeans is much higher than in government departments; that of the Asians, if not so high as the Europeans', is still higher than in the government positions; while for the Africans the salary-scale is about the same.

Also it should be borne in mind that very few Africans can have even this income as an office employee of the government or commercial organisations. Those who earn their living as wage-labourers have a remuneration much lower than that of a clerical employee. And the vast mass of them live as peasants on the cotton economy of the Protectorate, from which they derive, as will be shown in due course, an income less than that of a wage-labourer.

Very briefly, such is the economic differentiation between the three communities in Uganda today.

6. Tension between Communities

Obviously, therefore, the Africans are not satisfied with their present position. As will be mentioned in the course of this study, their discontent was manifest from the very beginning of British rule in Uganda; but in the early years it was sporadic and spontaneous. There were local outbursts against the colonial rule and in favour of political activities for furthering the cause of the separate peoples (officially described as "tribes"). After 1921 a united all-Uganda attitude against British policy was apparent when a proposal was taking shape to amalgamate the Uganda Protectorate with the other states in British East Africa for the eventual emergence of the East Africa Federation, an institution to turn Uganda into a White Man's country where the "white" minority would perpetually dominate over the "black" majority like it does in British Central Africa today. This all-Uganda resistance to British policy became stronger in the course of years.

The Uganda Africans also began to express clearly their attitude to British policy in the internal affairs of the Protectorate. For instance, it was the traditional custom of the Baganda that after the death of the Kabaka (the traditional ruler) his chief queen remains a widow and looks after his remains. But, contrary to the sentiments of the people, and even though British policy was said to acknowledge the customs of the local people, the Namasole or the Queen Mother of the present Kabaka was remarried. Referring to this incident the African Labour Party in Uganda wrote in a memorandum to the Commission of Inquiry on the 1945 "Disturbances" in Uganda that in 1941 the Bishop of Uganda told the leader of the party[34]:

> "It was H. E. the Governor who issued the marriage licence to Namasole and commanded me to join her in matrimony. . . . . Marriage is not the affair of God, but of Governments. The Governor was able to give the order because marriage is a political affair, and a concern of the government."

Whether marriage is an affair of god or is a social institution is not the subject matter of discussion. It need not also be discussed whether it is good to continue with the traditional custom that the Queen Mother of the reigning Kabaka must remain single for the rest of her life. What is, however, of importance to note here is that the charge that such a marriage was dictated by political moves suited to the needs of a foreign government is certainly grave and indicates the strong criticial attitude the Uganda Africans had developed in the course of years against the foreign rulers.

Besides such a critical attitude of the Uganda Africans against particular measures taken by the colonial rulers a united desire for the freedom of the whole of the Protectorate also began to be emphasised in the post-war years. Probably for the first time this was clearly stated in 1944 when in a publication entitled *Buganda Nyafe* (Buganda our Mother) the tension between the communities, especially between the Africans and the British, was openly stated. This book, which the authorities noted to have had a "very wide circulation" and was therefore promptly suppressed[35], stated unter the heading: *Great Britain is still fond of Slaves*:

> "But when Britain suppressed the existing slave trade it seems that she devised instead the plan of buying slaves but bought them instead in their native country....
>
> "The slave trade was abolished in one way and then it was introduced in another but the old slave trade was better because only a few were taken away and those who remained home lived in peace, but the modern slave trade includes children and women and English laws put yokes round our necks even though we think we are in our native home. Here in Buganda we have not yet suffered much on account of this slavery but the time is near when we shall be deprived of our land and settle on European estates, when we shall suffer bitterly from slavery just like our comrades in Kenya, South Africa and elsewhere. Our country will not fail to be in trouble and such trouble will be brought about by our own countrymen who have got tender hearts and who act on bribes and some of them are well known but beware of them when they rise to high posts such as the Katikiroship [the Prime Minister of the Buganda Native Government]."

The authenticity of the above statement, which refers to further alienation of land to serve British colonial interest at the expense of that of the Baganda Africans will be proved later. It may however be stated here that such alienation of land from the African people to the anti-people interests within the society and to the foreigners started with the annexation of Buganda in the Uganda Protectorate under the Uganda Agreement of 1900. Details of this process will be found in Chapter IV, where it will also be described how in this process the foreign rulers have been well-assisted by the previous feudatory ruling class which emerged under new circumstances as the native bureaucracy in African society.

To continue with the growth of anti-colonial feelings in Uganda, although the above-mentioned book refers mainly to Buganda, the sentiments expressed therein were much more widespread as the succeeding incidents indicated. Nor did this feeling remain limited to a few declarations like the above. In January 1945, serious "disturbances" broke out. The working class of Uganda although not large in number, played a significant role in the movement. There were widespread strikes for increased wages throughout the Protectorate, culminating, on January 15, in a general strike in Kampala which embraced even the African police. So big was this movement that army and police had to be brought in from Kenya to suppress it. The colonial government fired on and killed many workers, arrested hundreds of Uganda Africans, and deported their leaders. Nevertheless, they were forced to grant certain wage increases. It is of further interest to note that although the Commissioner of Inquiry into the 1945 "Disturbances" gave his opinion that only a few African politicians had engineered the trouble for their personal benefit, he noted in his report[36]:

> "There also existed dissatisfaction among the 'Bakopi' (the peasant class) because they had no or no proper representation in the Lukiko (the Buganda Native Parliament) and so had no say in the Government. For the same reason what may be called the 'intelli-

gentsia' also had a grievance. They, the educated businesmen, teachers, African medical officers, clerk class, etc., have no representation whatever and some of them who gave evidence spoke bitterly on this. There is, due to the spread of education, an ever growing clerk class. Some of them cannot find employment and some of them feel that their education has made them fit for any job however responsible and difficult and are unable to understand why the higher posts are not available to them."

Thus, the Commissioner himself admitted that dissatisfaction was widespread in all classes of the people, and therefore it could not have been engineered just by a handful of politicians for their personal gain. Even so, the Colonial Government appear not to have taken any positive measure to remedy the cause for such discontent. On the contrary, the plan for the alienation of African land went on in full swing. To remove any opposition to the scheme (which existed even among the African officials in the kingdom of Buganda), Wamala, the Chairman of the Lukiko (Buganda Parliament) and the Prime Minister of the kingdom of Buganda, was declared insane by three British doctors, for he had violently objected to this scheme in the Lukiko. The fourth doctor, Mukibi, who was also deputed by the British Government to examine Wamala in order to find the reason for such unwarranted behaviour of the Prime Minister, was a Muganda* African. He found that Wamala was in complete possession of his senses. The upshot was that while Wamala was forced to resign from his post, Mukibi was ordered to be transferred to an unfavourable spot in the Protectorate, and, if he objected to this, to be deported.

The Prime Minister of Buganda, appointed in place of Wamala, was assassinated on the 5th of September 1945. Immediately, and without any trial, the Colonial Government ordered the deportation of Wamala, Prince Suna (whom the Africans described as the "Prince of the Common People"), and fifteen other African chiefs and leading men of Buganda. Wamala and Suna died in deportation. It has been alleged that they were not properly cared for in the place they were forced to live.

Meanwhile, the newly-appointed Prime Minister passed the Land Act which had now an additional clause that any landowner who would violate the action of the government to acquire land would be penalised by a big fine and imprisonment for six months.

The relation between the Africans and the British worsened continually as the appendix to this book, written jointly by some Baganda students of the Makarere College, will explain. And this discontent did not remain confined only to Buganda. Also in other parts of the Protectorate the people learnt from the happenings in Buganda and began to express their united opposition to such a rule of repression and terror.

Simultaneously, with the rise of nascent bourgeoisie within the African community (see Chapter V), some of them began to protest that in the present state of affairs they were being deprived of economic opportunities. In this protest their anger was directed first to the Asians. For the Asians supply them with goods as retail traders and are also often their immediate boss in the government and commercial organisations. Also, to them the Africans bring their cotton crop for ginning. The Asians thus forming the middle layer of the society in Uganda between the Europeans and Africans became the immediate target of the latter's anti-colonial move. Also, as will be explained later, in view of the growing discontent of the Africans with their present state of existence, the ruling power so engineers its policy as to present the Asians as the scapegoat to the elemental wrath of the Africans. Probably for these reasons, the League of Uganda Citizens declared in 1948[37]:

"A similar game of enmity (like the deprivation of the power of the Buganda Parliament by the Colonial Papers No. 191 and 210) is followed against economic opportunities of the

* Muganda is the singular for Baganda.

natives. All trade and commerce are in the hands of the foreign people, the Agents of the British government. This symbolises how the natives (are) supposed to be inferior."

The scientists and administrators, however, did not consider that a fundamental change in the situation was necessary. Lucy Mair, an anthropologist of repute, had stated[38]:

"African societies have so far produced few, if any, sociologists. Consequently, the demands made by Africans, however intelligent, cannot be taken as decisive in determining the lines along which African development should go."

The administrators also did not consider the political tension in Uganda to be very serious. Whitley, the Commissioner of Inquiry into the 1945 "Distrubances", noted[39]:

"Buganda seems to be the only part of the Protectorate which is badly infected with the political virus. ... I have had before me many earnest patriotic Baganda honestly anxious for their country's good. All that seems to be required is that those who have been using their undoubted abilities to hinder Government should now use them to assist, and put aside their private political ambitions to work together for the common weal."

These "patriotic Baganda" are mainly the native officials and their associates whose social basis will be discussed later. Whitley now suggested that, as before, the Government should depend on them in order to maintain the colonial rule. The Governor of Uganda, Sir John Hall, also shared the same view. He noted in 1946[40]:

"The ignorance of the Uganda African, and the suspicion which is largely bred by ignorance, represent faults which, as civilian and army educationists have proved, are definitely curable, provided that education is not allowed to bring in its train, as it has done elsewhere in Africa and Asia, the political distempers and social malaise which in turn breed even more harmful suspicions and discontent."

Political discussions and movements were therefore expressly banned all over the Protectorate and in 1947 Elspeth Huxley even more frankly despised the slogan of *Africa for Africans*. She noted[41]:

"We run a race with time: on the one hand our good intentions, our needs and our resolve to re-make and enlighten; on the other, the natural and gathering impatience of the half-educated, fed on the vapour of our own philosophy, to be done with an alien ruler. ... It is perhaps a weakness that our aims are the aims of idealism."

But while the foreign rulers and their allies went on showering platitudes, and harping on the ignorance of the "primitive" Africans and their own laudable "aims of idealism", the Africans in Uganda did not give up their aspirations for a better life. Humbly they approached the colonial authority, but they were determined to attain their object. On the 24th January, 1949, the President of the African Labour Party sent a memorandum to the Chief Justice on "The Reconstruction of the Present Social Order". In this memorandum he wrote[42]:

"Your rank is high and respectable: you are trusted and honoured by all people and the government. But a heavy burden is on your shoulder. You are responsible for the peace of the country and welfare of the people. You are the father of Justice, which means 'father of every body and every thing'. But I have fear in my mind in regard to the previous acts of the government in Uganda. The Uganda Government is self-assertive, ignores the British proposals [for the Colonial Development and Welfare Act of 1940]. They seem to abuse their position in the Colonies and could produce troubles by creating enmity, ill-feelings. Therefore I came unto you on my knees searching for a remedy, and to ask you to interfere in this matter as a matter of importance to preserve the peace of the country."

Complaining of "Colour Discrimination", he wrote further:

> "Any destiny for a native (is) being closed up, sat upon, sandbagged and tinned like a dried fish by the Government, the British Government, our nurse and trustees."

To some this may appear as a puerile outburst; but the deep sense of oppression which had stirred up this emotion should be taken into account while reading the above lines.

Indeed, this upsurge did not remain only on paper. Again serious "disturbances" broke out in Uganda in April 1949, just four years after the first, proving that animosity of the Africans towards colonial rule was growing all the time. This upsurge was much more widespread, better organised and stronger in character than the previous one.

The Colonial Government retaliated with a resolute suppression of the movement. Again a good number of patriots were killed, and hundreds were arrested. The Federation of the Bataka, which was developing in Buganda into a national organisation, had already been banned in May 1948. Now the Uganda Farmers' Union, a mass organisation of the African people established to look after their economic needs, was also closed down. Yet the fire in the hearts of the people went on burning. The Colonial Government therefore took further steps. In a memorandum on the 1949 "Disturbances", sent by the African Labour Party to the Inquiry Commission, it was imputed that the Public Relations and Social Welfare Department of the Uganda Government was full of spies and that this organisation was meant not to *serve* the people but to "sense" public feeling. How true was this allegation of sensing public opinion by spies has been confirmed by the admission of one of the leading architects of Uganda's colonial administration in the post-war years. This will be seen in a following page. For the situation descussed here, it is not an exaggeration to state that the Protectorate was now turned into a prison house for the patriots and man-hunt went on for the progressives. Moreover, in 1950 a Bill was introduced in the Protectorate declaring that any person organising a strike in an "essential service" could be sent to prison for a year and fined £ 250, and that anyone supporting the strike could be imprisoned for six months or fined up to £ 50. It will be realised from a previous discussion that the stipulated amounts of fines were nearly ten and two times, respectively, the average annual salary of an African employee in the government offices, and is thus much more with respect to the wages of a worker who in the government employment (i.e. receiving comparatively high wage-rates) earned in that year about £ 2 a month. Still, in the same year, police, armed with rifles and clubs, had to be called out against some 1,000 African strikers.

The people's movement went on developing. In 1953 it flared up again when the question of East Africa Federation was presented by the Secretary of State in a speech at the East African Dinner Club. The three-fold demand which the peoples of Uganda then presented to the government has been noted earlier. Their movement had by this time become so strong that even the present king of Buganda and his ministers openly began to give at least a partial expression to the people-s demand in regard to the date of independence of Buganda. Indeed, the situation was very critical for the foreign rulers. It came to such a pass that the king refused to co-operate wiht the Colonial Government without consultation with the Lukiko — the Buganda Native Parliament.

The upshot was that the king was deposed and he was deported to England without any pretence at legal formalities. Regarding this deposition and deportation, Mr. Lyttelton, the British Colonial Secretary, argued that the Kabaka (the king of Buganda) had failed to "co-operate loyally" under the terms of the Uganda Agreement of 1900. But Lord Munster, the Colonial Under-Secretary, was forced to admit in the British Parliament that this Agreement gave no right to depose the Kabaka, much less to deport him, and stated that the action was taken under the Emergency Powers Act of 1939. Evidently, the fact that Mr.

Lyttelton made no reference to this Act in his report to the House on 30 November, 1953, the day a state of emergency was declared in Uganda and the Kabaka was sacked and deported, serves to demonstrate that the legality of the action was of no significance to the British Government.

The movement of the people, of course, could not be checked by the deportation of the king. In spite of several attempts the Colonial Government failed to seat a loyal "king of Buganda" in place of the banished one. Moreover, martial law which was imposed immediately upon deposition of the Kabaka and which was lifted after several months had to be reimposed again *sine die*. And, Her Majesty, the Queen of Great Britain, during her sojourn in Uganda in the first part of 1954, had to change her venue of reception from Kampala, the headquarters of the Buganda Native Government, to Entebbe, the headquarters of the Protectorate Government, where very few Africans other than the government employees live.

Such then are some indications of the growing discontent of the Africans towards colonial rule, which is permeating even into the African official circle. As will be seen from Chapter V, this discontent is likely to develop into such a powerful united struggle in the near future that it would be most difficult for Britain to maintain her position also in this part of her colonial empire besides the "troubles" she is facing in the neighbouring country, the Kenya Colony, as well as in the Malay Peninsula in Asia, in Cyprus in Europe, in Guiana in America, etc.

7. Why This Tension?

How can this growth of discontent among the Africans be accounted for? In this respect, while the great majority of British social scientists and administrators simply eulogise the beneficient role of the British in Africa and deplore the ungrateful attitude of the Africans, there are a few scientists and administrators who, at least on some points, have lately taken a rather critical view in regard to the consequences of imposing the so-called "western civilisation" on the African peoples. For instance, Malinowski noted[43]:

> "In Africa we have had of late a number of studies on the broad movement which is taking place as from tribal conditions towards the partial westernization of the Natives. The anthropologist is becoming increasingly aware that the study of culture change must become one of his main tasks in field work and theory. ... As a humanist, he ought to be aware that in this process [observations on culture change] there are involved human interests and passions that are still largely under the full control of agents of the active Western Civilisation. This control has not always been scientifically enlightened by a knowledge of all the facts at issue. Even now we must ask: are the changes in Native societies such as to bring about a common existence of harmonious cooperation; or must they lead to temporarily suppressed but powerful forces of coming disruption, upheaval, and historical catastrophe on an unprecedented scale?"

But such cases of "maladjustments" have been regarded either as "mistakes" in the mode of approach of the White Men or as an "error" in timing the course of change in African life. To continue with Malinowski[44]:

> "There is a moral obligation to every calling, even to that of a scientific specialist. The duty of the anthropologist is to be a fair and true interpreter of the Native. ... He ought to be able to make clear to traders, missionaries, and the exploiters what the Natives really need and where they suffer most under the pressure of European interference. There is no doubt that the destiny of indigenous races has been tragic in the process of contact with European invasions. ... The Native still needs help. The anthropologist who is

unable to perceive this, unable to register the tragic errors committed at times with the best intentions, at times under the stress of dire necessity, remains an antiquarian covered with academic dust and in a fool's paradise."

"The whole functional approach is based on the principle of the plasticity of human nature and of the possibilities of cultural development. It contains also a warning: the tenacity of custom and the difficulty of transforming the African family into a Christian one, or the indigenous system of administration into something approaching the Western ideal; are explained by the complexity and dovetailing of all human activities. The general rule that slow, gradual, and well-planned transformation can achieve what haphazard, fortuitous muddling through would only confuse is thus a moral to be drawn from the functional analysis of culture."

E. D. Tongue, an experienced administrator, was also of the same opinion. In "an exposition of purely personal impressions and conclusions, formed during twenty years of close and intimate contact with a variety of tribes in Uganda", he noted in 1935[45]:

> "At present we are witnessing, and some of us are directly assisting in, the imposition of a twentieth-century civilization of a highly individualistic type upon an eighth-ninth century communal organization of society. What are the reactions on the native mind? He is possessed of great potentialities — I am speaking of my own impressions only —; shrewd and full of practical common-sense, he has not yet reached the stage when abstract ideals can make much impression on him. He is at present like a child flung into the society of adults, and expected to react as an adult. He is being given all the privileges of individual liberty without any preparation to enable him to realize the personal responsibilities that accompany it, with the unfortunate result that liberty to him too often means licence. . . . Only by a thorough and a sympathetic understanding of African sociology and of native psychology can we hope to succeed in our endeavours. Western civilization has taken centuries to develop, and the future civilization of Africa cannot be evolved in a single generation."

Although such a critical attitude takes account of the fact that everything is not as satisfactory as could be desired, it fails to explain why the situation is getting "worse" day by day. There can be no doubt that individually some belonging to the ruling class in Uganda are genuinely interested in the welfare of the Africans who at the time of their contact with Europeans were undoubtedly at a lower stage of social development. But, irrespective of the fact whether the Africans by themselves could not have traversed the stages of development travelled earlier by the European societies, these scientists fail to realise that whatever may be the intentions of some members of the ruling class it is the *central aim of the colonial system*, and the policy emerging in consequence, which in the ultimate analysis decide the fate of the Africans and thus give rise to their discontent.

Changes were of course necessary in the Uganda societies after the British appeared. So, as will be explained in Chapter IV, some "improvements" were brought about in Ugunda as by-products while consistently serving the British interest only. But such changes could not but have a tremendous effect on the people "from the splintering and collapsing fabric of tribal and feudal society"[46].

For example, the foreign rulers wanted to create a "native" intelligentsia. Lugard, a British "pioneer" in Uganda, noted in 1893[47]:

> "Not only, I think, that may we hope in the present that subordinate officials for the administration of Uganda may be supplied by the country itself, but in the future we may even draw from thence educated and reliable men, to assist in the government of the neighbouring countries of East Africa generally, replacing the West Coast Africans, Parsis and Goanese, which I have elsewhere advocated."

But is it not natural that the African intelligentsia, for whichever reason it was created, would desire a progressive and prosperous Uganda run by the Uganda Africans themselves? Can they be disuaded from this goal by being told that they are not fit for self government, when they with their developed outlook see for themselves how things are working in Uganda and when they can also read outside their office hours how the people in other countries are running their motherland, of whom also it was said that they were not fit for the task?

Likewise, with the advent of British rule in Uganda, a monetary economy had to be developed in order to sell British goods in Uganda. And that could not but make a fundamental change in the economic relations of a country living in the past mainly on a subsistence economy. Also, it had to affect the political life of the vast mass of the people living as tillers of the soil. For in the present-day economy of Uganda the peasants have to fight for a fair price for their crops, against high taxes and high prices of the commodities they must buy, —and for a progressive economic, social, and political life.

Then, if in British Uganda the natural wealth of the country had to be extracted under the auspices of foreign capital, it could not but produce skilled labourers in the native societies in order to mine the ores, build the railways, run the processing industries, and so on. But, as Kartun has very aptly remarked[48]:

> "... a man who can read the servicing instructions for a pneumatic drill can also read newspapers. He can read those things in the Bible that tell him he is as good as any man of any colour. He may, perhaps, read the Communist Manifesto.
>
> A man who can write his name on a work contract can also communicate with his fellow workers. He can form a trade union, run a political party. And so a leadership emerges for the young national movement in Africa."

Hence, again without going into the question as to whether the Africans themselves could not have improved their life, the question that appears foremost is to what direction, in which forms, and how changes were introduced in the societies, and whose interests they served. Did the foreign rulers fulfil their responsibility in this or in other colonies? This is the basic question that has to be answered. But instead of getting to the bottom of the problem, the authorities quoted before appear to have sought the answer for the root cause of the "troubles" in the socio-psychological superstructure of society, which however important cannot over-ride the basic economic relations of production, as the present study will indicate.

It will be shown in the following pages that just as "fortuitous muddling through [in economic and other spheres] would only confuse" the issue, so an attempt to accomplish a "slow, gradual and well-planned transformation" of "the African family into a Christian one, or the indigenous system of administration into something approaching the Western idea" is bound to end in failure as long as the present-day relations exist between the Africans and the British in the basic economic organisation of society, and in the political organisation called upon to maintain such economic relations. This is the point the above authorities appear to have overlooked.

However, in spite of the fact that he based his conclusions on an abstract "value-judgement" of *western* civilisation, probably Tongue was getting nearer to the truth than many other scientists. He noted in his exposition referred to above[49]:

> "Closer contact with Europeans of all types, particularly during the hostilities caused by the extension of the Great War [1914—18] to African territories, has resulted in a definite loss to our prestige. While more than ever appreciating our superiority in material civilization, our command over the resources of nature, the native is under no delusion

as to our short-comings, and no longer regards us as superior moral beings. This loss of respect has encouraged him to believe that he can, and will attain an equality with us in every respect at a not too distant date. He is no longer prepared to take us on trust; this has to be borne in mind when dealing with him and his interests."

There were, of course, more basic reasons for the Uganda Africans losing trust in their British rulers, as it will be explained in due course; but without going into such a discussion here it may be of interest to note that if such was the effect of the contact during the First World War, one can imagine how it has been intensified during and after the Second World War, when the Africans went beyond their home territory (as far as Europe, India and Burma) and on return very few of them could be absorbed in positions suited to their training and experience during the war. The 1947 Colonial Annual Report on Uganda noted in this connection[50]:

"Recent figures received from the military authorities show that some 79,000 men were enlisted from Uganda in all branches of the Forces ... (of which) 1,950 men, including 11 Europeans and four Asians, lost their lives during the period of hostilities. ... The Central and District Employment Bureaux have continued to find employment for ex-soldiers, and although records show approximately 3,000 men (less than 4 per cent of the total number enlisted —RKM) to have obtained employment by this means, it is known that a large number have found employment with commercial undertakings by their own efforts. The majority of ex-soldiers returned to the agricultural pursuits in which they were engaged before joining the Forces. ... It was, of course, realised that the ex-soldier was bound to find it difficult to slip back easily into the village existence he had found adequate to his needs before the war, but it was hoped that he would be enthusiastically co-operative (!) and participate in plans for the general improvement of African village communities."

More will be said later on this abject economic servitude of the Uganda Africans. It is, however, worthy of note here that whatever may be its relative value, even such mild politico-economic explanation of the situation, as given by Tongue, was not acceptable to many, although, as the foregoing extracts indicate, their relevance to the occasion could not be completely ignored by Whitley, Hall (the Governor), and other administrators and advisers. Therefore, Whitley, while reporting on the 1945 "Disturbances", recommended setting up a Special Branch of the Intelligence Bureau of the Criminal Investigation Department. He suggested[51]:

"One of the principal duties of [the Special Branch] would be to keep the authorities thoroughly posted as to what was going on amongst the people, sense public opinion, and try to obtain the earliest and fullest possible information of any subversive activities."

The Commissioner of the Inquiry, Whitley, complained that, "like the Assistant Director, Security Intelligence", he had also "encountered a wall of silence" from the Africans he met. He therefore asked for close vigilence on the native people. Moreover, as a positive measure to stop recurrence of such "troubles", he suggested "encouraging the formation of social clubs especially in the out-districts ... [in which] politics should be rigidly banned"[52]. But the attitude of distrust of the Africans towards the British rulers had become so much stronger in the next few years that even an organised department with profound experience in running things in other colonies (such as in India) failed to sense the situation in 1949. The prophylactics of social clubs also failed miserably. The 1949 "Disturbances" were of much greater intensity than the former and also involved many more people. And then, in spite of further repression and special laws and ordinances, came the climax in 1953 when even the king and the ministers of the kingdom of Buganda expressed their support, however partially, for the demands of the Uganda Africans.

Evidently, the scientists and administrators have not been able to assess the situation correctly. To what extent their complacency was due to wishful thinking, or their lack of understanding, may be of interest to examine. But it cannot be doubted that the situation is indeed much more serious than can be explained by "mistakes" and "errors". "Mistakes" may occur accidentally without any serious consequence in the long run provided they are rectified immediately when they are found out; and "errors" can also be eradicated by paying local attention. But when an inherent weakness in a system formulates itself into an idea, and that idea grips the masses, it becomes an irresistible force. In short, reality has forced the foreigners — intellectuals and administrators — to detect the disease of discontent among the Uganda Africans. But the diagnosis for a cure has yet to be made. For they have failed both in judging the intensity and the character of the tension between the communities, so that the measures they used to stop it were futile.

These scientists and administrators have not learnt from their own sad history. For instance, in British India similar precautionary measures were taken by setting up a Special Branch of the Intelligence Bureau of the Criminal Investigation Department and terrible repression was let lose whenever the Indians raised their head against colonial rule. But that could not stop the development of the Indian national movement. In due course it was victorious. And now a similar scene is being enacted in Uganda, in the Kenya Colony, and in all such places in the world where a foreign ruling class sits on top of society and perseveres to keep the local people down by showing its iron heel.

It is therefore the duty of a social scientist to explain the situation in Uganda by taking into account the forces working in society, so that due measures may be taken for establishing a harmonious life. This is the task the author has undertaken in this study, although it is a vast problem and he may not be able to do full justice to all the points involved in the discussion. Furthermore, lack of up-to-date information may vitiate the analysis in some respects. This is particularly true for every part of Africa today, for in this continent life is advancing so rapidly that what is topical today is history tomorrow. Yet, in spite of all such difficulties and the inherent limitations of the author in treating such a big problem, it is no doubt a task worth undertaking. And he hopes that further comments on, and criticisms of this study will enable the world public to understand the *problem of Uganda* more clearly. Only a beginning is thus made here.

To conclude then this discussion on why there is ever-growing tension between the communities in the Uganda Protectorate, the author is of the opinion that the incomplete understanding of the present-day life in Uganda is the result of a superficial analysis of the effects of the contact between the three communities (mainly, of course, of the Europeans or British and the Africans), and of the consequent evolution of the social structure. A society evolves by the resolution of the contradictory forces working within it. A study of the present situation in Uganda must not therefore be limited to describing the apparent improvements in the life of Africans, because of the introduction of a cash economy through *money crops* like cotton, and of their "happy life" as a result of an efficient administration which has eradicated slavery, brought "peace" in the country, and is now ostensibly looking after the needs of the "people". A scientific analysis should go deeper into the problem, and explain what are the contending forces working in society and who *really* gain by the resolution of these forces which superficially appear to increase the wealth of the African people. Obviously such a study needs both a historical analysis of the situation and an analysis of the present socio-economic structure of Uganda which has evolved out of the historical development. The historical survey should indicate what was the situation before the contact between black and white, and how the contact has resulted in the present state of affairs. The socio-economic analysis of present-day Uganda will show the interactions between the units of the socio-economic structure of the changing society, and will indicate where these interactions are likely to lead the people in their economic, social and political life.

The following is a brief account of an analysis on these lines.

REFERENCES

1 Hailey, Lord — "Native Administration in the British African Territories", Part I, His Majesty's Stationery Office, 1950, London, p. 80
2 *ibid.*, p. 25
3 *ibid.*, p. 45
4 *ibid.*, p. 45
5 *ibid.*, p. 45
6 *ibid.*, p. 60
7 *ibid.*, p. 60
8 "Uganda 1953", Colonial Reports, Her Majesty's Stationery Office, London, 1954, p. 120
9 *ibid.*, pp. 123—124
10 *ibid.*, p. 122
11 *ibid.*, pp. 122—123
12 Speke, J. H. — "Journal of the Discovery of the Sources of the Nile", Everyman's Library, J. M. Dent & Sons Ltd., London, 1922, p. 25
13 See, for instance,
(a) *loc. cit.* 12;
(b) Stanley, H. M. — "Through the Dark Continent or the Sources of the Nile", Sampson Low, Marston, Searle & Rivington, London, 1890
(c) Stanley, H. M. — "In Darkest Africa", Sampson Low, Marston, Searle & Rivington, London, 1890;
(d) Baker, S. M. — "The Albert Nyanza", Macmillan, London, 1866;
(e) Lugard, F. D. — "The Rise of Our East African Empire", William Blackwood & Sons, Edinburgh & London, 1893;
(f) Johnston, H. H. — "The Uganda Protectorate", Hutchinson, London, 1902;
(g) Cunningham, J. F. — "Uganda and its Peoples", Hutchinson, London, 1905;
(h) Hobley, C. W. — "Eastern Uganda", 1902;
(i) Jackson, F. — "Early Days in East Africa", Edward Arnold & Co., London, 1930;
(j) Roscoe, J. — "The Baganda: An Account of Their Native Customs and Beliefs", Macmillan and Co., Ltd., London, 1911;
(k) Roscoe, J. — "The Bakitara or Banyoro", Cambridge at the University Press, 1923;
(l) Roscoe, J. — "The Banyankole", Cambridge at the University Press, 1923;
(m) Roscoe, J. — "The Bagesu and Other Tribes", Cambridge at the University Press, 1924;
(n) Roscoe, J. — "The Northern Bantu", Cambridge at the University Press, 1915;
(o) Roscoe, J. — "Twenty-Five Years in East Africa", Cambridge at the University Press, 1921;
(p) Roscoe, J. — "The Soul of Central Africa", Cassell and Company, Limited, London, 1922;
etc., etc.
14 For example, the writings of Roscoe
15 See, for instance,
(a) Hailey, Lord — "An African Survey", Oxford University Press, 1945;
(b) *loc. cit.* 1
(c) Huxley, Elspeth — "The Sorcerer's Apprentice: A Journey Through East Africa", Chatto and Windus, London, 1948;
(d) Coupland, R. — "East Africa and Its Invaders", Oxford at the Clarendon Press, 1938;
(e) Coupland, R. — "The Exploitation of East Africa: 1856—1890", Faber and Faber Limited, London, 1939;
(f) Mair, L. P. — "An African People in the Twentieth Century", Routledge & Sons, London, 1934;
(g) Mair, L. P. — "Native Policies in Africa", 1936;
(h) Mair, L. P. — "The Place of History in the Study of Culture Contact", a contribution in the "Methods of Study of Culture Contact in Africa", Memorandum XV, International Institute of African Languages and Cultures, 1938;
etc., etc.
16 "Cmd. 3234", Publication of His Majesty's Stationery Office, London.
17 *loc. cit.* 13(f), vol. I, p. 298
18 *loc. cit.* 15(c), pp. 199—201
19 *ibid.*, p. 365
20 *loc. cit.* 15(f), p. 286
21 Malinowski, B. — "The Dynamics of Culture Change", Yale University Press, USA, 1945, p. 161
22 *loc. cit.* 15(e), pp. 487—488

23 For instance,
(a) *Colonial No. 228* — "Information on Non-Self-Governing Territories", Colonial Office, His Majesty's Stationery Office, London, 1948;
(b) Hall, J., (Governor of Uganda) — "Some Notes on the Economic Development of Uganda", Foreword to E. B. Worthington's "A Development Plan for Uganda", Entebbe, Uganda Protectorate, 1949;
etc., etc.

24 *loc. cit.* 23(a), p. 34

25 *loc. cit.* 23(b), p. iii

26 *loc. cit.* 15(c), p. 243

27 *ibid.*, p. 240

28 *loc. cit.* 8, p. 17

29 Calculated from "Estimate 1948", Budget Estimate of the Uganda Protectorate Government, Entebbe, Uganda, 1948

30 Calculated from *loc. cit.* 29

31 Calculated from *loc. cit.* 29

32 Calculated form *loc. cit.* 29

33 "Memorandum on 1948 Draft Estimates", Publication of the Uganda Protectorate Government, Entebbe, Uganda, 1948

34 Extract from the copy of the unpublished memorandum received by the autor from Mr. D. K. Sekuma, the leader of the African Labour Party in Uganda

35 Whitley, N. H. P. — "Report of the Commission of Inquiry into the Disturbances which occured in Uganda during January 1945", Entebbe, Uganda, 1945

36 *ibid.*, p. 12

37 Sekuma, D. K. — "Matter of the Moment", Official Circular of the League of Uganda Citizens and African Labour Party, Kampala, Uganda, 1948; copy of the circular given to the author by Mr. Sekuma.

38 *loc. cit.* 15(f), p. 288

39 *loc. cit.* 35, p. 2

40 *loc. cit.* 23(b), p. v

41 *loc. cit.* 15(c), p. 365

42 Copy of the Memorandum given by Mr. D. K. Sekuma to the author

43 *loc. cit.* 21, pp. 2—3

44 *ibid.*, pp. 3—4, 8

45 Tongue, E. D. — "The Contact of Races in Uganda", British Journal of Psychology, Vol. XXV, Part 5, January, 1935, p. 364

46 Kartun, Derek — "Africa, Africa!", Lawrence & Wishart, London, 1954, p. 5

47 *loc. cit.* 13(e), vol. ii, p. 650

48 *loc. cit.* 46, pp. 4—5

49 *loc. cit.* 45, p. 363

50 "Uganda 1947", Colonial Annual Report, His Majesty's Stationery Office, London, 1949, pp. 82—83

51 *loc. cit.* 35, p. 28

52 *ibid.*, p. 28

CHAPTER III.

UGANDA BEFORE THE BRITISH

In that early dusk of a distracted age,
when God in scorn of his own workmanship
violently shook his head at his primitive
efforts,
an impatient wave snatched you away,
Africa,
from the bosom of the East,
and kept you brooding in a dense enclosure
of niggardly light,
guarded by giant tress.
There you slowly stored
the baffling mysteries of the wilderness
in the dark cellars of your profound
privacy,
conned the signals of land and water
difficult to read;
and the secret magic of Nature invoked
in your mind
magic rites from beyond the boundaries
of consciousness.
You donned the disguise of deformity to
mock the terrible,
and in a mimicry of a sublime ferocity
made yourself fearful to conquer fear.
You are hidden, alas, under a black veil,
which obscures your human dignity
to the darkened vision of contempt.

[From the poem entitled "Africa" by Rabindra Nath Tagore, published in Bengali in 1937 and in English in *Poems*, Visva-Bharati, Calcutta, 1946, pp. 158—159.]

What were the stages of social development the peoples of Uganda had reached before their territory came under British rule? This is a question not of mere historical interest, for today when the Uganda Africans are asking for self-government, it is often declared by the foreign spokesmen that an essential hindrance to attaining such an objective is the peculiar development of the African societies in pre-British Uganda. As noted before, this viewpoint was aptly put forward by Elspeth Huxley when she asked the world public: "Can one change natures moulded over centuries by their environment in the wink of an eye? Can one race pass to another its own experience?"[1] Obviously, the underlying idea behind such statements is that primitive and peculiar societies which have been "moulded over centuries by their environment" are so very *different* from that of *modern* societies, especially those belonging to the so-called Western Civilisation, that they have first to be *re-made*, even though the "half-educated" Africans, "fed on the vapour" of Western philosophy, are demanding self-rule with "gathering impatience"[2].

In this context the question that should first strike one who is interested in the *problem of Uganda* is what were the stages of social development the Uganda Africans had reached before the advent of the British. Did their societies remain static over centuries? Or, if they underwent changes, was the direction of such changes very different from that attained by peoples in other parts of the world? In short, was it likely that in the normal course of history the Uganda Africans could eventually have emerged to the same stage of social development arrived at by other "civilised" peoples, even though the tempo of changes might have been slow among them? Or, on the contrary, were their societies so "moulded over centuries" and thus produced such aberrant forms and functions that Britain had to take upon herself the responsibility of governing Uganda in order to "civilise" the "tribes" and to re-make their societies? And until then, as the British Government have often declared, did they have to remain in African territories as Trustees of these backward peoples who are "not yet able to stand by themselves under the strenuous conditions of the modern world"[3]?

This is the question the author intends to examine in this chapter. Unfortunately, the material available for such a study is not large. Yet, as the following pages will show, it may be considered sufficient to lead to the following important conclusions:

(1) None of the "tribes" of Uganda (as they are described in official and some scientific literatures even today) remained static in the pre-British days.

(2) The direction of changes and the stages of social development of these people were the same as have been observed in other parts of the world including the cradle of Western Civilisation.

(3) Most of these "tribes" had travelled beyond the tribal stage of a *class-less* society distinguished from other such societies by the exclusive possession of a territory, a name, and a dialect, and governed by the gentile organisation and kinship ties.

(4) In the remaining "tribes" also the tribal stage was fast disappearing by the time the British appeared on the scene.

(5) In the major part of the territory, later declared as the Uganda Protectorate, the people were living in semi-feudal kingdoms.

(6) In some of these areas there were also incipient signs of further development from the feudal stage.

However, before discussing whether or not the Uganda Africans were in a *tribal stage* when the British arrived on their territory, it is necessary to clarify what is meant by a *tribe* and why a particular definition has been adopted in this study for such a discussion. This necessity lies in the fact that not only every scientific term must have a precise meaning as otherwise its significance in analysing a phenomenon cannot be fully appreciated, but also because this insistence on exact definitions has a practical relevance to a sociological analysis of the Uganda Africans who are very loosely described as "tribals" even in some current scientific and government literatures.

As regards the theoretical necessity to define a *tribe*, even to some renowned sociologists, and especially to those who remain satisfied with "operational" definitions of societal phenomena and institutions, this approach may appear as unnecessary and perhaps they will also declare that it is not possible to give precise definitions at the present stage of our knowledge in social science[4]. But the author is inclined to the view that precise definitions of scientific terms are absolutely of fundamental importance in any scientific research, and it is more so in any study of society, for without such definitions it is impossible to make a proper analysis of a society in the course of its development or to make a comparative study of it with other societies.[5] How the lack of a precise definition leads to theoretical confusion, or to contradictory conclusions in any specific study, will be illustrated below. It will also be explained at the end of this chapter how the lack of interest in, or a deliberate avoidance of, defining a *tribe* in the perspective of social development of a people has facilitated the prosecution of sinister propaganda against the Uganda Africans. There cannot be any doubt, therefore, that a proper definition of the term *tribe* is of both theoretical and practical interest for this and such other studies.

It should however be admitted that the definition of any phenomenon or institution, as given on the basis of the existing stock of knowledge at a particular time, may not be final or true for ever. It is of course possible to improve upon such a definition or replace it by a better one in the light of additional knowledge gained from further researches on the subject; but at the same time one should not lose sight of the fact that in order to make it useful any study made at any particular time must abide by exact definitions of the phenomena or institutions involved in the discussion. Therefore, in a sociological study, as the present one is, these definitions must be justified in accordance with the existing state of knowledge in social science. A short discussion on what is meant by a *tribe* and why a particular definition has been accepted by the author is thus very relevant at the outset.

1. What is a Tribe?

The definition of a tribe as given above by the author may be a matter of dispute for, exact or not, today there are many definitions of a tribe as depicting a form of social organisation. But is there any one available other than what the author has accepted for the present study, which would give a better characterisation of a people in the course of development of their particular society? To examine some well-known authorities in this connection, the definition of a tribe is given as follows in the standard book for the British anthropologists, viz. *Notes and Queries on Anthropology*, which was revised and rewritten by a Committee of the Royal Anthropological Institute of Great Britain and Ireland in 1951[6]:

> "A tribe may be defined as a politically or socially coherent and autonomous group occupying or claiming a particular territory."

But would not such a definition also signify the Irish as a tribe, the French as another, and so on? In that case, what is the significance of the *tribal* stage of human society and how can a so-called "primitive" tribe be distinguished from a "civilised" community or folk or nation?

There are also other definitions of a tribe, which are equally unsatisfactory as not being precise and therefore leading to confusion. There is, for instance, the definition given by Professor George P. Murdock of Yale University at the International Symposium on Anthropology in the United States in 1952. He stated[7]:

> "A cultural species will be called simply a 'culture', and the people who bear it a 'tribe'."

Besides giving such "operational" definitions of *culture* and *tribe*, he noted[8]:

> "We therefore propose to define a culture as including all local cultural variants exhibited by communities within a particular geographical area which speak mutually intelligible languages and have essentially similar forms of economic adjustement."

And, probably being conscious of the limited value of his definitions, he further elaborated[9]:

> "This conception of a culture, and of the tribe as its social correlate, coincides very closely with actual anthropological usage as this has developed through general consensus rather than explicit definition."

Professor Murdock's labour thus leaves one as much in the dark as one was in before. For the definitions do not specify the attributes of what he means by "culture" or by "tribe", except that they involve *some* people.

In practice also it is seen that some very reputed anthropologists land in confusion because of the lack of a precise definition of a *tribe*. For instance, Professor Robert H. Lowie speaks of the Baganda (Ganda, according to his terminology) having "a well-knit political organization along with the refinements that accompany such features"[10], and further notes while discussing the evolution of "classes" in a society[11]:

> "In East Africa peasants subjugated by herders became a lower class. Here we even find a curious analogy to European feudalism: the chief of the ruling pastoralists, now king of a sizable population, claims in principle all the subjects' livestock as a medieval monarch claimed all their land."

It is well known that the Banyoro (Bakitara), Banyankole, and Baganda societies were characterised by this form of subjugation of Bairu or Bahera agriculturists by Bahima pastoralists*; yet Professor Lowie characterises Bairu as a tribe, Bahima as another, and Ankole and Ganda as still others[12]. Is this not a curious neglect of examining peoples in the course of their social development ?

On the other hand, as early as in 1877, Lewis Henry Morgan gave the definition of a tribe from his researches into the Indian tribes of America, which has not yet been superseded by a better and more precise definition. The following are the main characteristics of a *tribe* as stated by Morgan[13]:

> "I. The possession of a territory and a name.
> II. The exclusive possession of a dialect.
> III. The right to invest sachems and chiefs elected by the gentes.
> IV. The right to depose these sachems and chiefs.
> V. The possession of a religious faith and worship.
> VI. A supreme government consisting of a council of chiefs.
> VII. A head-chief of the tribe in some instances."

Friedrich Engels, in 1884, further elaborated on this definition and discussed the tribal stage of society in the process of social development. After giving an account of the tribal characteristics of the Iroquois of America, he continued[14]:

> "I have given Morgan's account of it in some detail because it gives us the opportunity of studying the organization of a society which as yet knows no *state*. The state presupposes a special public authority separated from the totality of those concerned in each case; . . .
>
> And this gentile constitution is wonderful in all its childlike simplicity! Everything runs smoothly without soldiers, gendarmes or police; without nobles, kings, governors, prefects or judges; without prisons; without trials. All quarrels and disputes are settled by the whole body of those concerned — the gens or the tribe or the individual gentes among themselves. Blood revenge threatens only as an extreme or rarely applied measure, of which our capital punishment is only the civilized form, possessed of all the advantages and drawbacks of civilization. Although there are many more affairs in common than at present — the household is run in common and communistically by a number of families, the land is tribal property, only the gardens being temporarily assigned to the households — still, not a bit of our extensive and complicated machinery of administration is required. Those concerned decide, and in most cases century-old custom has already regulated everything. There can be no poor and needy members of the community — the commu-

* Details of this process of subjugation of the indigenous peoples and the evolution of class societies with the break-up of tribal organisations in Kitara, Ankole, Buganda, etc., will be discussed in course of this chapter.

nistic household and the genes know their duties towards the aged, the sick and those disabled in war. All are free and equal — including the women. There is as yet no room for slaves, nor, as a rule, for the subjugation of alien tribes. . . .

This is what mankind and human society were like before class divisions arose. And if we compare their condition with that of the overwhelming majority of civilized people today, we will find an enormous gulf between the present-day proletarian and small peasant and the ancient free member of a gens.

This is one side of the picture. Let us not forget, however, that this organization was doomed to extinction. It never developed beyond the tribe; the confederacy of tribes already signified the commencement of its downfall, as we shall see later, and as the attempts of the Iroquois to subjugate others have shown. What was outside the tribe was outside the law. Where no express treaty of peace existed, war raged between tribe and tribe; and war was waged with the cruelty that distinguishes man from all other animals and which was abated only later in self-interest. The gentile constitution in full bloom, as we have seen it in America, presupposes an extremely undeveloped form of production, that is, an extremely sparse population spread over a wide territory, and therefore, the almost complete domination of man by external nature, strange and incomprehensible to him, a domination reflected in his childish religious ideas. The tribe remained the boundary for man, in relation to himself as well as to outsiders. The tribe, the gens and their institutions were sacred and inviolable, a superior power, instituted by nature, to which the individual remained absolutely subject in thought, word and deed. Impressive as the people of this epoch may appear to us, they differ in no way one from another, they are still bound, as Marx says, to the umbilical chord of the primordial community. The power of these primordial communities had to be broken, and it was broken. But it was broken by influences which from the outset appear to us as a degradation, a fall from the simple moral grandeur of the ancient gentile society. The lowest interests — base greed, brutal sensuality, sordid avarice, selfish plunder of common possessions — usher in the new, civilized society, class society, the most outrageous means — theft, rape, deceit and treachery — undermine and destroy the old, classless, gentile society."

How was the "power of these primordial communities" broken? As Engels noted, it was achieved by means of —

"differentiation in wealth affecting in turn the social constitution by creating first rudiments of a hereditary nobility and monarchy; slavery, first limited to prisoners of war, but already paving the way to the enslavement of fellow members of the tribe and even of the gens; the degeneration of the old tribal warfare to systematic raids, on land and sea, for the purpose of capturing cattle, slaves, and treasure as a regular means of subsistence. In short, wealth is praised and respected as the highest treasure, and the old gentile institutions are perverted in order to justify forcible robbery of wealth. Only one thing was missing: an institution that would not only safeguard the newly-acquired property of private individuals against the communistic traditions of the gentile order, would not only sanctify private property, formerly held in such light esteem, and pronounce this sanctification the highest purpose of human society, but would also stamp the gradually developing new forms of acquiring property, and consequently, of constantly accelerating increase in wealth, with the seal of public approval; an institution that would perpetuate, not only the newly-rising class division of society, but also the right of the possessing classes to exploit and rule the non-possessing classes. And this institution arrived. The *state* was invented"[15].

What are the essential characteristics of the institution of the state, and how does it differ from the gentile organisation of tribal society? To quote Engles again[16]:

"The state differs from the old gentile organization, firstly, in that it divides its citizens *according to territory*. . . .

The second is the establishment of a *public power* which no longer directly coincided with the people organized as an armed force. This special, public power was necessary, because an independent armed organization of the population became impossible with the division of society into classes. . . . This public power exists in every state; it is not only composed of armed men, but also of material accessories such as prisons and institutions of coercion of all kinds, of which gentile society knew nothing. It may be very insignificant, almost infinitesimal, in societies where class antagonism are still undeveloped and in out of the way places, as was the case at certain times and in certain regions in the United States. It becomes stronger, however, in proportion as class antagonisms in the state become more acute, and as neighbouring states become larger and more populated. ...

In order to maintain this public power, contributions from the citizens become necessary — *taxes*. These were absolutely unknown in gentile society; but we know enough about them today. As civilisation advances, these taxes become inadequate; the state makes drafts on the future, contracts loans, *public debts*. . . .

In possession of the public power and of the right to levy taxes, the officials are now organs of society standing *over* society. The free and voluntary respect that was accorded to the organs of the gentile constitution does not satisfy them, even if they could get it; being the vehicles of a power that is being divorced from society, respect for them must be enforced by means of exceptional laws which make them specially sacred and inviolable. The pettiest police officer in the civilized state has more 'authority' than all the organs of gentile society put together; but the most powerful prince and the greatest statesman, or general, of civilization may envy the humblest gentile chief for the spontaneous and undisputed respect that was paid to him. The one stood in the midst of society, the other is forced to assume a position outside and over it.

The state came into existence owing to the necessity of curbing class antagonism; but having arisen amidst these conflicts, it, as a rule, is the state of the most powerful class, the class which rules in economics, and with its aid become also the class which rules in politics, and thus acquires new means of holding down and exploiting the oppressed class. Thus, the state of antiquity was above all the state of the slave owners for the purpose of holding down the slaves, as the feudal state was the organ of the nobility for holding down the peasant serfs and bondsmen, and the modern representative state is an instrument for the exploitation of wage labour by capital*. . . .

In most of the historical states, the right of citizens are, besides, apportioned according to their wealth, thus directly expressing the fact that the state is organized for the protection of the possessing class against the non-possessing class. This was already seen in the Athenian and Roman classification according to property. It was seen in the medieval feudal state, in which political power was commensurate with the amount of land owned. It is seen in the electoral qualifications of the modern representative states. The political recognition of differences in wealth is by no means essential. On the contrary, it marks a low stage of state development. The highest form of the state, the democratic republic, the form of state which under our modern conditions of society is more and more becoming unavoidable necessity, and in which alone the last decisive struggle between proletariat and bourgeoisie can be fought out — the democratic republic officially knows nothing of property distinctions. In it wealth exercises its power indirectly, but all the more surely."

* Engels continued:

"There are periods in which, as an exception, the conflicting classes balance each other so nearly that the public power for a time gains a certain degree of independence of both, seemingly as a mediator between them. Such was the case with the absolute monarchy of the seventeenth and eighteenth centuries which held the balance between the nobility and the burghers; such was the case with Bonapartism in the First, and still more in the Second Empire, which played off the proletariat against the bourgeoisie and the bourgeoisie against the proletariat. The latest performance of this kind, in which ruler and ruled appear equally ridiculous, is the new German Empire of the Bismarck nation: here capitalists and workers are balanced against each other and equally cheated for the benefit of the depraved Prussian cabbage junkers."

It is thus seen that Morgan and Engles described the tribal stage of society as the lowest stage in social development in which *classes* had not developed and the societies governed by gentile organisation and kinship ties were distinguished from one another by the exclusive possession of a territory, a name, and a dialect. Later *classes* evolved at a certain stage of economic development of society, and the institution of the state was established to keep the dominating class in power. With this the tribal stage of society was gone for ever; the people entered into a class-society. This is how Morgan and Engels envisaged the primary stage in the evolution of human society.

Today, however, not only the name of Engels is virtually unmentionable even in some scientific circles, but Morgan is also practically tabooed by a large number of social scientists. For, as Professor Leslie A. White has rightly remarked[17]:

> "The forces of antievolutionism, which fought Darwinism so vigorously, entrenched themselves upon the sociological field after being driven from the field of biology. In America the attack upon Morgan and social evolutionism has been led chiefly by the anthropologist, Franz Boas, and his disciples. One of the latter has gone so far as to delare that the idea of social evolution is 'the most inane, sterile, and pernicious theory ever conceived in the history of science'."

It is therefore seen that even though he is still considered as a scientist, Morgan was first attacked and then ignored by a vast number of reputed anthropologists and sociologists. Also, "Morgan has been so grossly misrepresented by some of his critics that it is difficult to believe that they have even read his works"[18].

No doubt, there is no point in defending everything that Morgan wrote; but at the same time one cannot overlook the fact that although a schematic unilinear evolution of society is not taken seriously today, the concept of evolution as depicting the stages of social development in human history is not disregarded by the general run of social scientists[19]. And in this connection it is worthy of note that, as has been acknowledged, ,,Morgan was particularly interested in the unilinear aspect of cultural evolution, but he recognised, of course, its multilinear aspect as well"[20]. Even so, Morgan's viewpoint is seldom discussed with genuine scientific impartiality. This is probably due to the fact that, among other reasons, the acceptance of Morgan's researches by the Marxists was a very important factor in his losing favour with many social scientists. As it has been cogently remarked[21]:

> "*Ancient Society* came to the attention of Karl Marx, who regarded it as a great work and was much impressed with it. He planned to write a book dealing with Morgan's contributions but was prevented from doing so by his death. The book was eventually written by Frederich Engels, Marx's lifelong friend and co-worker, under the title, *Der Ursprung der Familie, des Privateigenthums und des Staats* (1884). *Ancient Society* became a Marxist classic. This fact has unquestionably had much to do with the change in attitude toward Morgan that has taken place since his day."

Yet, although put in somewhat different ways, the definition and characterisation of a *tribe*, as given by Morgan and Engels, are still found to be accepted by many reputed sociologists and anthropologists who in other respects may not agree with them. Professors R. M. MacIver and Charles H. Page, for instance, explained in their *General View of Social Evolution*[22]:

> "The functional interdependence of the groups and organizations of an advanced social system is almost totally lacking in primitive society. The main divisions of the latter — families, clans, exogamous groups, totem groups — are segmentary or compartmental. It may have a fairly elaborate system of ceremonial offices, and a more elaborate system of kin-distinctions than is characteristic of an evolved society. But there are few groupings or categories into which, for the practical purposes of co-operative living, the members

fall. The kin-grouping is usually predominant and inclusive. To be a member of the kin is *ipso facto* to share the common and inclusive rights and obligations, the customs, the rituals, the standards, the beliefs of the whole. There are, of course, certain 'natural' groupings, particularly those of age and sex. There may be prestige groups, perhaps a simple system of classes or castes, though these latter are not found under the most primitive conditions. There may be some rudimentary occupational distinctions, but the division of labour is narrow and usually follows 'natural' lines, such as that between the sexes or between the older and the younger. The great associations do not yet exist. There is no separate organization of religion — still less of religions: there are no schools, no distinct cultural associations; there is little specialization of economic productivity and exchange. The only clearly associational groups, other than temporary partnerships in trading ventures and so forth, are usually 'secret societies', not specifically functional, and the very fact that they are 'secret' is significant, implying that the group has not yet found a way to incorporate them effectively within its unity.

The undifferentiated character of primitive society is seen in the prevalence of a simple form of communism. The kin is a larger family and exhibits something of the communistic character of the family. The tribe devises a system of participation in the booty of the chase and the products of the earth. Where private or family rights are admitted, it is in the right of use, not in the ownership, of the land. Even what are to us the most intimate or personal of rights were then rights pertaining to the blood brotherhood. The lending of wives to tribal guests, . . . may be regarded as a mode of admission to the 'freedom' of the tribe. . . .

Such a communism typifies the simple solidarity of an undifferentiated community. The differentiations as exist are based on the natural distinctions of youth and age, of man and woman, of different aptitudes such as that for leadership, and on a few socially acquired distinctions, such as the inheritance of ceremonial office or of magical lore. The myriad aspects of differentiation belonging to a civilized society are latent. The divergent interests, aptitudes, capacities which may appear in rudimentary forms have no opportunity to develop within the restricted range of the communal life. The social heritage is too rude to afford them selective stimulation. The mores appropriate to that narrow heritage tend to be repressive of such differences, as endangering the solidarity of like-mindedness, the only solidarity of which the group as a whole is yet capable."

Professor A. L. Kroeber also appears to agree with Morgan's definition of a tribe[23]; and Professor White categorically stated that the distinction that Morgan made between *societas* and *civitas* was "one of his greatest contributions to sociological thought". To quote Professor White in full[24]:

"Morgan surveyed the entire span of human history and saw that, with reference to social organization, it is divisible into two great categories: 'ancient society' (or 'primitive society', as we tend to call it today) and modern civil society. He called the first type of society *societas*, the second *civitas*. Morgan's distinction of these two great types of human society and his description of each constitute one of his greatest contributions to sociological thought.

Ancient society, the *societas*, was founded upon 'persons and upon relationships purely personal'. Civil society, the *civitas*, was founded upon territory and upon property. Thus, in primitive society a person's status and role in social life were determined by his personal relationship to other members of the society. And, since in almost all instances, all members of a *societas* regard themselves as related to one another, a person's role in primitive society is determined and regulated by ties of *kinship*, as well as by merely personal ties. A *societas* is thus a community of kindred. In a *civitas*, on the other hand, a person's status and role are determined by his place of residence — deme, township, state etc.— and by his relationship to property. The relationship between persons is thus impersonal.

> The members of a *societas* are kinsfolk, the members of a *civitas* are citizens. The clan and tribe are the characteristic units of ancient society. The political state is the form characteristic of the *civitas*.
>
> Ancient society is democratic, and all men are free and equal. In civil society there are class distinctions: masters and slaves; lords, vassals, and serfs; kings and subjects; or the rich and the poor."

Furthermore, how useful is Morgan's definition of a tribe to the present study is proved by the fact that even Professor A. R. Radcliffe-Brown who is very much against Morgan and, in the words of Professor White, ,,made the remarkable 'discovery' that Morgan was not even a *social* evolutionist but merely a *petit bourgeois* with a simple faith in progress, measured by the provincial standards of his home town, Rochester, New York"[25], stated[26]:

> "Indeed we may agree with Morgan, that the passage from lower forms of civilization to higher forms such as our own was essentially a passage from society based on kinship to the state based on political organization."

It is thus seen that the definition of a *tribe* as accepted by the author is in accord with the process of social development of mankind; — the tribe representing the lowest stage at an undifferentiated state of society. Therefore, accepting this definition it can be examined in the following pages whether the Uganda Africans were in a *tribal* stage when the British appeared in their territory or whether they had moved farther from this stage. And if they were beyond the tribal stage, it can be further examined as to which stage they had reached at the time of their contact with the British. In this way, a scientific analysis of the Uganda societies can be made instead of loosely labelling the Uganda Africans even today as "tribals", and thus giving a "scientific" scope for subversive propaganda against them. Simultaneously, even though ignored or circumvented by many scientists, the validity of the definition of a *tribe* as accepted by the author will be proved or disproved by its usefulness in analysing the Uganda societies and thus helping one to understand the *problem of Uganda*. In the following pages, therefore, the "tribes" of Uganda, at any rate the most important of them, will be treated one by one in order to prove the conclusions made at the beginning of this chapter.

2. Kitara or Bunyoro

For a long time before the advent of Europeans the major part of the present boundary of the Uganda Protectorate was under the kingdom of Kitara, the name Kitara being later replaced by Bunyoro[27]. Roscoe, the first ethnographer who made a scientific study of the Bakitara or Banyoro people, noted[28]:

> "The country of Kitara, which is now generally known as Bunyoro, was at one time the largest and most powerful of all the autocratic kingdoms in the lake region of Central Africa."

This kingdom at that time spread well beyond the present bounds of Bunyoro. With regard to the importance of this kingdom and its people, the explorer Samuel Baker has thus quoted the impression of Speke, the first European to visit this kingdom, in his book *Albert Nyanza*[29]:

> "Formerly Unyoro, Uganda, and Uddu were all united into one vast kingdom called Kitara, but this name is now only applied to certain portions of that kingdom."

A study of pre-British Uganda may therefore start with an account of this kingdom.

Little is known of the indigenous people who later formed the bulk of the subjects of the kingdom of Kitara. However, Roscoe's researches have brought to light that before the kingdom of Kitara was formed among these Bantu-speaking agricultural Negroes, there were "several distinct tribes differing completely from each other in type and dialect and belonging to different parts of the country"[30]. They "were not united as a nation, but formed a large number of independent tribes composed of loosely connected clans, chiefly agricultural, living by a rude cultivation of the earth which sufficed to grow food enough to keep them alive. The clans were totemic and the marriage system was exogamous, but marriage ties did little to form any kind of bond between clans or tribes"[31]. They "lived in small detached communities or villages, formed each of some twenty to forty huts, sometimes enclosed by a rude fence or stockade of growing trees but more often unprotected. They had no settled form of government, but each village had its chief, to whom the people looked to settle all local differences and act as arbitrator in matters which concerned the whole community. In their personal affairs, the members of each clan looked for advice, direction and counsel to the head of the clan"[32]. It thus appears that these people probably lived in a tribal state based on gentile organisation.

However, even though they were in a tribal state, *social division of labour* had taken place to such an extent that different clans had different vocations. Roscoe noted that these "aboriginals were not all agricultural, for it is in their clans that we find all the artisans of the country"[33]. He also noted: "Among the clans which composed each of these tribes distinction must also be noted, for the iron-workers were a different people from the wood-workers and from the salt-makers, and each of these types, again, differed from the purely agricultural labourers"[34].

It is worthy of note that even at such a remote time in the past these people were appreciably advanced in material culture. Roscoe was much impressed by their knowledge of iron-working, and believed that it was here that this industry was first invented in the world. To quote Roscoe[35]:

> "Years ago the native, in some unknown way, became acquainted with the use of metals, and from this land went forth in early ages a knowledge of ironworking and smelting which revolutionised the world's methods of work. ... When and how the people began to use iron we cannot discover and they themselves can give no satisfactory account of the growth of the industry, but they have gained a knowledge of the value of different ores, the hard and the soft, and they produce from the amalgamation of two a metal of good quality. ... They build furnaces for smelting and make and use bellows; they know what wood to use to get best charcoal for their purpose, the amount of heat required, and the time needed for smelting operation. We have advanced so far from the methods of these primitive workers that we are inclined to overlook the country and the people who revealed to the world the secret of iron-working; but here we find the smith, with his stone hammer and anvil and his charcoal fire, using the methods adopted by his ancestors who were the fore-runners of our iron industry."

These indigenous people were also well-versed in the art of pottery-making. Roscoe noted that "Potters in Bunyoro are far more skilful than most of their craft in this part of Africa"[36]. These aboriginals also possessed knowledge of graphite and petroleum and made use of them[37]. And they were noted for salt-making. Roscose noted that when "the Kingdom of Kitara was at the height of its power it possessed two important centres of the salt trade", where the salt-works were situated[38].

It is true that Roscoe's investigations also refer to the period after the establishment of the kingdom of Kitara, but the fact that these industries were restricted only to these aboriginals indicates that they were not in a very low stage of development even prior to that period, and that in course of time internal changes in their society could have gone much beyond the tribal stage. History, however, had ordained a different course of development for them.

As Johnston has estimated, "some two thousand years ago, or more or less, there wandered into Central Africa Hamitic pastoralists from the direction of Ethiopia and the White Nile"[39]. These invaders, probably belonging to the Galla stock[40], and now generally known as the Bahuma or Bahima aristocracy, subjugated the indigenous people by means of their superior military strength, and "united all these scattered tribes and formed one kingdom under a pastoral king"[41]. And with this the tribal organisation of the indigenous people was lost for ever.

Evidently, the rule over a subjugated people is incompatible with the tribal organisation. In the latter no distinction can be made between rights and duties in the internal structure and in the external relations problems are solved by war leading to annihilation or absorption, but never subjugation. The domination of the indigenous people by the Bahima aristocracy thus completely changed the character of society.

The question that strikes one at once is what was the need for the Bahima aristocracy to subjugate the indigenous people, and likewise why did not the latter leave the place and go elsewhere instead of accepting the life of subjugation. The answer to the first question is evidently the fact that the subjugated people were a source of tribute and great advantage to the Bahima. Thus Roscoe noted[42]:

> "The pastoral people refrain, as far as possible, from all manual labour and expect the agricultural clans to do their menial work for them, such as building their houses, carrying firewood and water, and supplying them with grain and beer for their households."

True, the technique of production of the aboriginals was primitive. The instruments of production were simple. They lived on hoe-cultivation, and the surplus they produced by rudimentary agriculture was not great. Yet it was substantial enough to provide the comparatively smaller number of pastorals with beer and grain. Moreover, their labour could be exploited through other forms of services as noted above. The stronger invaders therefore did not want to exterminate the indigenous people. On the contrary, when the Bahima first imposed their rule on these people whom they called Bahera, meaning serfs, they "looked on these agricultural people as their slaves or serfs; and a chief had many of these serfs who settled on his land, and worked for him"[43]. "They were despised by the pastoral people, not because of their poverty, but because of their mode of life, for, in the eyes of a cow-man, anyone who ate vegetable food and cultivated the land, or worked at anything not connected with the cows, was low and mean"[44]. So great was the distinction between the Bahima and the Bahera in those days that "vegetable food was taboo" to the former ones[45].

Whether or not the indigenous people were slaves of the invaders in the beginning, is not clear from the account furnished by Roscoe of their past life. But that they were under semi-serfdom is evident from his writings. To quote[46]:

> "The serfs, however, were not slaves, for they were not bound to particular chiefs; they were free to move to other parts of the country and serve other chiefs without giving their former masters any indication of their intentions; nor indeed was a man bound to apply for permission to the chief in whose territory he meant to settle, though as a rule, he would present himself to his new chief and tell him of his desire to serve him. He was sure of a welcome from the chief, to whom each serf meant another labourer and an addition to his wealth, for besides doing building for him and perhaps herding his goats and sheep, each serf paid him a yearly tribute of grain and beer; this was not a compulsory tax but was regarded as a voluntary return to the chief for the land occupied. If a chief put a

serf in charge of his goats and sheep, he paid him a proportion of the young for his work. A serf was not limited to a definite amount of land, he might cultivate as much as he wished, and there was no restraint upon him as regards the possession of wealth, for he might accumulate large herds of goats and sheep."

Cattle, however, were sacred, and the Bahera were not allowed to tend them. Moreover, as Roscoe has noted, for many generations "these serfs were not expected to keep cows, and were liable to be plundered if they did so"[47]. The serf-like position of the Bahera under the Bahima was thus enforced.

For the indigenous people also there was probably no other possibility left for survival. Because the entire country was overrun by these invaders they could not move to some other place. Nor could they take refuge in some less accessible hills or mountains as the country did not provide them with such. They therefore accepted the fate of remaining as serfs of the invaders. They attached themselves to the pastoral overlords, and although allowed to move anywhere they liked, seldom did so. In the words of Roscoe: "In this attachment to the land we see the remains of the original tribal system under which these agricultural people, who are now amalgamated under one king, dwelt in isolated and probably hostile groups."[48]

Under the circumstances the institution of State developed in order to curb class antagonism and to hold down the indigenous people for the benefit of the invaders. Public Power was developed for this purpose, and to maintain this public power taxes and tributes were levied upon the common people. Simultaneously in possession of the public power and of the right to extract taxes and tributes, the officials as new organs of the society stood *over* the society. The tribal organisations disintegrated under the impact.

Within the established kingdom the ancient tribes might have remained, in the main at least, in their old territories and have spoken dialects of their own, but henceforth they represented only shell of the previous tribal organisation. "The country was divided into some ten districts."[49] For "public purposes" the people were henceforth divided not by their tribes or clans, but by territory. Politically the people became mere attachments to the territory, and the tribal structure became quite useless for political purposes. Neither could the people anymore decide their common affairs, nor could they select or depose their chiefs. Only matters of little importance, such as observance of peculiar religious rites and ceremonies were left to them. And instead of tribal authorities, there developed special public authority separated from the totality of those concerned in each case. A central administration was established which considered all the important affairs previously conducted independently by the tribes as common affairs but henceforth decided by the centre. Tribal democracy became a thing of the past; the institution of State developed on the ruins of the previous tribal societies.

The state apparatus of this newly developed serf owning society took the character of absolute monarchy. The demand for a strong leadership of the state was unquestioned. A strong hand was necessary in order to prevent disintegration of the *class society* through internal dissessions as well as to protect it from external aggression. Probably this was also suggested by the nature of the life the pastorals were used to lead. The Bahima were a fighting people. Protection of their cattle, raids and counter raids had infused them with a military discipline and obedience to a strong leadership. Now this leadership could be used to protect their interests over the dominated indigenous people and maintain discipline among themselves. And this leadership took the form of absolute monarchy, for this appears to have been the most suitable centralised authority for the Bahima. It forbade any internal strife among the invaders each time the leader died. There could be no competition for leadership except in the royal line.

The leadership had of course to be strong and remain always strong. Therefore, it was unbecoming for a king to grow weak or become invalid or infirm. "In olden times it was unusual for the king to lie ill for any length of time; he would call his medicine-men and from him obtain a cup of poison which he drank and which ended his life in a few minutes."[50] Then again, every time a king died, the succession was decided not merely by the wish of the deceased king or of the chiefs, but primarily by proving superior strength and qualities of leadership. Hence, with the death of a king fighting broke out between the contending princes until only one of them remained in the kingdom to claim the throne; others having been killed or gone into exile. Further competition for leadership was ruled out during the lifetime of the reigning king. Thus Roscoe has noted[51]:

> "There was always at least a general understanding as to which prince a king wished to reign after his death, and that prince was ready to assume the place. Most of the chiefs would stand by this prince, by which the others knew he was the chosen of his father, for this was the only way in which he was distinguished from any other who might claim the throne. He had to prepare for trouble, for rarely did a prince ascend the throne without opposition from at least one brother, and he had to be prepared to meet any claimant in battle and prove his right to the throne as the stronger man, the victor in battle. A prince had always the assistance of the chief who acted as his foster-father from his birth, and the smaller chiefs, the herdsmen, and the agricultural peasants took sides and joined the prince whom they favoured. There was a general rush to arms; fighting began at once and continued until one of the rivals was killed, when all his followers submitted to the victor and became his men. It seldom happened that more than two princes fought for the throne, the others would look on and accept the result of the combat. Sometimes, however, several would claim it, and whatever the number of rivals might be, the fighting would not end until only one of them was left alive. Had Kabarega died as king of his country under this regime instead of being deposed by the British, there would have been much fighting, for the princes who might have laid claim to the throne numbered about a hundred."

In this way the selection from among the princes decided the most able leader of the Bahima, while the institution of monarchy did not allow the strife to spread far and wide within the ruling class.

The king thus appointed because of his valour and evidence of strong leadership "was an autocrat, holding absolute sway. He was on a level with the gods, in fact was almost a god himself, and, though he might appeal to the gods for advice and confirmation of his judgements, his word was final; he had power of life and death and no man might question his decisions"[52]. The monarch owned the whole country. "All the cows in the country belonged to the king, and he had the right to take any cow from any of his subjects, just as he could take any man's daughter and make her his wife"[53]. The king could demand anything from his subjects. "There was no direct system of taxation of land though state labour, generally in the form of building for the king, had to be supplied when required."[54] This labour came from the Bahera. Moreover, although "there was no regular system of taxation, the king imposed other taxes than those of state labour, and these were collected in kind", namely cows from the pastorals and grain and beer from the agriculturists[55].

Under the king the state apparatus spread out over the entire kingdom by means of a hierarchy of chieftainship, starting from the immediate chiefs of the king in the capital to the district-chiefs and sub-chiefs. At the centre was a body of chiefs known as the Sacred Guild with whom the king consulted on most matters. "Two chiefs, *Bamuroga*, who was a member of the Guild and had charge of the king's tombs, and *Munyawa*, the head of the royal clan, were the most powerful of all the chiefs and during the interval between the death of one king and the accession of another, these two governed the country."[56] Besides these supreme chiefs, each of the "some ten districts", in which the kingdom was divided, had "its pastoral overlord who was directly responsible to the king"[57]. Under the district-chiefs, "there were numbers of smaller chiefs appointed by the king"[58]. Besides them there were also other pastoral chiefs in each district "who were independent of any official in the district"[59]. In this hierarchy of chiefs judicial function descended up to the district-chiefs, although "every one had the right of appeal to a higher authority, and, in the last resort, to the king himself"[60].

As obvious in a society where the institution of the state has developed, none of these chiefs were appointed by the people through the ancient tribal councils, nor had these councils any authority to depose them. The chiefs held their office at the king's pleasure, in return for maintaining "peace" in the country, efficiently collecting the king's taxes and tributes from the mass of the people, guarding the king's estates in cattle in the district concerned, satisfactorily supplying state labour whenever desired, and in case of warfare successfully mobilising the men of the district into the king's army[61].

All these chiefs belonged only to the Bahima aristocracy. As noted earlier, they extracted tribute from the mass of the people, the Bahera, and compulsorily exploited their labour-service free of charge. They also possessed large herds of cattle. Even the district chiefs "possessed large herds, which were not necessarily confined to their own districts, for pasture land was common to all and anyone might pasture his cows where he chose"[62].

The other members of the ruling class who were not appointed as chiefs also maintained large herds of cattle, numbering as much as several hundreds. The Bahima were absorbed mainly in "the care of the cows", and lived on cattle-rearing and tributes of grain and beer from the Bahera. "In the old days cows furnished the pastoral people with the essentials of existence and the means of purchasing anything else they desired. Spears, knives, milk-pots, water-pots, and other articles were obtained in exchange for meat, butter, or skins, or by selling sheep and goats, of which the pastoral people possessed flocks and herds, kept for them as a rule by some of the agricultural serfs. Salt, which they considered necessary for the health of the cows, was obtained by bartering meat, butter, or goats."[63]

Slavery existed in the society, but mainly in the form of domestic slaves. In this serf-owning society the ruling class did not find any particular necessity to use slaves for production. Thus Roscoe noted[64]:

> "The possession of real slaves was, however, universal throughout the country, for even the poorest man might capture one or more during some raid or battle and afterwards be permitted by the leader to keep one. The wealthy people bought slaves when they needed them and a man's heir inherited his slaves along with other property. Domestic slaves were regarded as superior to those who were used as labourers, and the domestic slaves whom a man inherited were regarded almost as members of his family and were not sold. A man might even marry a slave woman, and if she bore him a child she became free and was accepted by his clan, though her children might only inherit his property if he had no child by any other wife. The ordinary slaves, especially those who had been captured in battle, were bought and sold like cattle, and a man might kill a slave just as he might kill one of his cows and no one would question his action."

It is obvious from the above that long before the Europeans had come to Uganda, her people like those in the kingdom of Kitara had travelled far beyond tribal organisation. Not only had the institution of the state developed and the king and his appointed chiefs ruled the kingdom, and not only was the society distinctly divided into the classes of well-born and the despised ones, but the common people also who previously belonged to various tribes were now in the process of being welded into one mass under a centralised monarchy and one type of economic organisation. The language of the people, Uru-Nyoro, had developed with several dialectal forms. Johnston in his scientific study of the Bantu and semi-Bantu languages in Africa, noted that this language had eight dialectal forms[65].

Further changes were also taking place in the society. Semi-serfdom of the indigenous people was in the course of being abolished, and the sharp difference between the Bahera and the Bahima was in the process of being obliterated. Property-differences began to occur both among the aristocrats and the common people. While among the former some began to lead a life of poverty with fewer cattle and consequent shortage of milk and meat, "some of the agricultural people became wealthy and showed signs of ability, or became servants to the king and in that capacity rendered him some special service"[66]. Soon these differentiations led to significant changes in the legal status of the serfs. "Some generations ago a king introduced a new status among these agricultural people: any man who showed ability and rendered special service to the king was rewarded by being made a free man, *Munyoro* (pl. *Banyoro*). This freedom entitled the men to marry any woman of the pastoral clans who was willing to form such an alliance; a wife was not difficult to find, for there were many daughters of herdsmen whose lot in life among their own people would always be hard, while if they married Banyoro it meant that they would become persons of wealth and ease, though they had to face the fact that they were marrying into the despised class."[67]

It may be that at the beginning the free Bahera were despised by the Bahima aristocrats, but in course of time this became such a regular feature in the society that the neighbouring people, the Baganda, gave in derision the name of *Bunyoro* to Kitara which means a "nation [a country] of freed slaves"[68]. And this name Bunyoro "superseded the real name of Kitara and is now commonly used for the country"[69]. How extensive this change in the status of the Bahera became is evident form the fact that, according to Roscoe's findings, among these people there were 48 totemic groups of the pastoral aristocrats, 43 of the agricultural serfs, and 24 of the "freed" people[70].

Changes in the status of the Bahera were reflected also in other aspects of their social and political life. While previously "these serfs were not expected to keep cows, and were liable to be plundered if they did so, it was only in later years that this restriction was removed and they began to acquire cows, which they used chiefly for the payment of marriage fees"[71]. Furthermore, these "freed slaves", in the words of Roscoe, could also become district-chiefs like the Bahima. "The king at times appointed an intelligent member of one of these agricultural clans to govern a district, . . ."[72].

While the Bahera were thus being brought on to an equal footing with the Bahima, the latter were also gradually equating to the ways of living of the other people. Roscoe has noted that "for some years now the pastoral people of Kitara, who were the ruling class, have abandoned the strict milk-diet on which they used to live, and vegetables, plantains, and millet have become regular articles of food among them"[73]. Moreover, as among the Bahera, the women of the pastorals had begun to cultivate millet, plantains, maize and potatoes "to supplement the food supply"[74].

In this process of assimilation of the ruling class and the indigenous people, intermarriage introduced between them "some generations ago" played a very significant role. For the sons of the union between a Bahera man and a Bahima woman were free to marry girls either of the agricultural or of the pastoral families, "while cow-men of the pure stock were willing to marry the daughters"[75]. The result was a closer amalgamation of the previous division of society between aristocrats and commoners. Roscoe noted[76]:

> "Such marriages as these have in course of time brought about a lowering of the pastoral type, an ever-growing laxity in the observance of the milk-customs, and the introduction among the pastoral people of a vegetable diet."

In this way in the period just before the advent of the British in Uganda, another new set of relations was in the process of being established in place of the previously established division of society into aristocrats and commoners. This form of relationship, determined by an individual's wealth and property, was cutting across the previous classes determined by birth. Acquisition of wealth rather than the nobility of birth was begining to determine the social position of an individual; the society began to be divided into rich and poor, rather than into high or lowly born.

Another new situation that was developing in this period was due to the advent of Arab traders in the kingdom which, although felt not so strongly as in the kingdom of Buganda, was influencing the king and the chiefs to indulge in monopoly trading. The kingdom of Kitara was opening up to the world, and this process obviously had to have an important effect on the previously parochial society. In exchange for slaves and ivory the Arabs brought to the Banyoro rulers guns, ammunition, and other goods, and thus trading profit became a lucrative source of wealth to the latter in addition to the regular tributes obtained as previously from the mass of the people.

Before the Arabs appeared, the market in Kitara was confined within the kingdom. Superimposition of the pastoral invaders on the indigenous agriculturists had created the conditions for regular exchange. As explained before, trading relations between the Bahima and Bahera had developed within the kingdom, but the market did not spread appreciably beyond the frontiers. Only a small surplus of iron hoes and salt were traded with neighbouring peoples, like the Lango and Banyankole. But henceforth external trade in slaves and ivory began to develop on a large scale. The "primary accumulation of slaves" became "a state monopoly operated by the rulers who dealt with Arab traders as middlemen in the chain of distribution"[77]. In short, the rulers, headed by the king, were developing into merchants, and the society of Banyoro was stepping into a new world.

And while new social and economic relations were thus evolving in the society, the kingdom of Kitara was narrowing down only to the region where the Banyoro lived. Oberg noted[78]:

> "Although it is said that the Bahima were once united and held sway over a great empire called Kitara, in historic times this territory was split up into kingdoms stretching from Bunyoro through Toro and Ankole to Ruanda in Belgian territory. The kingdom of Buganda also belongs to this classification, . . . "

The kingdom of Ankole, however, had separated a long time before. According to the legend current among the Banyankole, Ndahura, "who is said to have consolidated the Kitara Empire", sent his son Wamara to rule in Ankole[79]. Wamara after a series of misfortune left the country with his clan-members, but Ruhinda, a young member of the clan, and his mother, were persuaded to stay; and this Ruhinda "became the founder of the Abahinda dynasty, which rules Ankole to this day"[80]. The kingdom of Buganda had also separated form the kingdom of Kitara a considerable time before the introduction of British rule in this part of Africa. When Speke for the first time went to this region, Buganda, which previously was a tributary state of the kingdom of Kitara, had developed into an independent state and a serious rival to the mother kingdom[81]. Toro had also virtually become a separate kingdom, although governed by a descendant of the kings of Kitara[82]. The kingdom of Kitara was thus more and more representing mainly, if not only, the Banyoro people.

It is also interesting to note that when Johnston made the pioneer study in Bantu and semi-Bantu languages he had noted[83]:

"The Nyorω language and sub-groups (2, 2a—g) ranges over a considerable stretch of country from the north-eastern extremity of the Bantu domain (at the juncture of the Victoria Nile with Lake Albert), to the south-west corner of the Victoria Nyanza and the Bukerebe archipelago in the southern waters of the great lake. This language, which with our present knowledge we divide into eight dialectal types, has been especially the speech adopted by the various branches of the aristocratic Hima type, in most cases the ruling people and cattle-keeping aristocracy of the Nyorω tribes."

But with the disintegration of the Kitara Empire, two of the eight dialectal forms, namely Uru-Toro and Uru-Hima, which were being spoken in two separate kingdoms of Toro and Ankole, developed in course of time into separate langauges[84]. The other main dialectal form of the language, namely Uru-Nyoro, remained as the language of the Banyoro people. Further to note, this *language* has no dialectal forms; the dialects Guy-gu and Kyopi, noted separately by Johnston[85], being according to Guthrie synonymous with Uru-Nyoro[86].

All these facts tend to indicate that before the advent of the British in Uganda the Banyoro people had developed into a nationality, and that, with the spread of a *national market* in Bunyoro (which the growing class of "merchants" could have eventually undertaken), in their normal course of development they would have developed into a nation without any foreign intervention.

3. Ankole

Besides the kingdom of Kitara, the kingdom of Ankole was also of a long standing in Uganda, and comprised a large area of the territory. As mentioned before, in this region also the institution of the state had developed in ancient times and the tribal life of the indigenous people was put to an end. Oberg's researches give us a glimpse of the aboriginals of Ankole living in a tribal stage long before the advent of the British. "Some time in the dim past this region was occupied by Bantu-speaking, agricultural Negroes. The rainfall, though scanty, was sufficient to permit a fairly even distribution of the population over the country, scattered thinly in the drier plains of the east, but more densely in the hilly regions of the west"[87]. At this time an advanced group of Bahima was also living in Ankole. "The Bahima lived in eastern Ankole with their cattle while the Bairu (Bahera) tilled the soil in the west. In those days the Bahima had neither king nor chiefs, but important men in the clans settled disputes"[88]. Thus in those days both the Bahima and Bahera were living independently in separate regions of Ankole, and neither of them subjugated or dominated over the other. Both were in their tribal stage, and so there was no need for "king or chief"; clan organisation was the notable feature of the tribes. Scope for exchange had of course developed between the pastoral and the agricultural tribes, but that did not make either of them superior to the other. "There were no wars between the Bahima and the Bairu in those days, each living in his own section of the land and trading beer and millet for milk and butter."[89]

"Then suddenly a strange people appeared."[90] These were the militant Bahima headed by Wamara. The invaders in course of time either brought the existing Bahima under control or absorbed them, and made the Bahera their serfs. The institution of State developed, and, as noted earlier, the Abahinda dynasty began to rule the kingdom of Ankole until the British appeared.

The tribal stage of the cattle-men was gone for ever. "No longer were the Bahima cattlemen free agents, united in extended families and loosely knit lineages and clans; they were now also members of a political group. If the Bahima were going to further their interests as Bahima, they had to organize and act together as Bahima. At bottom this new relationship was based on Bahimaship — upon race and cattle-ownership. But this special political bond had to be created, had to be consciously entered into. It involved leadership, co-operation, submission to authority. It gave rise to kingship and the dynastic principle, the organization of military forces and chieftainship. In short, it welded the Bahima into a state, the nucleus of the Banyankole kingdom"[91].

The agricultural people also no longer remained in their tribal stage. For the same reasons as noted for their brethren in the kingdom of Kitara, the Bahera of Ankole accepted their serf-like position in the newly-formed class society. "The pastoral people divided the land up into districts" and "the agricultural people, who cultivated small portions of it, were despised and regarded as serfs. They could cultivate any land wherever they wished, but they were expected to do any menial work required by the pastoral people of the district in which they settled and to supply them with grain and vegetable food should they require it."[92]

Moreover, there were several limitations of Bahera rights. "The Bairu were not permitted to own productive cows. For services rendered to the Bahima, they were sometimes given barren cows and bull calves. These cattle the Bairu either kept for making marriage payments or slaughtered for food. If a Mwiru [singular for Bairu] did have productive cows in his possession, any Muhima [singular for Bahima] could take them away from him"[93]. Also there was a strict prohibition of inter-marriage between the Bahera and Bahima. Furthermore, the Bahera were barred from military service. They "thus lacked the military training and discipline necessary for effecting any change in their status"[94]. And finally the Bahera had no political status, not even the right of blood revenge in regard to the Bahima. "The right of blood revenge which was exercised by the extended families of the Bairu among themselves could not be extended to the Bahima. If a Muhima killed a Mwiru, the extended family of the murdered man could not claim blood revenge, although it sometimes was able to exact compensation through the agency of the Mugabe (the king). The Bahima, on the other hand, could avenge the death of a kinsman if he were murdered by a Mwiru without consulting the Mugabe."[95] In short, the Bahera of Ankole had a status similar to their brethren in Kitara before a new course of change was introduced in the latter.

Kingship and the state structure were also the same in Ankole as in Kitara. The king must be strong and remain ever-young. Therefore, as soon as the king's "wives and followers observed signs of weakness, the Mugabe was given a poison which brought about his death"[96]. And, as in Kitara, two rules governed the choice of the successor to the king. "First, the new Mugabe must be in the royal line; second, he must be the strongest of the last king's sons. Patrilineal descent fulfilled the first requirement. The second depended upon some method by which the strength and courage of the Mugabe's sons could be tested. Primogeniture and favouritism, both important factors in the selection of a successor in the extended family of the commoners, also played their part in the royal family, but were overbalanced by the political and ritual demands of kingship. The Bahima demanded that the strongest of the king's sons should be their leader and that the test should be one of war. The brothers must fight among themselves until one of them alone remained alive in Ankole to claim the drum and the Mugabeship."[97]

Also, the king was an absolute despot. "The position of the Mugabe was exalted, his authority supreme, his leadership all-embracing."[98] "The legal status of the Mugabe gave him the highest political authority. Appointments to office were ultimately in his hands, as was the decision for war or peace. . . . The Mugabe's legal status gave him also the position of supreme judicial authority. He had the right to punish individually by death, exile, beating, torture, and cursing. He could confiscate the cattle of any of his subjects. He could prevent the execution of his people by his chiefs for criminal offences and could override the judicial decisions of the kinship-groups. In disputes involving two lineages, the Mugabe alone could grant the right of blood revenge."[99] Finally, the king was above the law. "No man could take legal action against him, for there was no authority higher than the Mugabe before which he could be accused, tried, and sentenced."[100]

The state structure developed into a hierarchy of chiefs, enjoying their exalted position at the pleasure of the king. "After the Mugabe, the most powerful man in the country was a chief who held the title of Nganzi, or 'favourite'."[101] He was the Prime Minister and Chief Justice, the last court of appeal being that of the king himself. The power of Nganzi "in the land and his influence with the Mugabe were great, and he often acted as the Mugabe's representative and judged cases of appeal from the jurisdiction of the chiefs"[102]. "The Nganzi was always a wealthy man, for he was constantly receiving presents of cattle and land from the Mugabe."[103] He "had estates in various parts of the country where large numbers of peasants lived and worked for him, and he had great herds of cattle which were pastured all over the land under his herdsmen"[104].

Besides the Nganzi and the Royal Pages, who also wielded great authority, there were the district-chiefs Mukungu (pl. Bakungu) appointed by the king; and other chiefs, such as Enyana, (the Keeper of Royal Cows), Abanga and Engangula (the warrior-chiefs)[105]. There were also subordinate chiefs in each district. "There was no animosity between the Mukungu and the subordinate chiefs in his district, but the latter were quite independent and only acknowledged him as their superior when some disputes arose among them and required authorative settlement."[106] In addition to these district-chiefs and subordinate chiefs, "there were a number of chiefs who were known as Bagalagwa. These were men who had been pages in the service of the Mugabe and who had grown too old for such posts. To such men the Mugabe would give estates, cows and serfs. The Bagalagwa in each district were under one of themselves and he, in difficult cases, appealed to the Mukungu of that district, who in his turn might refer the matter to the Mugabe"[107]. Like in Kitara, the district-chiefs were also invested with judicial power, and they retained their position so long as they diligently fulfilled the wishes of the king. The king always gave a district chief "a present of from one to three hundred cows which became his personal property and were used for the food of the chiefs and his household"[108]. But, although "these cows were a gift to the man and he regarded them, as well as any others he might possess, as his own property, the Mugabe might deprive him of them all if he saw reason to do so"; and no man could sell or exchange cows outside Ankole "without the king's permission"[109]. Similar was the position of all other chiefs, conditional to the king's pleasure.

The above facts show that long before the British appeared on their territory the Banyankole were far beyond tribal organisation. The institution of the state with a centralised *public power* had developed and the society was sharply divided into the ruling class of noble pastorals and the agricultural serfs.

However, in the pre-British period of their history the Banyankole did not progress to the extent the Banyoro did. Roscoe noted[110]:

> "Until recently the people of this region were comparatively unknown, though they were one of the earliest of those tribes who invaded the Lake Region and subdued the small and isolated village communities of negroes who were the original inhabitants of the land.... Evidently their nomadic habits, combined with their complete disregard of everything unconnected with cattle, prevented their making much impression either on the surrounding countries as warriors or on their own country as reformers.... Even the Mugabe, by which title the native ruler is now known, had originally only a roughly built bee-hive hut, with a stockade round it forming an enclosure in which cows were kept at night; the ground was always slimy with the animals' droppings, but this and the smell

of the cows were considered wholesome and pleasant. There was no courthouse, for the Mugabe's councils were held under the shade of some large tree; his main interest in life was guarding and improving the condition of his large herds; and his food was milk, which he drank in great quantities, and beer, which he drank, often to excess, at night."

To such an extent was this cult of the cow developed by the pastoral ruling class that the "cow was the standard by which all prices were regulated"[111]. Moreover, as Roscoe has noted[112]:

> "Contrary to the usual custom of kings and of the tribes surrounding Ankole, the king does not reckon his greatness by the area of his kingdom nor by the number of his subjects nor yet by the amount of land cultivated, but by the cattle he possesses."

It is also worthy of note that because of the dominance of pastoral life among these people they did not take to cultivation of any crop before the introduction of British rule. They "set no value on their land except for grazing purposes"[113]. Land was, however, an indirect source of income to them, "for the agricultural people who resided upon a chief's estates might be called upon to work for him, and they kept him supplied with beer, grain, and vegetable food"[114]. The distinction between the pastoral ruling class and the agricultural serfs thus remained very strong among the Banyankole.

Why was it so? The answer probably lies in the fact that conditions of agricultural production were not so good in Ankole as in Bunyoro, and that the Bahera of Ankole were not well-versed in arts and crafts. Roscoe has noted[115]:

> "The land is not so good for cultivation as in Bunyoro or in Uganda [that is, in Buganda], though there are a few agricultural people who eke out a miserable existence upon scanty crops of millet and a few plantain they grow."

Then again[116]:

> "These peasants cultivate the smallest amount of land possible for their existence and adopt the most primitive methods of cultivation, so that the results they obtain from the land can scarcely be said to be a satisfactory test of what the produce would be if it were properly tilled and given a fair trial."

And, while the state of productive forces in agriculture was less developed in Ankole than in Bunyoro, the country "was well-suited for cattle-rearing, and the good climate, excellent pasturage, and plentiful water made it possible for them to continue the nomadic life led by pastoral people, few of whom had any settled dwelling places"[117]. Pastoral life therefore remained the dominant note of Ankole.

There was also the point that although certain industries, like carpentry, iron-working and pottery-making, were pursued only by certain clans of the agricultural people as in Bunyoro, these were not so well developed as in the kingdom of Kitara. Hence the market in these products was not very extensive and hardly ever spread beyond the borders of Ankole. The parochial attitude of the people therefore remained as before, and there were not sufficient demands in the society for a change in the serf-like position of the Bahera.

It is also worthy of note that although slavery as an institution was present in the society, it was purely domestic slavery. Indeed, the slaves could be so much assimilated in the family of their masters that "if a man married his slave as his first wife and she gave birth to a son, he would inherit even though the father had later married a free woman and by her had a son"[118]. Moreover, slavery was very much restricted only to the top stratum of the ruling class, and slave-trade as a lucrative source of income to the king and the big chiefs appears not to have developed so much in Ankole as in Bunyoro. "Slaves were owned by the Mugabe, the leading chiefs, and the wealthier cattle-owners, and they consisted entirely of Bairu captured in raids made upon neighbouring kingdoms. . . . From all accounts, slavery was restricted to the very wealthy and slaves were restricted in numbers. Only those individuals who had sufficient surplus wealth could afford to keep slaves."[119] Furthermore, "no member of the nation may enslave another member; they are free born and therefore cannot enslave one another"[120].

Thus, from all aspects there was very little scope in the society for the development of new relationships beyond the already-formed class society determined by the nobility of birth.

Yet the society of Ankole did not remain altogether static. Though not at such a rapid rate as in Kitara, changes were taking place in Ankole and their direction was also the same as in Bunyoro. Although, unlike in Kitara, none of the Bahera were legally freed from serfdom, the king "might for some special reason grant a portion of land to one of them as his own estate"[121]. Some Bahera were thus made into landholders, although in a limited sense, for land essentially belonged to the clans. The social position of some Bahera was also enhanced by the union between a Bahima man and a Bahera woman. Oberg noted[122]:

> "In spite of the prohibition of intermarriage, miscegenation took place. A class of half-castes arose known as Abambari, whose status, although not clearly defined, was not always that of the Bairu. An *omwambari* whose father was a chief often came into the possession of cattle and was recognized as a man of importance, if of uncertain status."

The fact that a special name had to be coined for these "half-castes" tends to indicate that this was not at all an uncommon feature in the society. It is of further importance to note that such "half-castes" were produced mainly in the families of Bahima chiefs, so that the probability of the Bahera attaining property and a social position was more than if marriage was allowed indiscriminately between the Bahera and the Bahima. Oberg noted[123]:

> "Bairu concubines were especially common among Bahima chiefs and gave rise to a class of half-castes known as *Abambari*. From the point of view of legal status, the Abambari were classed as Bairu, but personal consideration often modified the strict rigour of the rule. A Muhima chief or cattle-owner without sons by a Muhima marriage would make an illegitimate son his heir. In time entire lineages were formed from such unions."

It is also worthy of note that the strictly inferior position of the Bahera in Ankole was further undermined by allowing some of them to join the king's army. Oberg wrote[124]:

> "It is also said that the present Mugabe's father established a Bairu band of warriors in order to counteract the determined effort of the Banyaruanda to conquer Ankole."

This was some time before the introduction of British rule in this territory.

While the position of the Bahera in the society was thus raised in the course of time, *differentiation* within the Bahima ruling class was leading to some of its members equating to the position of the Baheras. This appears to have been taking place in the economic and legal spheres, though not in the field of social and religious rights. Within the class in which "a man was considered poor or wealthy according to the number of his cattle"[125], eventually an impoverished stratum had developed which possessed very few cattle or none. These poor Bahima came to be employed by their wealthier compatriots on a basis similar to that granted to the Bahera. To quote Roscoe[126]:

> "Those pastoral people who owned large herds of cows employed as herdsmen men of the pastoral stock who either possessed no cows or had not enough to support a wife and family. Such a man would become the servant of some rich cow-owner, who supplied him with a number of cows for his own use and for the support of his wife.... These herdsmen were quite free to leave their masters without any warning."

Thus, although distinction by wealth and property had not yet seriously cut across the previously-formed classes by birth qualification, it would be wrong to say that these features were completely absent from the Banyankole society.

To sum up then, like the kingdom of Kitara, the kingdom of Ankole also represented features of a semi-feudal society. Although strict observance of distinction between the pastoral ruling class and the agricultural serfs remained a hindrance to the Banyankole developing much further before the British appeared on the scene, yet it should not be doubted that changes, even though incipient, were in the direction of a progressive development of the society. One would therefore be inclined to think that, given the normal course of development, these people also would have by themselves passed on to a higher stage of civilisation. This would result from (i) the ever-growing contradiction between the super-imposed Bahima and the dominated Bahera; (ii) the tension generated between the wealthier and impoverished strata within the Bahima ruling class, and (iii) as in Kitara, with the eventual opening-up of the country to world-trade while maintaining sovereignty and independence. Such, however, was not the course of history.

4. Toro

Although until the introduction of British rule in Uganda Toro was only a tributary state of the kingdom of Kitara, here also a class society had developed a long time previously. The society of the Batoro, the people of Toro, was formed in the same way as in Ankole or Kitara, that is, by the imposition of Bahima aristocracy on the indigenous agricultural people. Henceforth, as in Bunyoro and Ankole, the agricultural people were called Bahera or serfs by the pastorals. "Toro, according to tradition, is a part of Bunyoro."[127] Only at the beginning of the British period of Uganda's history, when the territory was controlled by the Imperial British East Africa Company, was Toro completely separated from Kitara and made into a kingdom by itself under the kingship of Kasagama, descendant of a prince of Kitara who had previously defied the king of Kitara and whose son had virtually proclaimed his sovereignty over this territory. Roscoe has noted[128]:

> "This Kasagama was descended from a prince of Bunyoro who, on being sent by the king to collect tribute from his subject peoples on Luenzori, rebelled and refused to return to Bunyoro. He baffled the armies sent against him, rejected all efforts to reconcile him to his king, and lived as a rebel in Toro, gaining power over the tribes there. His son, Kasagama's father, declared himself the ruler of Toro and assumed kingly powers over the whole country, which was thus entirely lost to Bunyoro."

This happened at the time of disintegration of the kingdom of Kitara when some of its feudatory chiefs in the tributary states came into a position to estrange themselves from the sovereignty of the Mukama, king of Kitara. A little later, when the Imperial British East Africa Company was lording over Uganda, Kasagama rendered "valuable assistance" to Captain Lugard, the representative of the Company, "and it was in return for this help that Toro was made an independent state with Kasagama as its first ruler"[129].

However, although Roscoe described the people of Toro as belonging to a tribe, they were far from a tribal stage. A long time before, with the domination of the Bahima over the indigenous people and the territory being brought under the rule of Kitara, a class society of pastoral aristocrats and agricultural serfs had developed. Probably because they belonged to a tributary state, the plight of the Bahera in Toro was even worse than their brethren in the kingdoms of Ankole and Kitara. Roscoe has noted, for instance[130]:

> "The agricultural clans of Toro are reduced to a serfdom which is little more than slavery; they are given the heavy work of building, road making, and of supplying plantains and corn for beer-making; they also supply food for domestic slaves attached to household."

It is also interesting to note how these people were being assimilated in the *milieu* of Banyoro. According to Roscoe, the indigenous people of Toro were "more likely to have affinity with the dwarf tribes of the Congo than with their taller neighbours who are pure Bantu"[131]. Probably they spoke a dialect different from those of Banyoro. But, having been brought under the orbit of the kingdom of Kitara, they had not only lost the basis of their tribal existence but also their peculiar speech. Roscoe noted[132]:

> "Whatever the vernacular of the people in the past, they have now acquired a dialect of Lunyoro (the language of the Banyoro), though it is far from pure, when compared with the language spoken near the Bunyoro capital."

Johnston in 1922 described the language of the people of Toro as a dialectal form of the language of the Banyoro[133].

Now with the definite establishment of a kingdom over them the characteristics of the kingdoms of Kitara and Ankole soon emerged among the people of Toro. "Chieftainship in the form of that in Bunyoro developed after Kasagama became the king under the auspices of the British"[134]. The characteristics of the society in Toro were thus similar to that of Ankole and Kitara, with the difference that probably as a mark of time here the rigidity of pastoral life was less apparent and soon the distinction between the Bahima and Bahera was less visible than in Ankole. Roscoe noted that the king Kasagama "still follows most of the pastoral customs, though he has no sacred cattle nor does he adhere to any of the more stringent milk rules of the Bunyoro kings"[135]. And Oberg wrote[136]:

> "From reports given to me by the natives of Toro and from Roscoe's account of the Bakitara, it appears that the Bairu-Bahima amalgamation had preceded much farther in these kingdoms than in Ankole."

It is also worthy of note that in 1948 Guthrie classified Toro as a separate language and not as a dialect of Nyoro[137]. It thus appears from the available material that, reflecting the course of development in the mother kingdom of Kitara, the society of Toro was quickly approximating the stage of development arrived at by the Bunyoro society.

At present it would of course be pure conjecture whether without British intervention the people of Toro would have eventually been assimilated in the society of the Banyoro or whether the Batoro would have developed into a separate nation, or at any rate a nationality, as they have done now. But one thing is clear; that is, the Batoro were not in a tribal stage when the British came, and that the course of change they were undergoing was not essentially different from that of the Banyoro or the Banyankole, or, for that matter, from other societies in the world.

5. Busoga

Like Toro, Busoga was a tributary state first of the kingdom of Kitara and later of Buganda. It first came under the influence of Kitara when some pastorals of highest birth were implanted from Bunyoro as the overlords of the Basoga. "Later the Baganda began to encroach upon much of the land which belonged or was subservient to Kitara; and by degrees the whole of Busoga, with the exception of one district, came under the rule of the *Kabaka*, or king of Buganda."[138] The history of Busoga in the pre-British period is interesting, for here again one sees how another "tribe" of Uganda had travelled beyond the tribal stage before the advent of the British, and was pursuing the same course of change as the other people.

The people of Busoga "were of the same class as the peasants of Kitara or Bunyoro, with broad noses and thick lips. At one time they formed a collection of small independent family groups, whose only connecting link was the intermarriage rendered necessary by the laws of clan exogamy"[139]. They lived mainly on agriculture. "The people said that they brought millet and plantains with them when they came into the country, but they learned to grow and use other food by contact with neighbouring tribes. Plantains were the most important crop."[140] Among them land was owned by clans, "and the head of the clan was known as the *Mutaka* or freeholder. This land could never pass from the ownership of the clan, and a clan member who was granted land by the head of the clan could not be deprived of it. There was always plenty of land, and a clan member who desired to possess some had merely to apply to the head of the clan, who would allocate to him a certain defined portion"[141]. The Basoga also possessed cows. "In Busoga each man was the real owner of his cows and no one had the right to deprive him of any."[142] It is interesting to note that the Busoga cows appear to be similar to those claimed to have been possessed by the indigenous agriculturists of Ankole before they were dominated by the Bahima. Roscoe noted[143]:

> "The cows of Busoga were smaller than those in most parts of Uganda; they were black or black with white markings, and had a small hump and short horns."

One should compare this with what Oberg wrote in connection with Ankole. To quote[144]:

> "There is a story among the Bairu that long ago they owned cattle but that these cattle were taken from them by the invading Bahima. Some veterinary officers in western Uganda believe that this is true and that the cattle which the Bairu owned were of a different breed from the present-day Ankole longhorn. This belief they base upon the existence of shorter horned stock upon the fringes of Bahima country, as, for instance, the cattle of Bakiga of Kigezi."

Cattle-rearing, however, was not the main occupation of the Basoga, although division of labour was noticeable between sexes in regard to herding of cattle and cultivation. In Busoga there "were no large herds, for even the chiefs had only a few animals; but every peasant, however poor, kept at least one cow, while some had as many as ten. . . . Men or men and boys herded them, and sometimes they were looked after with the goats and sheep by children; but women were not allowed to herd or to milk, though they might help to drive the cows into the houses at night or to let them out in the morning."[145] While some assistance from the women was thus given for the rearing of cattle, the men also helped the women in agriculture by clearing the land, "removing the trees and scrub, after the women planted", and also by giving "some assistance in the later work, such as pruning the trees"[146].

But although agriculture was the main source of livelihood of the people — "For many years Busoga was looked upon as the finest plantain-growing district in the Uganda Protectorate"[147] —the state of productive forces was poorly developed. "The only instruments used in agriculture were the large knife for pruning the plantains, and the hoe."[148] Industries also were not so well developed as, for instance, in Bunyoro. "Canoe-building was one of the important industries of Busoga, and at one time the Basoga canoes were the best and largest on Lake Victoria"[149]. This was when Busoga was a tributary of the kingdom of Buganda. Other industries however were rather poor. "Smithing was seldom attempted, for all iron goods, spears, hoes, knives, etc., were purchased from the Bakitara, who brought them into the country and exchanged them for goats and sheep."[150] Pottery was also far inferior to that made in Bunyoro. "Pots were made by the women", but these "were thick and heavy and little attempt was made to decorate or polish them"[151].

Such then were the life and living of the people of Busoga who apparently had lived in a tribal stage before they were brought into the orbit of the kingdom of Kitara. "In later years, however, the Bakitara, or Banyoro, became overlords of the country, and the scattered family groups were united into districts governed by chiefs."[152] This was the end of the tribal life of the Basoga. No more were the people demarcated by their clan areas and clan authority of the Mutaka. The country was now divided into districts, and the "paramount chief, or *Mwami*, in each district" became "the head of the clans in that district, the office being hereditary. He was regarded as the owner of all the land, and could grant to any man a piece of ground which would become the freehold of his family"[153]. Not only was the authority of the clan-heads thus seriously undermined, but henceforth the people had very little say in the matter of selecting their chiefs. "These chieftainships were hereditary; but disputes often arose as to which son should inherit, and such questions were settled by the king of Kitara."[154] The district-chief was also invested with judicial powers. "He tried any cases which were referred to him from the lesser chiefs, and might refer them in his turn to the king of Buganda"[155], that is, after Busoga was forced to transfer her allegiance from Kitara to Buganda.

The Mwami was not the only authority imposed on the Basoga after they were dominated by the king of Kitara. A state structure similar to that in Bunyoro developed in Busoga. "The first great chief was said to have been a prince of Kitara, called *Kitimbo*, who came with his brothers *Musali* and *Ngobe* from Kitara and ruled Busoga. His descendants, who succeeded him and were chiefs of western Busoga with the official title of *Kitimbo*, . . . were considered of greater importance than all other chiefs."[156] "The mother of such a chief was known as Namasole, and possessed land in her own right, where she had complete power, trying cases through her own prime minister or *Katikiro*. No other woman in the land possessed property."[157] Such was the power held by the imported chiefs and their families.

"The second grade of chiefs were known as the *Bakungu* and they, with the third grade, the *Bamitale*, formed a council to assist the superior chiefs when required. The *Bamitale* were appointed by the district chiefs or *Mwami*, who had the right to appoint any man to one of these posts, regardless of clan, and to depose him. Usually another man of the same clan as the former holder succeeded to such an office, but this was not a binding rule. Below the *Bamitale* was yet another rank of chiefs who were known as the *Wakisoko*."[158] Thus a host of chiefs was planted on the Basoga to keep them under the control of the kingdom of Kitara.

It is worthy of note here that in Busoga domination by these chiefs did not lead to the emergence of two sharply distinguished classes of pastoral aristocrats and agricultural serfs. This was probably so because not many pastorals migrated to Busoga with the chiefs. There is also the point that Busoga was a suitable region for agriculture, so that the agrarian economy prevailed even after domination. However, although the class division of society was not so widespread as in Ankole, Toro, or Kitara, it nevertheless developed between the chiefs and commoners. The former lived on the tributes of animal and vegetable food obtained from the latter, while in all matters remaining answerable not to the people of Busoga but to the court of Kitara and later of Buganda.

Indeed, this domination by a foreign power was the most serious problem for the Basoga. It affected their life in many ways. Roscoe noted[159]:

> "The effects of the rule of Kitara are still plainly visible, for the people follow in many ways the customs of the Bakitara. For example, at puberty they extract the six front teeth in the lower jaw, and many of their ceremonies, especially those connected with death and mourning, are similar to those of their former overlords. Their language, too,

was much influenced by Lunyoro and when contact with Buganda became close, the combined effect was such that nothing is now left to show what the original tongue was."*

Subservience of the Basoga to the foreign power was reflected not only in their social and religious life, but its most devastating effect was in their political subjugation which also resulted in a steady drain of the economic wealth of the country. One of the effects "of this subservience to a neighbouring kingdom was that the sons of chiefs were sent at one time to the court of the king of Kitara and later to that of the king of Buganda. This was not only a method of educating the sons of the chiefs, but it ensured that these boys would succeed to the chieftainships of their own country and rise to high positions there"[160]. In effect, it meant large amounts of gifts and bribes, which evidently in the last analysis came from the people of Busoga, to the top stratum of the ruling class of the imperial country. "The Basoga never entered Uganda [Buganda] without bringing large numbers of goats and cows as presents, and they often brought slaves too. The king frequently demanded cattle and women of them, and thus obtained much wealth. Whenever a principal chief in Busoga died, there was much quarrelling as to who should be his successor; and these disputes brought wealth to Uganda, whither each aspirant came to have his claims investigated."[161] Further to note, each "of the greater chiefs of Busoga had a small estate in Buganda, which was looked after for him by a steward. He was thus enabled to visit Buganda and remain there for some considerable time, living on the products of his own land"[162]. Such was the bond between the Busoga chiefs and their imperial masters; a bond which was stronger than that between them and the common Basoga, even though the latter supplied all the wealth to them.

Besides this indirect exploitation, the domination by the kingdom of Buganda in later years was felt directly by the Basoga. Busoga "was a mine of wealth, yielding slaves, cattle and ivory"[163], und therefore "for years the Baganda looked to Busoga especially for their supply of domestic slaves"[164]. Besides slaves there were other demands from the king of Buganda. "Under the rule of the king of Buganda, the Basoga were not subject to any fixed taxation, but messengers were sent into the country each year to collect a tax, and the king, if he found himself in need of animals or slaves, would send into Busoga and demand to be supplied with what he required[165]. In short, Busoga was a veritable wishing-tree for the imperial rulers. Furthermore the tax-collectors sent to Busoga from Buganda "seized the opportunity and demanded as much as they liked, keeping the surplus for themselves"[166]. Thus the general mass of the Basoga people were exploited not only by the chiefs who were planted on them and by their colleagues who emerged in the class society after the institution of the state had engulfed this territory; and, more over, they were exploited not only by the imperial authority alone, but also by the lackeys of the Buganda rulers. There-

* To a certain extent the language of the Basoga still remains a problem. But the problem itself suggests how the tribal dialect had undergone considerable modifications and alterations in the course of the changing history of the Basoga from the time their tribal life in all essentials was at an end. Both Johnston and Doke considered the language of the Basoga as a dialect of Lu-Ganda, the language of the Baganda. (cf. Johnston, *loc. cit.* 39, vol. II, p. 2; Doke, C. M. — *Bantu Modern Grammatical, Phonetical, and Lexicographical Studies since 1860*, International African Institute, London, 1945, p. 10). But Johnston further noted (*loc. cit.* 39, p. 20):

> "But Lusoga has not been studied to anything like the same extent as Luganda, and when it is set forth as amply and as clearly, it may prove to be entiled to the rank of a distinct language rather than a dialect. Moreover, when Lu-kena of the great marshes to the north of Busoga is examined, it may be found to be so distinct from Lu-ganda, yet so affiliated with Lu-soga that it would have to be regarded as the type of a distinct language in the Nyanza group of which Lu-soga was an outlying dialect strongly impregnated with Lu-ganda influence."

Guthrie in 1948 classified Lu-soga as a separate language (*lot. cit.* 84, p. 77).

fore when the British appeared in Uganda they received spontaneous help from the Basoga, for the latter hoped "that the British would free them from the yoke of Buganda"[167].

Here again an African people, who were no longer in a tribal stage and were intent on getting rid of their greatest obstacle to further development, might have developed to a higher stage of civilisation in the normal course by freeing themselves by their own efforts from foreign domination. But the advent of the British put them on a different course of history.

6. Other Regions of Uganda

In the remaining parts of the territory now included within the Uganda Protectorate, but excluding the kingdom of Buganda (of which a detailed account will be given at the end of this chapter), various other "tribes" were living from olden times. Unfortunately very little is known about these people, and especially about their social organisation in the period when they were free. But it appears from whatever account is available that just before the advent of the British most of them were either in a higher than the tribal stage, or were in the process of disintegration of their tribal system. Among the Banyaruanda, living at the Uganda-Congo border, the institution of the state had developed as in Ankole and Kitara and a class society of hereditary chieftainships under a king and commoners had definitely emerged. But apart from the Banyaruanda (and the Baganda), the general characteristic of the remaining peoples of Uganda was that among them the State had not emerged out of the domination of the indigenous people by invading hordes. It appears, on the other hand, that changes towards detribalisation and evolution to a higher form of organisation were taking place *within* these societies.

The essential features of such changes were (i) emphasis on territorial organisation of society at the expense of clan organisation, (ii) growth of centralised authority — the harbinger of the eventual emergence of *public power*, and (iii) the development of age-classes in some places, as a special institution under peculiar conditions, in order to replace the gentile organisation of tribal societies.

In most of these societies the territorial organisation by village-unit was becoming rather more important and beginning to influence the life of the people, especially in economic and political matters. Here, politically, the people were being divided not so much by their clans as by the areas they inhabited. Thus, not membership of a body of consanguinei but place of domicile was becoming the deciding factor of society, like the classification of the peoples of Ankole and Kitara and Toro and Busoga by the newly-demarcated districts which they inhabited and not principally by their clans and tribes.

Simultaneously, among some people the customary election of chiefs from one family or clan had led to hereditary succession, which was evidently first tolerated, then claimed and finally usurped. It then became a custom for the chiefs to receive honorary tributes from the people in return for protecting them or leading them in a war or raids. And in some places the supreme military commander or the highest chief began to aspire for despotic power. Tribal democracy, in which there is no distinction between the duties of the people and privileges of the rulers (in fact, no perpetual distinction between rights and obligations for all members of society, as chiefs were elected and deposed by the body of the clans or by the tribe as a whole), was thus being superseded by real governmental authorities; and the foundation was led for the eventual emergence of royalty and hereditary nobility from *within* a society.

Finally, among some people war or organisation for war were becoming regular functions in the life of the people. Acquisition of wealth by systematic raids on neighbouring people to capture cattle, treasure and slaves was becoming a regular feature of the societies. Consequently, in some places age-classes evolved to uphold political and military power in place of the total authority of the clans. Politically the ancient structure of these societies was thus undermined and particular age-sets which represented the military power of society in particular periods and in various grades of responsibility began to dominate the entire community.

Most of these societies thus took the character of *military democracy*. Military, because war and defence were now one of the prime concern of the people; and democracy, for the structure of tribal democracy still persisted, although it was now perverted in order to justify forcible robbery of wealth. The institution of the state had not, however, developed yet, which would of course have to follow logically in order to perpetuate the newly-rising class division of society as well as to stabilise and protect the right of the possessing class to exploit and rule the non-possessing class. Military democracy was thus in effect the transitional stage between the tribal and state organisation.

An attempt has been made in the following pages to show, from whatever evidence is available, how the above changes were brought about among the remaining "tribes" of Uganda (except the Baganda), or, at any rate, among the most important among them. It should however be noted that because of lack of information some of the sparsely populated areas of Uganda have had to be left out of the present discussion. It is also worthy of note that this tentative study may have to be drastically revised in the light of new and additional information about these people. Nevertheless the following account may be useful in initiating further discussion and research on those people of whose life and living conditions in the pre-British period of Uganda's history very little is yet known.

7. Acholi

Of those less important "tribes" of Uganda in the pre-British days, the Acholi and the Lango were perhaps the most important. Authorities, like Westermann, Schweinfurth and Driberg, are of the opinion that the Acholi and the Lango are linked with the Shilluk people, the former representing the most advanced guard of their southward migration and the Lango the next[168]. However that may be, the Acholi living in the northern part of Uganda (as well as in southern Sudan, of which no account will be given here) trace their descent from various other peoples who still live in the neighbouring regions. According to a survey conducted in 1949—1950, 43 per cent of Acholi *kaka* (clan or the biggest lineages) claim origin from the Luo, 17 per cent from the Karamojong, 14 per cent from the Lango, 9 per cent from the Madi, 2 per cent from the Bari, and so on*. It appears thus that the Acholi society was formed of various people settling down in one territory, taking up a common economic life of hunting and rudimentary agriculture, in the course of time speaking a common Acholi language and evolving a common Acholi culture, so that ultimately they considered themselves only as Acholi.

Probably at the beginning the Acholi lived under a strong patriarchal clan system. Even today the Acholi territory of the Uganda Protectorate is demarcated into sub-counties,

* This survey was conducted by Dr. F. K. Girling, at that time a research scholar of Oxford University. The writer collaborated in the survey in the initial stage to prepare the sampling design, the questionnaire-schedules and preliminary interrogation, and later to analyse some of the material collected from the survey. To the knowledge of the writer this very interesting work of Dr. Girling, which would throw considerable light on the life of the Acholi both in the pre-British as well as in the British period of Acholi history, has not yet been published. Needless to say, the opinions expressed here on the basis of this survey are entirely those of the writer, except where Dr. Girling has been specifically referred to.

parishes and villages according to the clans and lineages of the old inhabitants of these respective areas; the names of the modern administrative areas being synonymous with the old ancestors whose descendants were the people of these areas. A sub-county, or Jago, refers to the *kaka madit*, the clan or the biggest lineage, such as Payera, Patiko, etc. "Payera" in Acholi language means "of Ayera", that is, the descendants of Ayera who was the traditional head of this clan or basic lineage. A parish, or Mukungo, refers to a *kaka matidi*, a small lineage. Thus, Payera is divided into four Mukungu, the names of which refer to Ayera's four direct descendants who a few generations later had separated from one another and started separate lineages. Finally, a village, or Won Pacho, very often refers to a *kaka matidi totwal*, a very small lineage. Such are Ngotor and Buchota in the Mukungo of Ogum' in the Jago of Payera. These names again refer to the descendants of Ogum (who himself is a descendant of Ayera) a few generations later. At present smaller lineage units (formed by the sub-division of the *kaka matidi totwal*) can be found in a village, and people belonging to different clans or lineages are also found to have settled in one village. But the persistence of the names of ancient clans or lineages for these areas signify that long before British occupation these areas were held communally by different clans or lineages[169].

Differentiation in the society, however, took place in course of time, and hereditary chieftainship began to be the form of political organisation. Crazzolara is of the opinion that "the Acholi consist of a number of tribal groups, some of them very small, each comprising several kingdoms or chiefdoms; the chiefdoms are ruled by independent chiefs (*rwot*)" [170]. By the time these chieftainships came into existence, the great chieftains or Rwots were selected only from certain favoured linages, which Crazzolara described as Royal Clans but presumably meant favoured lineages[171]. The 1949—1950 survey elicitated the fact that today there are only some 40—50 such clans or basic lineages among the Acholi out of a total of about a thousand. This suggests how class-differentiation had taken place in the Acholi society even in pre-British times. It is interesting to note that "the royal lineage is called *lokal* (or *lokeer* or *lobito*) and the commoner clans are called *bon* or *lobon* or simply *lwak*, meaning the mass or bulk"[172]. It should further be noted that according to the survey mentioned above these royal lineages mostly claim their origin from the Luo. As suggested by Girling, this may be explained by the fact that the ruling class of Kitara had a common origin with the Luo and that during the civil wars in Kitara (for succession to the throne, as explained before) the defeated sections often fled to the Acholi country and eventually became chiefs[173].*

Such a selection of chiefs only from a few lineages evidently undermined the total authority of the clans or the biggest lineages after which different areas of the Acholi country were named. Henceforth, the people were no more under the control of their respective clans or lineages; on the contrary, chieftains selected only from the favoured lineages began to rule *over* them. With the growing differentiation in society, the old organisation was thus giving way to a new one.

However, the great chief or Rwot of an area, thus chosen from a favoured lineage, had not yet developed into an absolute monarch or a despot. The support of the people was still needed for raising a particular person to this authority, although the selection was restricted to "a reigning line of chiefs belonging to a royal lineage of the district" [174]. The ancient tribal authority was thus partially recognised, but the organisation was now perverted to suit new conditions of making raids and counter-raids on neighbouring regions and people in order to acquire wealth. Fights between chiefdoms were not of unusual occurrence. The Acholi society was therefore based territorially, and a great chief ruled over each of

* This suggestion may be disputed. Also the author does not know whether Dr. Girling still holds to this view.

such territories. "Girling states that a chiefdom usually consists of one or more villages, varying greatly in size, and includes the surrounding territory." [175]

Why was so much stress given to village organisation? Because war and defence were now the prime concern of the society. Therefore, "in former times villages were fairly large and were surrounded by stockades. Many were built on hills or rising ground and houses were packed closely together, the population being about 100—200. Nowadays, stockades are no longer so important as in the past when there was considerable inter-village raiding." [176]

Under such territorial demarcation various clans lived together. Crazzolara noted[177]:

> "One clan is never found wholly together in one unit; it may be divided up and dispersed in various district units. Such dispersed groups never had a common headman; only the single clan group within a unit has. Such a single clan group may consist of various colateral groups with their own headmen."

Thus it appears that no trace of clan authority was found when the European investigators began their investigation in modern times.

On the other hand, age-classes had developed in the society to suit the military demands of society. These age-classes of Acholi males from 15 years and upwards formed the fighting organisation of the people and were organised on a territorial basis[178].

And in such an organisation of society, the Rwot played the central role. He was "a very important factor in bringing about the cohesion of the various territorial parts of" a chiefdom[179]. Because of his position he thrived on tributes collected from the people. "The *rwot* is given gifts of meat and fish. Also the village pays the bridewealth for the rwot's chief wife, *dak ker*. The heirs to the chiefship and to the rain-making function must be born from this union." [180] The Rwot was not only the leader of war, but "very often an hereditary rain-maker"[181]. Besides usurping political power, the social position of the Rwot was thus enhanced, and as a rain-maker he claimed further tributes from the people. "The skins of all lions and leopards killed are considered the property of the rainmaker and are brought to him."[182]

Within a territory demarcated by a chiefdom a hierarchy of chiefs also developed. "The *jago*, a sub-chief, comes next to the *rwot* in authority. Girling states that the larger village units in a chiefdom are represented by the *jago* while the *ladit* is an 'important man' and is head of a small kinship group. The *jago* is the head of the dominant lineage of his village as well as the representative of the village as a whole. The dominant lineage to which he belongs is not necessarily of the same clan as that of the *rwot* whom he acknowledges."[183] Thus it is seen that the new organisation of society arose on the ruins of the old, of which the Jago as also "head of the dominant lineage of his village" was a feeble representative. In short, classes had begun to develop in Acholi society on the basis of military democracy.

The institution of the state however had not yet emerged. But there is evidence to indicate that just before the advent of Europeans the class-society of the Acholi was leading to the usurpation of despotic powers by the great chiefs, which was obviously the harbinger of monarchy and the formation of the state. This was particularly conditioned by the prior arrival of the Arab slavers and traders. To the Acholi also the world was opening up with the greed and lust of a new age. While the common people and the weaker chiefs suffered from the predatory raids of the Arab traders to capture slaves, some of the great chiefs (who had already consolidated their position to the extent of aspiring for the supreme leadership of all the Acholi by means of superior military strength) took advantage of the situation. The Arabs traded with them in guns and ivory and encouraged them to make raids on neighbouring people to capture slaves for themselves. Driberg noted that "the slave-traders' headquarters were among the Acholi at Fatiko" and thenceforth they conducted raids on other peoples' territory with the help of the Acholi; so that, "it is quite conceivable that the Lango would identify the Acholi with the slavedealers"[184]. Thus, just as the arrival of Arab traders increased inter-people warfare and brought great sufferings to the people, so it also enhanced the autocratic power of the great chiefs. By the time Europeans like Samuel Baker or Emin Pasha first visited the Acholi territory, their society was developing towards centralised political system under the despotic leadership of some very able *Rwots*[185]. Each of these Rwots wanted to be the monarch of all Acholi. Pre-requisites were thus created for the formation of State among the Acholi, but the introduction of British rule put them on a different course of history.

8. Lango

Among the Lango also the course of change was similar. Like the Acholi they also were in the stage of military democracy when the British arrived. In those days "war was the natural condition of things and peace an interlude"; hence, the most important political and social unit among the Lango was the village, of which the primary object was "for protection against attack and for purposes of offence"[186]. For this purpose the "villages were larger than in these more settled times, as the larger the village the greater the security on the one hand, and the more extensive their opportunities for plunder on the other"[187].

This territorial organisation of the Lango had sharply cut across their previous clan organisation. Under the stimulus of war and migration clans and even families were often widely scattered. Driberg found in the early years of British rule among them that "in one village at Kibuji sixty-seven men belong to the clan Agodya, while the remaining twenty-six men are members of seven different clans"[188]. That the village had become more important than the clan organisation of former times is evident from the researches of Driberg. He found that "property in land is held communally by the village, the land being held to include the grazing and water rights"[189]. He further noted that though clans have become much scattered and broken up, "indications are not entirely wanting to show that in the remote past land was held communally by the clan within the sphere of tribal occupancy, and that such clan-land was at the disposal of the individual members of the clan for their use so long as they required it. This clan tenure, however, is by now almost non-existent; but in a few cases, as at Abako, traces of it can still be found, a member of an alien clan settling in their property being required to give the head of the Jo Abako some small present, such as a bull or a goat. By this present he secures the privilege of tenancy and has equal rights in the usufruct of the land with the members of the clan, without, however, being absorbed into the clan itself. For all practical purposes, however, clan tenure may be disregarded, and the village is the actual unit of communal occupancy."[190]

With this growing importance of the village the most important person in the community became the Jago or the headman of the village. Although he was selected from the main clan of the village, he had to depend on the council of village elders for dispensation of justice; and this council obviously included members of various clans. Moreover, suited to the exigencies of the time, such councils also were "in reality courts of arbitration rather than of judicial decision, and there was no machinery existing for the legal enforcement of their awards"[191]. Hence, although law among the Lango operated with any certainty only within the limits of the village, the administrative or judicial body of village elders was no more than an informal gathering. The rule of the Jago also was not decisive. This was so probably because military force and fighting qualities were becoming the decisive factors in the society. Indeed, the impotence of this judicial body was particularly revealed in the case of intervillage disputes which were more often solved by inter-village raiding or declaration of war instead of coming to a peaceful settlement. Butt noted[192]:

> "In inter-village disputes much depended on the overall relations between the two villages involved. In a nearby friendly village a man might obtain some compensation for his grievance. In an unfriendly village or one under another *rwot* he would probably gain nothing without starting an inter-village feud."

And these feuds often took a very destructive character, causing several deaths and looting of persons, cattle, etc., as war booty.

Such, however, was not always the course of solution adopted by the Lango. In a still earlier period the tension between two villages was resolved by a whipping duel, if and when the elders of the two villages could not reconcile the two parties. In this duel the warriors of the two villages stood in front of each other with a felled tree between them, and whipped each other with eight-feet long whips "until an umpire who stood at one end raised a stick between them when he thought that they had punished each other enough"[193]. Later, however, internecine wars spread among the Lango, which, according to Driberg, they learnt from the Banyoro. The underlying motive of such internecine wars was evidently the desire of the warlords to obtain paramountecy over the entire people and of his participants to plunder the enemies' territory. Driberg noted[194]:

> "For more than the last hundred and twenty years Lango mercenaries have made periodical incursions into and beyond Bunyoro, assisting bickering factions, abetting rival claimants to the throne, or allied with the Banyoro in carrying war and fire into Buganda. During these raids and forays they were spectators of civil war, of villages striving on against another to the death; they saw women and children killed in the heat of battle or sacrificed afterwards to the lust of revenge. It is not surprising therefore that after many years what they saw beyond the Nile gradually changed their own view-point, until fired by these examples Akena (whose mother was a Madi captive) resolved to win for himself mastery over the Lango at the point of the spear. His struggles with neighbouring leaders were long and persistent, especially with Ogwali of Chegere, so much so that songs are still sung of them, and *yi pa ogwali ked' Akena* (the struggle between Ogwali and Akena) has become proverbial of any difficult situation."

Henceforth, "there was a perpetual state of war, unrestrained by chivalry or rules of conduct, with the very briefest interludes of peace[195]. From that time "internecine strife became the rule, and with the old order was jettisoned the chivalry which formerly prohibited the killing of women, children and cripples even during war. Men, much less women, could no longer go abroad from their villages by day unless accompanied by a band of armed friends"[196]. Looking-out for spoils of war and raids, and undermining the position of others became the prime concern of the Lango. Driberg noted[197]:

> "While fighting is always liable to break out, the favourite season for war is in the months of June and July, when the young crops are in the field, as part of the rewards of fighting is the destruction of your enemy's crops, with the result that he has subsequently to buy food or to go hungry. Women and children, cattle and goats, are the spoil of victory, but women and children are sometimes killed in the heat of battle. They are not sold as slaves to other tribes, but not infrequently change hands among the Lango in consideration for cattle. Their captors, however, often marry the women and adopt the children, especially the girls, who will subsequently bring in a good marriage dowry. Captives can nevertheless be ransomed by the defeated village, who will in all probability have suspected or been warned of the impending raid, and will have placed their livestock in the charge of friends several days' journey away."

This situation had a far-reaching effect on the Lango society. The people "were divided into a number of factions unrelated to their clan system, which were often hostile to each other and rarely worked in cooperation"[198]. The organisation of the society became essentially military. Driberg noted[199]:

> "Hunting and fighting were the pursuits of the Lango, the former providing them with meat and the latter with booty in the way of cattle, in which their hearts delight. To this end, that their youngmen might be worthy warriors, tobacco was only allowed to the old men, a solace to their unmartial existence. Similarly, the games of children were of such a type as to train the hand and eye to that swiftness and co-operation which are essential to a successful fighting man. In the same way the dance in which they voluntarily endured painful flogging was to develop courage and the power of endurance. Accordingly, just as all their young lives were moulded in such a way as to produce warriors, so their government was of such a nature as to suit a warrior people. Over each village was a *jago* or leader of a company, whose duty it was to lead his detachment under the *rwot* or leader of a column."

The office of the Jago was not necessarily hereditary. It depended on his ability in leading in battles and managing the village during the interludes of peace. Generally the son of the deceased Jago was elected to the post, but should this son "be evidently an impossible person, either on account of mental or physical capacities, the most suitable member of the family is elected. He holds his office for life, and his son would be the heir presumptive; the title would not revert to a lineal descendant of the old *jago*"[200].

As mentioned above, the principal duty of the Jago was to organise and lead "the fighting men of the community, regardless of which lineage they belong to, and so protecting the village from the attacks of other *jegdi* (plural of *jago*)"[201]. Consequently, the life-long aspiration of a Jago was to become a Rwot, who "was the nearest approach to a chief which the Lango knew, his sphere comprising a varying number of *jegdi*, possibly only three, but in exceptional cases as many as nine to ten"[202]. For this purpose the Jagos were always striving to obtain "a following in neighbouring villages if his lineage is particularly powerful and his ability in war especially remarkable"[203]. And a Jago who could thus extend his influence became the war leader for a district and was known as Rwot. The Rwot's "functions were almost entirely military, and though often a elected clan headman, he owed his position largely to military success, combining with other *rwodi* (plural of *rwot*) for larger operations under a *twon-lwak* or war-leader (lit. bull of the crowd)"[204].

This emphasis on the military qualities of the Lango was climaxed by the office of *twon lwak*. Driberg noted[205]:

> "The *twon lwak* was not elected, and his title depended on his success; for from being a *rwot* or even a *jago* he gained his position as the war-leader over a group of *rwodi* by his personal influence due to prowess in battle."

Thus the Twon Lwaks were in effect the military despots in making, who had practically no connection with the ancient social structure of the Lango but were now becoming the supreme political leaders, — "in fact becoming paramount chiefs of a large area"[206]. Suiting to the new demands of society, the Twon Lwak's "tenure was precarious, since just as success attached *rwodi* to his standard — the hope of booty being a compelling factor — so in time of failure they would gradually desert to a new and more successful leader. Thus we find many cases of men winning a wide reputation and much renown, in fact becoming paramount chiefs of a large area, suddenly losing their position and following almost entirely. Occasionally, indeed, they have maintained it to the end, possibly dying gloriously in attle, . . ."[207].

Although despotism was thus gaining ground in society, it was not yet stabilised. Thus, even though a Twon Lwak might have retained his prestige and position up to the end of his life and career, he would not have been automatically succeeded by his adult sons or brothers. They could inherit the office only when they have "already proved themselves to be worthy inheritors of the title, as was the case with Odongoja, who was succeeded by his brother Otwal. But such a succession is accidental: the son succeeds not by virtue of his sonship, but by reason of his own prowess"[208].

The upshot was obvious. In this transitional stage of the society towards the centralisation of authority and the final emergence of *public power*, the only aim of the Twon Lwaks was to outbid their rivals, so that one of them could finally stabilise his position in the society. Therefore, each "*twon lwak* was against his neighbour, endeavouring by the bribe of success to detach his neighbour's villages to his own standard, in order that he might *chamo piny duchu*, eat up the whole earth"[209].

In other words, the attempts of these big war-leaders was in the direction of overthrowing the stage of military democracy in the society and inaugurating the institution of the state. Accordingly, "we find that the history of the Lango is one of internal strife, war-leader vying with warleader — at any rate since the example of the Banyoro civil wars, for prior to that influence inter-village hostilities appear to have been contrary to the tribal consciousness, a fact which perhaps argues a greater degree of civil authority at that time in their chiefs and headmen"[210].

The above facts suggest that before the British arrived in their territory, the Lango had travelled beyond simple tribal organisation and were in the stage of military democracy. They had not, however, yet reached a higher stage of development. Their arts and crafts remained ill-developed, and as agriculturists also they did not command any particular attention[211]. Even their military skill did not make itself felt beyond their borders. And although the big warlords were aspiring for a supreme position in society, there was not yet any paramount chief for all Lango or a centralised monarchy. Driberg noted that "normally there is nothing to distinguish the chief's house either by size or position"[212]. This was probably so because the land of Lango had not yet opened up to the world; in many essentials the society therefore remained parochial. Things might have moved faster if the Arabs had come to terms with them just as they did with some great chiefs of the Acholi. In the end, of course, the Lango came in contact with the outside world through the British, and this made a profound change in their society. But, because of the nature of this contact, the course of change was very different from what could have been expected in a free Lango society. This will be discussed in the next chapter.

9. Iteso and Other Nilo-Hamites

Not much is known of the life of the Nilo-Hamites in Uganda before they came under British rule. But from what evidence is available it appears that these people also had advanced beyond simple tribal organisation. Among them also territorial organisation as the social and political unit had become more important than clan organisation, and the people were constantly at war with each other. Referring to the Nilo-Hamites in East Africa, that is, including also those who are now living in Kenya or Republic of Sudan, Gulliver noted[213]:

> "Turkana made war on all their neighbours with the exception of the Jie, with whom they occasionally allied themselves against the Karamajong and the Dodoth. Karamajong similarly made war on all their neighbours with the exception of the Dodoth, with whom

> they occasionally allied themselves against the Jie. Jie claim friendship with the Toposa, but since they have no common boundaries this would have been of little importance. Toposa and Donyiro did not fight each other, and are known to have formed an alliance against the Turkana. Toposa and Jiye were enemies."

Thus, war and organisation for war had become the prime concern of these societies, the purpose of which was obivous. "War was carried on more or less continuously by raiding parties, whose chief interest was the capture of cattle and other stock"[214].

And the result of such war-like organisation of the people was that one of the primary attributes to define a tribe among the Nilo-Hamites was found to be that "its component parts do not go to war with each other". Thus, while describing the "Political Organization" of the Nilo-Hamites (of whom the "Central Group occupies much of north-eastern Uganda"[215]), Huntingford noted[216]:

> "The investigator soon becomes aware of the fact that the Nilo-Hamitic languages have no word for 'tribe', for the words which are sometimes given this meaning turn out on examination to mean something different: e. g., the word *emet* which Hollis gives as the Nandi for 'tribe' turns out to be the name of an internal territorial division. Among the southern and central Nilo-Hamites a major, independent group of people comprising a number of clans and other sub-divisions forms a 'tribe', its members united by a common name, land, and language. All those not belonging to the group are 'foreigners' who do not count, and therefore there is no need for a word to define the group, beyond its name, which gives all the information needed. One does not ask a Nandi what his *tribe* is: the question is, what sort of man are you? The tribe is a cohesive unit, its members linked by social and ritual ties, and its component parts do not go to war with each other."

The imprint of the situation was visible in the social organisation. The clans were so much dispersed in different villages or territorial locations that they could not function as political bodies. Gulliver noted[217]:

> "The types of clans within the tribes vary from small territorial ones in Jie to large widespread ones in Turkana and Karamajong. It is most likely, however, that the latter are a development of the former, following the increase in population and the colonization of new areas. Jie is typified by a high concentration of a relatively small population, which has scarcely moved for many generations; Turkana and Karamajong are both typified by larger populations which have spread out in the last hundred years or so. Indeed, the Turkana say that formerly their clans were concentrated."

And then, referring to the "Restricted Political Function of the Clan System" among the Nilo-Hamite peoples, Bernardi wrote[218]:

> "All the Nilo-Hamites are organized into a system of dispersed patrilineal and exogamous clans. Descent is traced in the father's line, though it seems that the agnatic aspect is not everywhere stressed to the same extent. Clans are generally, but not always, subdivided, and it is usually these sub-clans that are exogamic. Nowhere is the lineage stressed as a distinct unit of the clan, nor is it ever suggested that the lineage provides the basis of the political structure. Indeed, the political significance of the clans is very limited. Generally clan-units have a corporate standing in matters concerning cattle ownership, inheritance distribution, the use of wells or other water sources, and the payment of murder compensation. Another aspect of considerable relevance is the present territorial dispersal of the clans. When clans are referred to as corporate political units, the reference applies to local groups of clan members. Clans seldom, if ever, act by gathering all their members from over the country. The occurrence of a whole clan in a single territorial section, from which some people have argued that, in the past, territorial organization was based on a clan pattern, is in fact very rare. The present dispersal of clans is, however, clear evidence

> that the political integration of Nilo-Hamitic societies is no longer based on clans. Therefore one has to say that, though the clan principle still provides a mythical and ceremonial link for clan members and a pattern for some forms of corporate activity, it lacks such cohesive power as would integrate the territorial, military, and juridical systems into a single political whole."

In other words, in the changed situation, some other institution had cut across the clan-organisation as the political authority. In most places it was the age-class system.

In a society where chiefdom as an institution had not developed to a decisive extent — "The Nilo-Hamites are conspicuous among African peoples for having no form of chiefship with centralized authority, or any other centralized machinery for political and administrative control"[219] — the age-class system was the agency which kept the society from bursting asunder. Referring to the "Integrating Power of the Age-System", Bernardi noted[220]:

> "The structure of the age-system takes living shape in the local groups. All the activity carried out in a corporate form at any level reflects the structural stratification of the age-system. Thus, though its actual shape is cast in a territorial mould, the system as a whole overflows all territorial limits and provides the basis for the unification of the territorial divisions."

This age-class system was not just a physiological classification of the members of a society. This was an institution by which the members of a society were classified into definite age-grades, and it thus provided "the Nilo-Hamitic societies with the machinery for organizing every form of corporate activity — social, military, juridical and ceremonial"[221]. The importance of this age-class system among the Nilo-Hamites, at a time when the authority of clans had declined and territorial organisation had gained ground, and when, while war had become the prime concern of the people, and no centralised authority had yet emerged, has been aptly stated by Bernardi, although, as is evident from the previous and the following extracts from him, he did not look at this institution from the point of view the present author is inclined to do. To quote Bernardi[222]:

> "Normally territorial divisions are integrated by some unifying principle, such as the institution of kingship, which, through its centralized power, gives cohesion to all territorial divisions and makes their members conscious of unity at a higher level. In societies with no centralized authority other agencies may produce the same cohesive effect. Among the Nilo-Hamites the age-system is such an agency."

Evidently, the people no longer remained under simple tribal organisation.

Thus, among the Jie the clans "are largely ritual groups, clan affiliation determining the course of ritual connected with marriage, fertility, and children"[223]. On the other hand, the age-class system "compelled a degree of social conformity, and could corporately award punishments, especially for ritual infringements"[224].

Among the Karamajong, another Nilo-Hamite people in Uganda, the hamlets "appear to consist of homesteads of mixed clan affiliation"[225], and here again the age-class system is perhaps the most important social and political institution of society.

Among the Dodoth, who now inhabit a county of Karamoja district of the Uganda Protectorate, hamlets were formed with "homesteads of different clan affiliations", for the clans had been earlier dispersed over the whole area inhabited by these people[226]. It is also worthy of note that, in the words of Gulliver, "co-operation within a hamlet follows the usual lines in the economic, social and ritual spheres"[227]. This suggests that territorial organisation and the age-class system had become the dominant features of Dodoth's life as among other Nilo-Hamites of Uganda belonging to Gulliver's "Karamajong Cluster".

Among the Iteso, the most prominent of the Nilo-Hamite peoples in Uganda, "the age-set (*asapan*) was undoubtedly the most important organisation of society in the past"[228]. This was the situation some time before the British arrived. And this was not all; some more significant changes had taken place in this society in the pre-British period. While there was no "political system based on clans and clan leaders", territorial organisation of society had become an established fact[229]. *Etem* "was the most important territorial grouping in Teso society"[230]. "*Etem* means a fireplace, the ceremonial hearth where members of all the different clans in a locality met to discuss religious and judicial matters, war and hunting"[231]. Thus, in all aspects of the people's life, the territorial organisation of society had the primary control; "the clan system (*atekar*) appears to have little other than a ritual significance"[232].

It is also worthy of note that probably further changes were taking place among these people in the pre-British period. Lawrence noted that the Iteso, "at the time of the Baganda occupation, were changing from a nomadic pastoral to an agricultural existence"[233]. This was the time when the British arrived, and this was also the time when while there "were no chiefs who had any extended authority" among the Iteso, "here and there a man might make himself the head of several villages"[234]. This was probably the appearance of despotic power in its incipient stage, and was thus the harbinger of a new situation among these people. But before it could establish itself in the society, the people were brought under British control, and their normal course of development ended thereby.

10. Bakiga

Of the rest of the "tribes" in pre-British Uganda, Bakiga is worthy of note. The Bakiga of the Kigezi country who lived in plains enclosed by hill ranges "of considerable height", appear not to have been seriously affected by the invasion of pastoral hordes, although their close neighbours were the Banyankole. When the British occupied this land, the inhabitants numbered "well over a million souls"[235]. "The Bakyiga were a wild race who in the past resisted all attempts to bring them into subjection"[236]. Among them there "was no supreme chief, but the tribe was divided into clans which were ruled by their own elders and lived completely isolated from each other; they were even hostile, for one clan would not associate with members of another and it was unsafe for a man to travel alone beyond the boundaries of his own clan land"[237]. "The people were of the Bantu stock and were mostly agricultural, though a few pastoral clans might be found on the lower slopes of the hills where the large plateaux afforded excellent pasturage."[238]

The state of productive forces in agriculture was, however, poor, and industries were not well developed. "There were men who had a certain rude knowledge of smithing. They collected their own iron-stone where it lay on the surface of the ground and smelted it, following the Ankole methods, which were rough and inferior to those of Kitara. The chief things made were hatchets, knives, and hoes, but warriors also required arrows, good axes, and one or two spears. Carpenters were few and much less expert than those of Ankole, and the articles they turned out were not in the least artistic in shape or well finished."[239] "The art of pottery was also in a very primitive state. The vessels were thick, brittle, and of a clumsy shape."[240]

Seemingly the Bakiga thus lived in a primitive tribal stage. But their life was not so simple as that. Territorial division had made sufficient in-roads in the society to make the village more and more the most prominent unit. "There might be several villages belonging to one clan, for men might build for themselves a little apart from the first village. If anyone wished to join one of these groups, he had to bring the leader a sheep in order to get permission to build in his village. Such a village was called *Ekilolero*, and the head-man was a *Mukungu*."[241]

Village-units were thus cutting across the ancient structure of society, and in course of time village-councils invested with judicial power were formed. Roscoe noted[242]:

> "As the members increased, men were chosen as elders of the village to assist the *Mukungu*, who, however, retained the right of final decision in all matters. No judgement of a case was valid unless it had been set before these elders of the village, and a man had always the right to refuse to accept any other means of trial."

The village headman was not only the supreme judicial authority, but henceforth he also began to collect tribute from the villagers. "When a man wanted land he applied to the head of the village for it, and an annual rent of a pot of beer was often imposed. Land thus granted was handed down from father to son and anyone who intruded on it or questioned the owner's right to it ran grave risk of being speared down on the spot."[234]

Indeed, the village eventually became such an important unit in the society that the "members of each village claimed as their own the sides of the hill in which their village was built, and any intrusion by strangers was fiercely resented and often led to strife and bloodshed"[244]. This territorial consciousness had become so acute among these people that when working in distant fields of a village — "for naturally those immediately surrounding the village were soon all occupied" — "men went in companies of three or four and took with them their weapons, keeping them near so that they might seize them in case of alarm or sudden attack"[245].

It appears that the desire to plunder was the cause of such inter-village attacks. Thus Roscoe noted in this connection that "if an attack was made the men were killed, the women who were with them were spared, but were taken captive and kept until ransomed by their husbands or clans"[246].

It should further be noted that since neighbouring villages might have been inhabited by members of the same clan, as is evident from the above description of the formation of new villages, such inter-village fights suggest that territorial consciousness had gained so much over clan solidarity that members of different villages but of the same clan could have used weapons against one another and taken away the women of the defeated side in order to collect ransom from "their husbands or clans".

It is also worthy of note that slavery as an institution had developed among the Bakiga, and slave-markets used to be held regularly. "Slaves used to be common, and were bought for a cow or a sheep."[247] Evidently, the society was in such a stage that human labour could yield sufficient surplus over the cost of its maintenance. No doubt, the purchase of slaves marked the rise of property differentiation in the society, for obviously not all people could buy slaves or buy a number of them. Further to note, currency and exchange had also developed among these people, and probably due to pastoral influence the cow was the recognised standard for it. Roscoe noted[248]:

> "Cows were of importance, not so much for their milk, which was looked upon as a luxury and not a necessity, as because they were the means of obtaining wives and of purchasing other things, for prices were fixed on the standard of the value of a cow. Cows were seldom killed except to celebrate some special occasion such as wedding, or in the event of an offering made to some ghost. Animals were too valuable to be killed to gratify cravings for meat."

Thus, in short, the Bakiga living in the extreme southwest of the present boundaries of the Uganda Protectorate were no longer in a simple tribal stage in the pre-British period.

From the little information available it seems that even if the tribal life had not yet been superseded by a higher stage of social organisation, it was at least in the process of disintegration.

11. Bakonjo

Among the Bakonjo also the tribal organisation was probably in the stage of desintegration. This small "tribe", numbering "only a few hundreds" and "inhabiting the eastern slopes of Mount Ruwenzori", was "purely agricultural" [222]. Industries were practically nonexistent, and the techniques of production were very poor. They used only a short hoe, and possessed "no knowledge of manuring or of rotation of crops" [250] Yet there is evidence to show that territorial organisation was cutting across the clan organisation.

The Bakonjo were "divided into a number of clans which followed the usual custom of clan exogamy" [251], but the tribal organisation was such that the "clans did not live apart but intermingled, forming villages of from three to twelve huts. When a man married and felt that he required more room, he left his own village and made a new clearing where he built his hut and settled. Other men joined him, and in a few years his settlement might have become a village of some eight to a dozen huts" [252].

In due course of time a village headman emerged who was invested with judicial power, and was also probably entitled to tributes from the local villagers. However, his power was not yet strong enough to go beyond that of "the superior chiefs of the clans and tribe". Thus Roscoe noted [253]:

> "The man who first settled in a place generally became the head-man of the village, and to him matters of local disputes and questions of land were referred, and he acted as the mediator between the other members of the community and the superior chiefs of the clans and tribe. Anyone who wished to cultivate a new plot of land had to get the permission of the head-man of his village."

So, even in a small "tribe" like the Bakonjo significant changes were taking place before he advent of the British, although they were probably still in the tribal stage.

12. The Minor Tribes

It is possible that like the Bakonjo there were some other peoples in Unganda in the pre-British period who were still in a tribal stage. But they were numerically insignificant and had very little influence in their regions. Even so it is interesting to note that, for instance, among the Alur (who dwelt mostly in the Congo but also extended into Uganda beyond her western border), although the clans formed "the basis of the social and political organisation", two distinct types of clans could be found, namely, "either *djuker*, noble or aristocratic clans, or *luak*, commoner or client clans" [254].

It is also interesting to note that many of these primitive tribes were forcibly cut-off from the general stream of life in their regions, and therefore they could not undergo the normal course of development. Thus, the Bakonjo, for example, was a hill-tribe without much contact with the people on the plains. Similarly, considering Bambwa, Bagishu and Basabei, in the pre-British period of their history, here one finds a special situation whereby driven out by a superior force these people moved to secluded mountainous regions, and, having thus cut themselves off from the general current on the plains, they lived like museum-pieces.

The Bambwa dwelt on the western slopes of the Ruwenzori mountains. Roscoe noted[255]:

> "They were a turbulent people and were never completely subdued, though in the past they were regarded as free-men under the king of Kitara. Their subservience, however, though nominally the head-man of a clan was accountable to his over-lord, only meant that from time to time small presents and submissive messages were sent to Kitara or sometimes to Toro."

For the rest, the tribe lived within itself. The tribe was divided into a number of clans. "Clan exogamy was practised, no man being permitted to marry a woman whose totem was the same as his own."[256] "Each clan had its own head-man, who not only settled the affairs of his clan but acted as intermediary when negotiations with the overlord in Kitara or Toro were necessary."[257] These clan-heads were evidently unlike the chiefs imposed on the indigenous people in the plains of Bunyoro, Ankole and such. "The members of a clan usually dwelt together, forming a small village, and all disputes were brought to the head-man."[258] Gentile organisation was thus the key-note of Bambwa society.

The state of productive forces in the society was also primitive. The Bambwa lived on rudimentary cultivation, and their industries were very poorly developed[259]. And it is evident, "from the precautions taken to secrete food and their preparedness for hurried flights up the mountain, that they were little accustomed to fighting"[260]. Therefore, when the pastoral invaders overran the plains, they took shelter in this inhospitable mountainous region to safeguard their freedom. From that time neither the outside world nor the uncongenial atmosphere they lived in stimulated them to change their tribal life, until the British appeared on their territory and made a drastic change in their course of life.

Like the Bambwa, the Bagisu lived a similar tribal life before the advent of the British. This, "one of the most primitive of the negro tribes of Africa", was "driven from the plains to the east of Mount Elgon by the attacks of the Masai and Nandi. To escape the ravages of these warlike tribes, the Bagesu fled to the mountain, only to find that on the lower slopes they were subject to periodical raids by the Abyssinians and those tribes who inhabited the borders of Abyssinia. They therefore made their way to the less easily accessible heights and seldom left the high valley and plateaux of the southern and western sides"[261].

Thus being forced to live in an inhospitable region, the Bagisu carried on a miserable existence based on a primitive economy. The people lived on rudimentary agriculture, for the mountainous region was not suitable for cattle-rearing. Therefore, "cattle-rearing is not a great feature in the life of the Bagesu or great aim with them. The more wealthy people keep a few cows, and a village may have from ten to twenty in it, though it seldom has fifty"[262]. Agricultural production was also limited both in quantity and quality, primarily because of the conditions they lived in. Roscoe noted[263]:

> "Plantains are grown freely, but they are not so prolific nor is their quality so good as in Busoga or Uganda; doubtless this is due to the cold nights and the altitude."

The Bagisu therefore also grew "small millet and sweet potatoes", to enable them "to have a supply of grain to fall back upon" when the stock of plaintains was exhausted[264].

The state of productive forces was also very primitive among the Bagisu. "The earth was turned over with hoes or digging sticks to a depth of some eight or ten inches, and the grain was sown upon this land without further preparation"[265]. So much for agriculture. Other industries were also equally poorly developed. "Both men and women were potters, but their work was very primitive and the vessels very irregular in shape."[266] "There were very few smiths and no smelters; indeed iron-stone was not known, and the smiths bought their metal ready for use. They made only spears and knives and rough ornaments. Spades and hoes the people bought ready made from Kitara and Busoga"[267].

On the basis of such a primitive economy, the Bagisu maintained a tribal life in which the gentile organisation was the decisive factor. "Clans and divisions were clearly marked and clan exogamy was enforced, while there was at the same time constant enmity between the clans, and, to a less degree, between the subdivisions of the same clan."[268] Therefore, "each clan keeps aloof from its neighbours and jealously strives to maintain its own rights and independence"[269].

In this tribal society, land was held communally by the clans. "When the Bagesu settled on Elgon, the different clans took possession of land on the upper slopes of the ridges and this land was meted out to sub-divisions and within these sub-divisions to the families. These plots of land might never pass out of the sub-division which had taken possession of them, but they might be given by the head-man to different members. When a member was given a new plot and cultivated it, it became his own and he could pass it on to his family."[270] In this way, while land was "regarded by members of a clan as freehold property "and was so "jealously guarded against any encroachment by fellow tribesman", the individual members of the clan enjoyed usufructuary rights over different plots for cultivation[271].

The gentile organisation of the Bagisu was most noticeable in their political life. Roscoe noted[272]:

> "The great source of weakness in this large tribe of Bagesu is their lack of a leader to bring together all the clans and form a united nation."

And thus, while "there was no tribal government", in the words of Roscoe, "each clan had its own head who settled serious matters within the clan and also dealt with disputes arising between members of different clans. Each village had also its head-man, who settled ordinary cases of disputes, but any difficult matters had to be submitted to the head or father of the clan"[273].

It is also worthy of note that tribal democracy was plainly visible in the way the people conducted their affairs. Roscoe noted[274]:

> "In any serious question affecting the clan there is a senior member of the clan who presides in council, but every initiated man and youth has a voice in the meeting."

In other words, it was the whole community which decided the issue.

It should further be noted that there was no sharp distinction between rights and duties in the internal structure of society. The heads of the clans and the villages received gifts from the people, but these were not obligatory contributions. To quote Roscoe again[275]:

> "There was no regular tributes given either to the head of the clan or to the head-man of the village, but both expected to receive gifts of beer about harvest time from the men under them."

Thus the Bagisu lived under tribal organisation, in a place far away from other human habitations, until the British occupied their territory.

Similar to the Bagisu were the semi-pastoral Basabei, who lived on the "north and north-eastern slopes of Mount Elgon"[276]. "Though not so fine either in feature or physique as the pastoral people of Ankole, they resembled the Negro-Hamitic tribes of the Lake Region in appearance, but differed from them entirely in their mode of life."[277] This was probably because by being compelled to live in the hills, they could not undergo the normal course of changes for a long time. Roscoe noted[278]:

> "The people said that they first lived in Sengweri, on the plains to the east of the mountain, and were forced by the constant raids of the Masai upon their cattle to betake themselves to the hill."

The hill-life stunted their development, and partly altered their mode of life. "The mountain homes not being favourable for the breeding of large herds of cattle, the people took to agricultural pursuits; and, as generations passed, milk fell into a secondary place in their diet and vegetables became the staple food."[279] But at the same time the productive forces in society did not undergo progressive development. Thus, although agriculture became an important occupation to these people, "they never used the hoe"[280]. Industries also remained in a rudimentary stage. "Pots were made either by men or by women, but they only made those which were used for cooking."[281] Other pots, like iron-goods, they bought from outside, as well as hoes and spears, "for none of the smiths were capable of making these"[282].

In such a secluded life, the gentile organisation remained the decisive factor in society. Among these people, "the clans are numerous, though the tribe is numerically small, not amounting to more than several hundreds"[283]. These clans formed the fundamental units of socio-political organisation. "The members of each clan lived in their own particular part of the mountain, and had their own head-man who was responsible for all affairs concerning the clan and its relations with other clans."[284]

It is interesting to note how democratic was this clan-leadership within the tribe. The headman of a clan "claimed no rights as ruler over the clan, for he was its Father, and not its king. His power was greatest in connection with the land, and it was to him that all disputes concerning the boundaries of cultivated plots or of clan land were brought"[285]. Evidently, unlike in the class-societies of Kitara, Ankole, etc., the representative of political authority was *within* the society; he did not rule *over* the people.

Under the clan-head was the village headman, who was, again, "looked upon as the head" by the people and to him they brought their problems. Thus he was also one *of* the people; and, moreover, in all cases the clan was the supreme authority. Roscoe noted[286]:

> "In each village there was one man who was looked upon as the head, often because of the site he occupied. To him the members of the village brought their difficulties and disputes, and all ordinary cases he settled himself. If, however, anyone refused to accept his decision or if a case concerned any person outside his own village, he would appeal to the Father of the clan."

Thus, being cut off from the general current of life in Uganda in those days and living in an inhospitable region, the pastoral Basabei also remained in a tribal stage like the agricultural Bambwa and Bagisu, until the British arrived on their territory.

Bakene, the Lake-dwellers, also lived a partly secluded life, and maintained their tribal organisation. Roscoe noted[287]:

> "The district to which the Bakene belong has necessarily a limited area, because their mode of life, together with their staple article of food, restricts them to the few stretches of water in which papyrus-grass grows freely; for that grass forms the foundation for their floating houses, and is also the chief material used in building them, while fresh water fish is their principal article of food. Hence the sources of the Nile and its feeders are the home of these people and their domain is one of water rather than of land."

Roscoe was of the opinion that the Bakene were previously connected with the Basoga, among whom however a class society had emerged, as explained earlier. To quote Roscoe[288]:

> "In customs, language and appearance these people are closely allied to the Basoga of the north-east, and there is a tradition that their forefathers came from Busoga."

But, having thus cut themselves off from the main-land and living at a very low stage of economic development (being interested only in fishing), the social organisation of the Bakene remained essentially tribal. Roscoe noted[289]:

> "There is no person responsible for the government of the whole tribe of Bakene; each clan has its headman to whom members look for advice and redress, should there be any dispute among the members."

Gentile organisation is thus seen to have been the main feature of Bakene society.

It is also worthy of note that the clan authority was democratically elected and the people had also the right to depose an inefficient clan-head. To quote Roscoe again[290]:

> "The headman is chosen by the clan when a vacancy occurs through death. When chosen a man holds office until death, unless he forfeits his right through vice, or in some way shows he is unworthy and incapable of fulfilling the duties. Such cases are rare: still, the clan has power to depose an unworthy chief."

It may further be noted that there were no compulsory tributes to be paid to the clan-heads. Roscoe noted[291]:

> "There are no taxes levied by the headman, but, when there is a case tried, the parties pay an amount of fish to the headman before the action is commenced; they sometimes pay as much as a goat."

It appears then that tribal democracy remained the key-note of Bakene life in pre-British Uganda, while their compatriots in Busoga were living in a class society resembling the feudal system.

To conclude then, only those, or mostly those people of Uganda remained in the tribal stage in the pre-British days who for some reasons or other maintained a secluded life in unfavourable regions, so that neither internal nor external forces could initiate significant changes in their societies leading them out of their tribal existence. And, at any rate, those people who remained in their tribal stage in pre-British Uganda were numerically small and had very little influence in their regions. On the other hand, those who really counted in Uganda in those days were either living in semi-feudal kingdoms or were on the way towards a similar stage of development.

13. Buganda

The foregoing pages have shown that life in Uganda was not static in the pre-British period of her history. In major parts of the territory now included within the Uganda Protectorate the course of change had led to the establishment of semi-feudal kingdoms. Also, as among the Banyoro, tendencies towards a further stage of development were evident with the advent of Arab traders and the opening-up of the country to the outside world. Such changes, as noticed among the Banyoro, were much more marked among the Baganda whose society in the nineteenth century was at a higher stage of development than any other society in Uganda.

The first British to arrive in this region in the later half of the nineteenth century found the Baganda living under a central monarchy and governmental machinery resembling the feudal system[292]. Lugard, who "was despatched to establish the influence of the Imperial British East Africa Company in Uganda in 1891", described the Buganda kingdom as representing a "barbaric civilisation"[293]. It should therefore be of interest to examine in some details the course of development the Baganda had undergone from the earliest times.

Before the institution of the state developed among these people and the Buganda kingdom came into being, the Baganda were probably living in a tribal society. Roscoe noted[294]:

> "Prior to Kintu (the first king of Uganda according to Roscoe's chronology of the kings of this territory — RKM), there were a few aborigines who dwelt in isolated communuties or clans; each clan was governed by its chief, who owned allegiance to no other chief. Kintu came either from the north or the north-east, and began at once to subdue and amalgamate these clans, and to form them into a nation."

Controversy exists between Speke and Roscoe regarding the chronology of the kings of Buganda and the exact time when the State was established in this society. Thus, the first king according to Speke, viz. Kimera, was the third king according to Roscoe[295]. Roscoe, however, could collect very little information regarding the two earlier kings, viz. Kintu and Cwa, according to his chronology. It also appears from his monograph on the Baganda that distinct changes began to take place in this territory after the arrival of Kimera. Furthermore, Roscoe noted that the name of Kintu is associated with the origin of the clan system among the Baganda, which obviously took place much earlier in time[296]. Also, Sir Apolo Kagwa, the noted Muganda, from whom Roscoe gathered much of the material he used in his monograph on the Baganda, stated[297]:

> "It is thought by every Muganda that we are descendants of Kintu, his successors having given rise to the princes and all the rest of his descendants to the common people. That is the reason for the old saying 'All Kintu's children did not perish'. However, upon inquiry from the old people, we are informed that it was not Kintu who was the father of us all but that all the clans came into Buganda with him . . ."

How different clans of the Baganda owe their origin to the attendants of Kintu, who had either come with him or joined him in Buganda, and how the "lands each clan originally occupied are still distinguished" in the Busudju county[298], "in which Kintu first made his official residence"[299], have been noted by Sir Apolo; and Ernest B. Kalibala, who translated Sir Apolo's book into English, further noted[300]:

> "A recent missionary report says that Kintu was a native of Southern Abyssinia, and that during the great tribal movement he happened to settle in Buganda."

It thus appears that after coming to this part of Africa from Abyssinia, Kintu and his followers formed the Baganda people by amalgamating the existing clans in this territory and those which came with Kintu. And henceforth centralised power of chieftainship developed with Kintu and later his son Cwa as the established hereditary chiefs. Later Kintu, his wife, and his children became the central figures in many legends of the Baganda[301], and Kintu and Cwa were also designated as "kings"; but it will probably be more accurate to say that both Kintu and Cwa were the precursors of monarchy in Buganda rather than being full-fledged kings themselves. For, although the authority of Kintu and Cwa had been established to such an extent in the Buganda society that there were representatives of different clans performing specific offices for them, as being their cattle-keeper, bird-trapper, weapon-maker, personal doctor and coffee-cook, carrier of drinking water, etc.[302], in those days Buganda was a distant appendage of the kingdom of Kitara and Kintu's and Cwa's rule was essentially of a hereditary chieftainship over a people undergoing the process of detribalisation. And it is worthy of note that in those days while the tribal society in Buganda was in the process of disintegration, it was effected by *differentiation* within the society and not by the imposition of a pastoral horde upon the indigenous agriculturists, as it happened in Bunyoro or Ankole. Later the process of differentiation was quickened when Kimera arrived in Buganda after the territory, in the words of Roscoe, "had an interregnum of some years from the death of Cwa to the crowning of Kimera"[303], and Buganda was separated fully from the kingdom of Kitara. Henceforth monarchy was established in Buganda and the institution of the state came into full bloom in course of time.

That monarchy in Buganda originated with Kimera is also substantiated by Sir Apolo's statement that "Kimera will always stand out in the memory of the people because it was he from whom the present royal household descended"[304]. Moreover, it is worthy of note that Kimera and his five successors to the kingdom of Buganda sat on a cow skin known as Ebiwu (until Mulondo, the seventh king from Kimera, had a Royal Chair made for him and his successors), and that this word Ebiwu "is of foreign origin, borrowed from the Luima language"[305].

About Kimera, all authorities agree that he came from Bunyoro or Kitara. Furthermore, Sir Apolo has specifically stated that Kimera's mother "was not a Muganda, but a Muima"[306]. Kimera thus brought pastoral blue blood in the royal line of Buganda, and also some customs of the pastoralists, one of which is noted above. It is also of interest to note that while the stories behind Kimera's arrival in Buganda are somewhat different according to Speke and Sir Apolo or Roscoe, Speke's account explains how the kingdom came to be named Buganda, which the British writers (like Speke) spelt as Uganda. It is, therefore, worth quoting in detail the information supplied by Speke on the origin of the "Kingdom of Uganda"[307]:

> "In the earliest times the Wahuma (Bahuma as it was spelt later — RKM) of Unyoro regarded all their lands bordering on the Victoria Lake as their garden, owing to its exceeding fertility, and imposed the epithet of Wiru, or slaves, upon its people, because they had to supply the imperial government with food and clothing. Coffee was conveyed to the capital by the Wiru, also mbugu (barkcloaks), from an inexhaustible fig-tree; in short, the lands of the Wiru were famous for their rich productions.
>
> "Now Wiru in the northern dialect changes to Waddu in the southern; hence Uddu, the land of the slaves, which remained in one connected line from the Nile to the Kitangule Kagera until eight generations back, when, according to tradition, a sportsman from Unyoro, by name Uganda, came with a pack of dogs, a woman, a spear and a shield, hunting on the left bank of Katonga valley, not far from the lake. He was but a poor man, though so successful in hunting that vast numbers of the Wiru flocked to him for flesh, and became so fond of him as to invite him to be their king, saying, 'Of what avail to us is our present king, living so far away that when we sent him a cow as a tributary offering, that cow on the journey gave a calf, and the calf became a cow and gave another calf, and so on, and yet the present has not reached its destination?'
>
> "At first Uganda hesitated, on the plea that they had a king already, but on being farther pressed for consented; when the people hearing his name said, 'Well, let it be so; and for the future let this country between the Nile and Katonga be called Uganda, and let your name be Kimara, the first king of Uganda.'"

It appears from the above that whatever may have been the origin of the "Kingdom of Uganda" (Buganda), this territory was formerly a distant appendage of the kingdom of Kitara and was inhabited by tribal people (later in the process of detribalisation) who had very little contact with the Banyoro and who belonged to the same type as the indigenous agriculturists in the whole of this area, Wiru of Speke being obviously indentical with Bairu of Oberg or Bahera of Roscoe and others[308]. Now with the establishment of a monarchy on their own territory, distinct changes were brought about in the life of the indigenous people of Buganda.

With the royalty rose a class of nobles. Besides the royal courtiers and royal pages, etc., there were many chiefs — arranged in a hierarchy according to their order of importance. The foremost of them was the Katikiro or the Prime Minister and Chief Justice. The next one was Kimbugwe who was in charge of protecting the king from spiritual agencies. Like Namasole, the king's mother, the position of these two chiefs was exhalted; all of them having personal estates in each district within the kingdom, and being exempted "from contributing labourers to keep up the royal buildings"[309]. Besides the Katikiro and the

Kimbugwe, there were ten more important chiefs in the Court who "administered the country, and rendered accounts for its management to the *Katikiro,* and through him to the King"[310].

Then there were the district-chiefs. A district-chief "was required to keep his official estates in good order, or he would be fined and deposed, but the benefit of any improvements he might make, or of the houses he might build, was reaped by his successor"[311]. Also, a "District-chief had to maintain in good order a road, some four yards wide, reaching from the Capital to his country seat; in some instances, as in the case of Budu, this road was nearly a hundred miles long"[312].

Next to the district-chiefs were the sub-chiefs who were placed under the authority of a district-chief and "worked for him"[313]. The sub-chiefs were, however, "independent of the chief in managing their own portion of land; they ordered their men to work upon roads, to bridge swamps, or to build for them; but they had to consult their District-Chief about matter concerning the State and State work"[314]. "Each sub-chief had to keep in good repair the road from the District-Chief's residence to his own residence; thus it was possible to reach all parts of the country with comparative ease. The rivers, owing to the growth in them of papyrus and grass, often formed large swamps, sometimes several miles wide, and it became necessary to make paths of raised earth through them, with bridges thrown over the actual streams."[315]

All these chiefs were entrusted with judicial function. The final court of appeal was of course the king, and the next below the Katikiro. Roscoe has noted[316]:

> "In each district there was a supreme court, at the District-Chief's residence or near it; to this court cases were referred from the sub-chiefs, when the parties tried were dissatisfied with the decision. Every chief, even a petty chief, with only a dozen followers, was able to hold a court and to try cases among his own people. The people could, however, take their cases from one court to another, until eventually they came before the *Katikiro* or the King."

In this class society the centralisation of *public power* was evident not only in the king being the supreme judge, but that all chiefs, including the sub-chiefs at the bottom of the hierarchy, were appointed by the king, and could not be deposed except by the king's consent. And, in any case, the will of the people to elect or depose their own chiefs was nowhere in the picture. As in the kingdoms of Kitara and Ankole, the class of nobles and chiefs held their power at the king's pleasure. Their office depended entirely on the will of the reigning king, and not of the people. "If the King wished to depose any person from a chieftainship, he could do so summarily; usually, however, he would trump up some charge against the man, imprison him, and then depose him, instructing the Council (Lukiko) meanwhile to nominate someone else. When a chief was deposed, he was not allowed to remove any property, but might only take his wives and cattle, provided that he could find them."[317] Similarly, "when a chief was promoted to a new office, he was required to give up all his possessions except the moveable goods; nor did he receive any compensation, either for buildings or for any improvements which he might have made"[318].

Indeed, as in the neighbouring feudatory states, authority in this society was so much centralised that on the "death of the king every chief, even the prime minister, went out of office, leaving the vacancies to be filled by the new sovereign. The old chiefs retained their titles, and were given small estates near the temple of the deceased king, who was never admitted to have died, but to have entered upon a new existence. The duties of these former first grade chiefs were now connected with the temple of the late king, and they were under the direct control of the dowager queen, who removed her residence so as be to near the temple of her late brother and husband and to manage the temple service and the estate connected with the temple"[319].

In all essentials the kingdom of Buganda represented a feudal society. The royalty and the chiefs lived on the tributes and free labour supplied by the common people. Centralisation of *public power* was felt by the fact that the "King alone levied taxes, some portion of which he gave to the District-Chiefs"[320]. The nature of taxation was as follows. "The amount usually demanded was a fixed number of barkcloths and one hundred cowry-shells from each peasant; of the smaller chiefs each paid a number of goats and also a few hoes"[321]. The tax was collected from each district by six representatives of the State, one each from the king, the queen, the king's mother, the Katikiro, the Kimbugwe, and the chief of the district concerned. When taxes have been collected from all districts, the chief of each district having "received a portion of the taxes for himself and for his sub-chiefs, the King took half for himself, while the *Katikiro* and *Kimbugwe*, the Queen, and the King's Mother also had their portions"[322]. Over and above that, "the King, the Queen, the King's Mother, the *Katikiro*, and the *Kimbugwe* kept the whole of what came from their own estates, in addition to the portion which they received of the taxes from the entire country"[323].

Free labour from the people was also exploited in various ways. "From time to time the King would send out special messengers to collect boys and girls for the royal enclosure; and each chief would have to supply a number according to the population of his district"[324]. "A similar method was adopted to obtain labour for building in the royal enclosure", each district-chief having "certain buildings in the enclosure allotted to him, which he was required to keep in repair"[325]. And the men, sent by the sub-chiefs through the district-chiefs for the building in the royal enclosure or to build streets in the Capital, had not only to contribute free labour for the monarch, but they had also to pay first a tribute to the king's overseer under whom they were to work. Failing this payment of tribute, they would be continuously fined until the dues were fully reimbursed. Meanwhile, of each of such workmen, "his wife, or some other member of his family, would be taken as hostage, until he should bring the necessary sum; the woman or child thus taken would he required to work for the chief during the time of detention"[326].

Besides supplying free labour to the king, the people had also to supply the same for the chiefs, in order to build roads, huts, etc. Here again they had first to pay tributes to the overseers in charge of such work, failing which they were fined and members of their families detained and made to work for the chiefs[327].

And while the people's labour was thus extracted for the benefit of the feudal lords and the kings, these rulers maintained their responsibility in upholding class power in the society in a hierarchical order; thus, the sub-chiefs were answerable to the district-chiefs, the latter to those higher in the scale, leading finally to the Katikiro and lastly to the king[328].

Besides the people in the metropolitan country, those in the tributary areas also "paid their tribute through the chiefs under whom they were placed, making their payments in cattle, slaves, ivory, cowry-shells, salt, hoes, etc."[329]. These people the king of Buganda had brought under his control either by snatching them away from the suzerainty of the kingdom of Kitara or by subdueing them after making frequent raids on their territories, so that finally they agreed to pay regular tributes to live in peace. In the course of time, some of them at least were partially assimilated among the Baganda.

To give a short account of these tributary areas of the kingdom of Buganda, "in the north was Busoga, which was a mine of wealth, yielding slaves, cattle and ivory"[330]. "Koki, to the south-east of Budu, was another tributary country, and was taxed each year in iron hoes and cowry-shells. Koki has had, from time immemorial, its own king and governing body (Lukiko); but being unable to withstand the raids of the Baganda, it became a tributary state. To the west there was the pastoral country Ankole, which though it was never subjugated, yet sent from time to time large herds of cattle, in order to keep peace with the kings of Uganda. To the south of Budu, the Kiziba country paid its yearly tax of cowry-shells and trade-goods, which came into that country from the south. Both the Kiziba country and Koki were ruled by the *Pokino*, who was the chief of Budu"[331].

The kingdom of Buganda thus flourished, and this was reflected in the ever-increasing feudal grandeur of the king and the overlords, and also to a certain extent in the improvement of the standard of living of the people. Indeed, in course of time, the kingdom of Buganda went much farther in development than the kingdom of Kitara, the next best developed kingdom in this area. Roscoe has remarked in this respect[332]:

> "In the early days of travel the remarkable spectacle of social progress and order which burst upon the gaze of the traveller when he entered Uganda (kingdom of Buganda — RKM), caused him the utmost astonishment. . . .
>
> He was prepared for a lower stage in morals and culture, and to find gentlefolk with the courtesy and manners of highly educated people was indeed a surprise. . . .
>
> "As a nation the Baganda are clean in their habits and most particular about their person, dress, food and general appearance; the floors of their houses are neatly carpeted with sweet-smelling grass, upon which the inmates sit, while a mat or rug is spread upon it for the visitor, . . . There are, it is true, among the poorer people some few homes where the wife is found to be untidy and the house dirty, but these houses are happily the exception, not the rule. These signs of cleanliness and order at once attract the attention of the most casual observer, and he further learns that the people are superior to their neighbours in almost every other respect."

Referring to the lay-out of the country he noted[333]:

> "The roads are striking, especially when compared with the narrow tracks in other parts of Africa; they are kept free from weeds and grass; even the private path leading to some small single hut from the main road is kept free from the growth of weeds and is at least three feet wide, while public roads from the capital to the dwellings of important chiefs are often from twelve to fourteen feet wide. Where a road crosses a swamp, a bank of earth is raised well above the water and four or five feet wide to form the path; openings are left in it where there is a stream; the bank at the opening is faced with piles driven into the mud and bridged with trees so that pedestrians can cross it without wetting their feet."

And he further remarked that all these were achieved in the kingdom of Buganda before the foreigners came there, viz. first the Nubians or the Arabs and later the British. To quote[334]:

> "These roads and bridges are indications of a superior people, and they strike the pedestrian, who has travelled from the coast through less civilised countries along narrow tracks, with agreable surprise; in districts even near the coast a tree falling across the path is left where it falls, it is no man's duty to remove it; so the traveller has to walk round or climb over it. These advanced qualities and marks of civilisation of the Baganda are not the result of any outside influence being brought to bear upon them, they are of much earlier date than the time of their contact with either the Nubian or the Arab.
>
> "A most interesting study which especially concerns the anthropologist is to account for the great progress which the people have made in social and political life; they have advanced from the degraded barbarity of primitive nations to a stage approaching very nearly to that of civilised government."

Why was Buganda better developed than the neighbouring regions? One of the reasons may be that the class society which emerged here was different from that in Kitara or

Ankole. In the latter kingdoms the sharp distinction between the pastoral overlords and the indigenous agriculturists was a hindrance to further development. Only at a late date was the resistance of the high-born among these people being superseded by differentiation in the societies according to the possession of wealth. But in Buganda from the beginning there was no such impediment. Here the class society was not formed by the imposition of large numbers of invading Bahima pastorals on the indigenous agricultural people. Here the pastoral blue blood flowed mainly, if not only, in the royal line; and the feudatory chieftainships which emerged in society with the development of centralised authority were evidently from within the mass of the indigenous agriculturists. Roscoe has noted that the Baganda "were a progressive agricultural nation"[335]. Among them, while "class distinction were marked, even though there was no blue blood among the chiefs", as described by Roscoe, "custom permitted any man of ability to rise to a chieftainship"[336]. He has further noted[337]:

> "The Royal Family alone were regarded as superior to, and as separate from, other classes in virtue of their birth. With this single exception, any member of a clan could rise to the highest position in the land, if he succeeded in making himself conversant with State affairs, and was brave in warfare and shrewed in council.... Class distinctions were chiefly determined by a man's abilities and by the bravery and skill which he displayed when in the service of the King or in that of some leading chief."

Thus, the scope for individual initiative and for being rewarded for individual qualities was much greater in this society than in Kitara or Ankole or Toro. And this was reflected in the daily life of the people as the foregoing extracts from Roscoe suggest.

There is also the point that because from the beginning class differentiation in this society was based not on birth-qualification but on the possession of wealth, and, following it, on the usurpation of political power to uphold their property, the relation between the classes was quite straightforward. Roscoe described the common people of Buganda as "freemen". They were indeed free to the extent that they were not despised and forcibly treated as inferior beings because of birth-qualification, like the Bahera in Kitara, Ankole or Toro. They could keep cattle, and even those "who went to live on the private lands of a chief were called his freemen (basenze); they worked or fought for him in return for the land which they tenanted"[338]. In peace-time, however, they could not attain great success and good positions, for when it came to the question of possession of wealth, their freedom suffered drastically at the hands of the property-owning class. For instance, referring to the possession of cattle, one of the most important forms of property in the kingdom in those days, Roscoe noted[339]:

> "The King numbered his cows by the thousand, and the chiefs also had large numbers of cattle; the peasants might possess one or two cows, but they would soon have excited the cupidity of the chiefs, and would have been robbed by them if they had possessed many."

During raids and wars, however, some of the masses could improve their position by exhibiting their bravery and skill. This was possible because unlike in Ankole for instance, the common people were not debarred from military service. On the contrary, this was one of the ways by which their labour was usurped by the king and the chiefs. These common people "held the land from chiefs, on condition of doing work for them, especially building work, and rendering military service"[340]. Even those who worked on the private lands of a chief, "fought for him in return for the land which they tenanted"[341]. And "by the bravery and skill which he displayed when in the service of the King or in that of some leading chief", one of these common people could attain important position and wealth and power.

Probably for this reason Buganda became the most powerful kingdom in this region. The upshot was that new areas were brought under its control which eventually enriched the country and added to the material welfare of the people. Thus, because of this straightforward development of classes in this society, social energy could be better utilised which in greater or lesser degree benefited also the common people.

Another, and possibly more important, reason for the spectacular development of the kingdom of Buganda, especially in the later years, was its early contact with the outer world; a contact which was more consistently maintained than even by the kingdom of Kitara. A significant introduction in the kingdom of Buganda which dates back to Kintu, the first king according to Roscoe's chronology, and which had a very important bearing upon the development of the external trade of the Baganda, was canoe-building. Roscoe has noted[342]:

> "From tradition we learn that the principal canoe-builders came to King Kintu from the north of the lake, that the Lung-fish Clan from that time onwards have been the chief canoe-men, and that one of their number has held the office of admiral (Gabunga). Many of the islands, with their numerous population, became subjects of the kings of Uganda, and were reckoned as part of the nation, though there has always been a marked line of separation between the Baganda and the Basese, as the people of the islands are called. No one has studied the marked differences between these peoples, so far as I am aware; and it is impossible to do so now, owing to the barrier raised by the terrible disease, sleeping sickness, which has been so prevalent on the islands and along the shores of the lake. Prior to the outbreak of this scourge, the islands were thickly populated, and they formed a source of revenue to the King, both by the fish which they supplied, and by the yearly tribute which they paid."

During the early years of British rule, the Basese became practically extinct due to the ravages of sleeping sickness which remained unconquered in the British period of Uganda's history, and therefore they had practically no role to play after British occupation of the kingdom of Buganda. But prior to that, the kingdom of Buganda could make very good use of them, for, with the development of trade with the Arabs, the canoes made by the Basese came to be of great service to the Baganda. Roscoe noted in this connection[343]:

> "As canoes were the principal means used for carrying on commerce with the nations to the south of the Nyanza, and also for connecting the main road from the east coast with the capital of Uganda, the Basese were of great importance to the country; and it was through their means that the Baganda influence had become so extended at the time when Arab traders first discovered the lake. The number of canoes kept on the shores of the mainland together with those on the islands, was very great, and a fleet of a hundred strong could easily be collected in two or three days. Each chief on the islands had to maintain a number of canoes ready for state-service, and he was liable to punishment if they were not kept in serviceable condition. Most of the men on the lake shore, as well as those on the islands, were skilled in the art of canoe-building, so that they were able on a journey to repair their canoes, if one of them met with an accident or leaked badly. There were, however, special men in each district responsible for the condition of all the large canoes, and these men made canoe-building a speciality."

The most developed feudatory state in this region thus ventured upon external trade, which, though not on such a large scale, was soon followed by the kingdom of Kitara mainly by the land-route.

Henceforth, foreign influence in the development of the kingdom of Buganda could be felt more and more. This influence, however, was on a commercial level with the outer world, and did not infringe upon the sovereignty or independence of the kingdom of Buganda.

King Kimbugwe, the eleventh king after Kimera, introduced leather sandals in the kingdom. Since then sandal-making became "an institution of long standing" in Buganda, and it "gradually became part of the dress of wealthy people"[344]. Roscoe is of the opinion that the sandals were copied from the Nubians whom the Baganda had visited in Khartoum on one of their early trade mission[345].

From the time of Katerega, the step-son and successor of Kimbugwe, changes in the kingdom of Buganda were also seen in the extensive practice of polygyny, which in those days obviously meant that the families had become prosperous enough to maintain large households. Not only the kings, as befitting their exalted position and ever-increasing grandeur, began to marry many wives and had concubines and "maiden servants" running up to hundreds, but the common people also "began to marry two or three wives apiece." Sir Apolo noted[346]:

> "The custom of marrying several wives is a relatively recent one. Formerly the king married only one wife, Kadulubale, who was responsible for the care of the entire palace. Later on two more wives, Kabedja and Nasaza were added. Kabedja took over charge of half the palace....
>
> "When the system of plural marriage became established, the queen always outranked the other wives, and Kabedja and Nasaza were immediately below her. The chiefs in their homes also had ranking wives, called by the same titles. It was not till the reign of Katerega that polygamy became popular, and the common people began to marry two or three wives apiece. Katerega and his chiefs decided to demand of the prominent and wealthy chiefs taxes in the form of goods, to be given to the wives of the king. This custom was not abolished until the reign of Mwanga (that is, the time when the British had appeared on the scene — RKM)."

The next important change was introduced during the reign of King Kyabagu, the twenty-third king after Kimera, who introduced "plates, cups and saucers, and glass into Uganda"[347]. This was in the eighteenth century when trading with the Arabs had apparently developed to a significant proportion. Kyabagu also "conquered the whole tribe of Busoga" and thus brought the attacks made by previous Baganda kings into a final conclusion[348].

After Kyabagu, his son Djundju subdued all of Budu and Kiziba; and then during the reign of his brother Semakokiro (who probably died in 1790) the Kingdom of Buganda progressed farther. Semakokiro was the first king who "forced his people to go about clothed; prior to his reign the people wore no clothes but small skins; the men wore them over the shoulders, and the women round the waist. Semakokiro also ordered the peasants to plant the barkcloth trees in their gardens, and fined them if they neglected to do so. At one time there were special kinds of trees from which the royal barkcloths were made; there were also men whose sole duty it was to keep the Court supplied with barkcloths; but the art of barkcloth-making was common property, and the humblest peasant had his trees, so that nothing but idleness prevented him from being well-dressed"[349]. Furthermore, Semakokiro "was the first King to introduce cotton-goods into the country, but during his reign their use was restricted to princes, and even chiefs were prohibited from wearing them"[350]. The king "had many huntsmen who brought in much ivory, with which he traded to the south of the Lake, buying in exchange cotton-goods and cowry-shells"[351]. It is of interest to note that during his reign commercial enterprises were undertaken on a larger scale, "which brought in wealth and his chiefs actually were obeyed when they suggested cessation of execution"[352].

The effect of trading with the Arabs was thus felt more and more in the kingdom. Arabs also taught the Baganda mat-making and soap-making[353]. "When cotton goods were brought into the country, it became necessary to have soap for washing them. The Arabs introduced the art of boiling down the fat of animals, burning the peelings of plantains, and mixing the fine ashes with boiling fat"[354]. The soap thus produced "was commonly used for washing cotton goods"[355]. "In later years it was also used for washing the human body, especially the head, though as a rule only water was used for that purpose."[356]

The contact between the Africans and the Arabs became increasingly closer, and changes in Buganda made rapid strides. During the rule of King Suna Kalema, the grand-son of Semakokiro, [1810—1852], for the first time the Arabs were admitted into Buganda to trade. The Arabs "introduced firearms and gun-powder in exchange for ivory and slaves"[357]. This king "made the neglect of certain sanitary conditions in the capital an offence punishable by death"[358]. During the reign of Suna Kalema (Suna II) "the wealthy men acted in a very ceremonious fashion. At meal times, their wives washed their hands with sponges and then dried them with dry sponges"[359]. The king also ordered that "anyone that wronged a young girl be put to death. This caused the people to be very careful"[360]. Suna Kalema changed and simplified some traditional customs[361], and, like his father Kamanya [1790 to 1810], also "added fresh territories to Uganda (that is, to the kingdom of Buganda — RKM)"[362].

After Suna Kalema, his son Mutesa Mukabya ascended the throne in 1852. Roscoe has described him as "the greatest of all the Kings of Uganda, not only in mind, but also in his progressive policy, for he was ambitious as well as able"[363]. He effected the most significant political changes in the pre-British days. Roscoe has noted[364]:

> "The first great change was made between 1854 and 1884, when King Mutesa instituted a standing army with a permanent General (Mujasi) and with sub-chiefs (Mutongole, pl. Batongole), or captains, over the troops; these captains were given estates in each district (Saza) in lieu of pay, and were exempted from any obligations to the chief of the district (Owesaza). The captains were responsible to the General (Mujasi), who in his turn was responsible to the Prime Minister (Katikiro) and to the King."

A navy was also formed with an Admiral, as noted before. Speke has noted[365]:

> "Fleets of boats, not canoes, were built for war, and armies formed, that the glory of the king might never decrease. In short, the system of government, according to barbarous ideas was perfect."

Moreover, "Guns were also brought into the country in numbers during this reign, and spears began to fall into disuse amongst wealthy people[366]. With this military power, the King reduced the surrounding people to order, "so that his rule extended over the whole of Busoga, over Koki, and over the Kiziba country to the south"[367].

The king also introduced measures to improve the standard of living of the people. "In Mutesa's reign the people were permitted to wear calico and all kinds of cotton goods, which up to that time had been restricted to royalty and favourite chiefs."[368] "Short breeches for men were introduced during the early part of Mutesa's reign; they were copied from some Egyptian troops who made their appearance in the country. When calico came to be commonly used, knickerbockers became universal for men and boys."[369]

It is significant to note that king Mutesa also effected changes in the social and spiritual life in the kingdom. According to Sir Apolo, "Mutesa did a great deal towards abolishing unnecessary old traditions"[370]. He simplified court etiquettes, invited the chiefs to dine with him, and reduced in many other ways the marked distinction between the king and his chiefs[371]. Previously, in the Court Room the king sat on a throne and the chiefs on the floor. But when Mutesa became the King of Buganda, "he ordered his blacksmiths to make several steel chairs, and the carpenter to make several wooden ones. These he placed in the council house for the use of his chiefs. There were four hundred of these, and half of them were ranged on each side of the house. The king sat in the centre of the house, the prime minister next to him, then Kimbugwe, and after him the Saza chiefs"[372]. Also Mutesa endeavoured to establish a cordial relation with all chiefs in his territory. There were some chiefs in Buganda and Sese who were not allowed to see the king, for "these chiefs were originally as much rulers as the king himself", and so, although "several tyrannical kings reduced their power to that of mere chieftainship", they were regarded as hostile to the king[373]. But Mutesa "called these chiefs together and advised them to visit his court and guaranteed their positions"[374]. Furthermore, as Roscoe noted[375]:

> "He changed the status of the gods, and reduced the power of the priests from its unique position, and though he never entirely broke away from heathen customs, his faith in them was very limited. When he was ill, he gave instructions that his body was to be buried, and not embalmed, and that no one was to be killed to accompany his ghost into the outer world. His jawbone was not removed, and no temple was built for him, but a large house was errected as his tomb."

Such changes during Mutesa's reign indicate that social progress was not lagging far behind economic development. It may not therefore be wrong to surmise that in course of time the Baganda as well as other Uganda Africans would have attained the stage of socio-economic development reached by other peoples of the world without any forcible penetration of "western civilisation" through colonisation. This would have been achieved through their close association with the rest of the world, but without losing their sovereignty and independence.

To continue with the account of socio-economic development of Buganda, while this kingdom prospered at an accelerated rate with the development of her external trade, a new relation was making in-roads into society. The king and the chiefs were turning into traders, and new ideas about social and spiritual life were also gaining ground in the society, as exemplified by Mutesa's reform measures.

Trading with the Arabs which was increasingly becoming an important occupation of the upper class of society was particularly felt in slave-trade. In Buganda, "the King's canoes went over the Lake to the Island Karagwe and to the mainland in the south, and brought over trade-goods, which were exchanged for ivory and slaves. The King ceased to send people to death at the slaughter-places, and sent thousands of slaves to the Coast instead"[376]. The Arab's trade in slaves which had its most important centre at Zanzibar on the eastern coast of Africa, to which the above extract refers, thus swelled in volume, and in course of time engulfed the whole of central Africa, and especially the kingdoms of Buganda and Kitara. Referring to such business organisation of the upper class of the society, mostly in Buganda as well as in Kitara, Thomas noted[377]:

> "In contrast with the southern Sudan and Nyasaland, this section of Africa was never a hunting ground for alien traders; for, under the rulers of Bunyoro-Kitara and Buganda, it had long enjoyed governments sufficiently strong and organized to secure it from such intrusions. Adopting modern phraseology, primary accumulation of slaves was a state monopoly operated by the rulers, who dealt with Arab traders as middlemen in the chain of distribution."

Indeed, in this profitable business it was not only the kings who were interested. The chiefs had also followed suit to capture slaves and hunt for ivory, and trade these commodities directly with Arab traders[378]. Many raids beyond the borders of their respective territories were instigated by the chiefs in the hope of capturing slaves. Wars were also encouraged with the same purpose in view. Thus, referring to the situation in Buganda, Roscoe has noted[379]:

> "Slaves were obtained by raids, or from wars made upon neighbouring tribes, or they were inherited from the owner's predecessors, or they were given in payment of a debt. As a rule slaves were foreigners, chiefly Banyoro and Basoga; Baganda who were slaves were treated with much consideration in their own country; they were men and women who had beeen sold by a relative in trouble, children who had been kidnapped, or who had been pawned to raise money in an emergency."

The situation was similar in the kingdoms of Kitara and Ankole in regard to domestic slaves. But in all these areas the slaves offered for sale to the Arabs were primarily those who were captured in raids and war and about whom the slave-owners therefore felt very little compunction.

The upshot was that, on the one hand, the weaker peoples who could be dominated by the powerful kingdoms, especially by that of Buganda, were always in mortal fear of being captured as slaves and sold to the Arabs, and, on the other, the same process brought prosperity to these kingdoms, and especially to Buganda, for this was the most powerful kingdom in this region and therefore could conduct this business of "primary accumulation of slaves" on the largest scale.

Indeed, so impressive was this development in this region that after some years of governing the Equatorial Province, which then included part of the northern areas of the Uganda Protectorate, Emin Pasha looked enviously at the "development set on foot" in the south. In a letter expressing his wavering between sticking to his moral scruples and to bringing prosperity to his province by joining in the slave-trade, he wrote[380]:

> "Development (in the south) is set on foot by external influences and is promoted by the inhabitants' love of commerce. Intercourse of the different tribes with one another as well as with the Arabs is resulting in constant efforts to open up new trade routes and to create and satisfy fresh wants. . . . The Egyptian territories in the north have remained in these respects very far behind. In spite of our occupation of the northern region, which has now lasted for many years, no progress has been effected, for the unfortunate system of well guarded parties and a monopoly in trade, up till now looked upon as the only protection against kidnapping and slave trading, has only prevented the natural development of our countries and has done little enough to abolish the purchase and sale of slaves. While in the south civilisation is now advancing and land after land is being opened up to trade, our fertile mountain districts lie fallow, and with folded arms we watch the advance of commerce from south to north when we ought to be up and doing and striving continuously to open up new routes for ourselves. Would it not be better to break lose from the old Utopian system to say farewell to philanthropic whims and to adapt the administration of these fruitful lands to the development of their resources and the supply of their needs?"

However, while Emin Pasha could not decide whether or not "to say farewell to philanthropic whims" and follow the "southern" system of lucrative slave-trade, the British appeared in Uganda and changed the entire course of development. The development of Buganda in its normal course was thus checked. Before the primary accumulation of capital by trading in ivory and slaves could be utilised to raise society to a higher level by improving the state of productive forces and thereby changing the *social relations of production**, or

* It should be noted that cotton, which under British rule became the backbone of Uganda's economy, had already been introduced in Buganda, so that changes in her economy could have come without British intervention. Moreover, as has been noted before, from the knowledge received from the Arab traders in other ways also production in Buganda had already been diversified. Hence, a demand for the development of the productive forces of society and consequently a change in the *social relations of production* (which could not but have a tremendous effect on the evolution of society) were inherent in the course of changes which were taking place in the Buganda society before the advent of the British.

before a distinct class of mercantile bourgeoisie could develop out of the society, the country came under British rule. "It was during Mutesa's reign that Speke paid his first visit to Uganda (in 1861), and later on Stanley (in 1875) also reached the country, and through his representation the Christian Missions were sent"[381]. And this was the prelude to the British occupation of the kingdom of Buganda, which soon spread all over what is known today as the Uganda Protectorate. The peoples of Uganda entered upon another course of history.

14. The "Tribes" of Uganda

To sum-up now the present discussion on the *social* development of the peoples of Uganda in the pre-British period of their history, it should be obvious from the preceding pages that the societies of Uganda "moulded over centuries by their environment" were nothing very different from those in other parts of the world. Of course, local conditions had left their mark on the societies in all aspects of the life of the people, just as they have done in all other societies in the world. But the general characteristics of social development of the Uganda Africans were neither peculiar nor incomprehensible. As in all parts of the world, once upon a time they lived under tribal organisation, but by the time the British set their foot on the soil of Uganda, most of them were beyond this stage. Moreover, the majority of the Africans were then in the feudal epoch of historical development, and tendencies towards further changes were also not altogether absent.

This point cannot be stressed too much, for it at once makes it obvious that even in those days to speak of the Uganda Africans, or at any rate the majority of them, as "tribals" did not fit in with the real state of things. It should also be borne in mind that from Speke onwards the British in Uganda first came in contact with the Baganda whose social development was the highest of all Uganda Africans. Therefore, one cannot resist the conclusion that to speak of the Uganda Africans as "tribals" was either a misrepresentation of precise scientific terminology and a neglect of the study of societies in their historical course of development, or it was a wilful distortion of facts.

Indeed, the second reason is becoming more and more forceful, for, whatever might have been done as a "mistake" in the past, even in the present-day literature of the British Government as well as in the writings of many social scientists, the Uganda Africans are described as "tribals". Even the latest official study of the Uganda Africans issued in 1950 has been entitled *Geographical and Tribal Studies, Uganda Protectorate.* And while even, now the Uganda Africans are described as "tribals", although capitalism, however slowly has entered into the Uganda societies during the last six decades of colonial rule and has made profound changes in the life of the people, the ethnographers and administrators who, in the earlier phase of British rule in Uganda, had gone among these peoples (such as the missionary Roscoe or the administrator Johnston), often described some of them as "nations". True, these authorities used the terms like *tribe* and *nation* rather loosely, just as some social scientists are still now doing with regard to the Uganda Africans, but, in those days, that might have been regarded as due to an unintentional confusion of the precise concepts behind specific terminologies. On the other hand, today the description of the Uganda Africans as belonging to several "tribes" has a sinister meaning attached to it, for it tends to give the impression that since the tribal peoples are supposed to live under a quaint social organisation, they have first to be "civilised" before their societies can be made amenable to such modern concepts as *democracy* and *self-government.* As noted before, a recent expression of such a view-point, couched in lyrical phraseology, has been found in Elspeth Huxley's announcements to the world public.

This argument is of course disputable in itself. But what may be of gerater importance to note is that by such arguments the British rulers attempted to confuse the mind of the world public, although from the beginning it was evident to them from the researches of Roscoe, Johnston, and others that there never was any question about changing the natures of the African societies "in the wink of an eye"; nor is there any point in *re-making* them before the peoples of Uganda attain self-government. Changes were of course necessary in those days and they are also necessary today. But there is no evidence that either then or now such changes had or will have to encounter a peculiar situation unknown in other parts of the world in their normal course of development. Nor is there any point in believing that such changes could not have been brought about by the Africans themselves without undergoing foreign domination.

Yet Uganda was occupied by the British to lift the Africans from "savagery to civilisation". Why did they come to this part of the world, and what has happened to Uganda since then? These are the questions which will be discussed in the next chapter.

REFERENCES

1 Huxley, Elspeth — "The Sorcerer's Apprentice: A Journeyt hrough East Africa", Chatto and Windus, London, 1948, p. 365
2 *ibid.*, p. 365
3 Governmental Papers published by His Majesty's Stationery Office, London: "Reports of the Commission on Closer Union of the Dependencies in Eastern and Central Africa", Cmd. 3234, 1929; "Information on Non-Self-Governing Territories", Colonial No. 226, 1948; etc., etc.
4 See, for instance,
(a) "An Appraisal of Anthropology Today", edited by Sol Tax, Loren C. Eiseley, Irving Rouse, and Carl F. Voegelin, The University of Chicago Press, USA, 1953;
(b) "Anthropology Today: An Encyclopedic Inventory", prepared under the Chairmanship of A. L. Kroeber, The University of Chicago Press, USA, 1953;
etc., etc.
5 See, for instance, Émile Durkheim's *The Rules of Sociological Method*, The Free Press, Glencoe, Illinois, USA., 1950, p. 34ff.
6 "Notes and Queries on Anthropology", Sixth Edition, Revised and Rewritten by a Committee of the Royal Anthropological Institute of Great Britain and Ireland, Routledge and Kegan Paul Ltd., London, 1954, p. 66
7 Murdock, George P. — "The Processing of Anthropological Materials", *loc. cit.* 4(b), p. 478
8 *ibid.*, pp. 478—479
9 *ibid.*, p. 479
10 Lowie, Robert H. — "Social Organization" Routledge & Kegan Paul Ltd., London, 1950, p. 294
11 *ibid.*, p. 341
12 *ibid.*, p. 409
13 Morgan, Lewis Henry — "Ancient Society", First Indian Edition, Bharati Library, Calcutta, pp. 113—114
14 Engels, Frederick — "The Origin of the Family, Private Property and the State", Foreign Languages Publishing House, Moscow, 1948, pp. 136—141
15 *ibid.*, p. 154
16 *ibid.*, pp. 242—246
17 White, Leslie A. — "Lewis Henry Morgan: Pioneer in the Theory of Social Evolution", in *An Introduction to the History of Sociology*, edited by Harry Elmer Barness, The University of Chicago Press, USA, 1948, p. 149
18 *ibid.*, p. 150
19 See, for instance,
(a) *loc. cit.* 4a, pp. 69—72, etc.
(b) *loc. cit.* 4b, pp. 313—326, etc.;
(c) Childe, V. Gordon — "Social Evolution", Watts and Company, London, 1951, Chapter 1;
(d) White, Leslie A. — "History, Evolutionism and Functionalism: Three Types of Interpretation of Culture", South-Western Journal of Anthropology, Vol. 1, pp. 221—248;
(e) White, Leslie A. — "Diffusion vs. Evolutionism: An Anti-Evolutionist Fallacy", American Anthropologist, Vol. XLVII, 1945, pp. 341ff.;
(f) *loc. cit.* 17;
etc., etc.

20 *loc. cit.* 4a, p. 71
21 *loc. cit.* 17, p. 150.
See also Julian H. Steward's *Evolution and Process* in *loc. cit.* 4b, p. 315; etc.
22 MacIver, R. M. and Page, Charles H. — "Society: An Introductory Analysis", Macmillan & Co. Ltd., London, 1953, pp. 594—596
23 Kroeber, A. L. — "Anthropology", Harcourt, Brace and Company, New York. USA., 1948, p. 281ff.
24 *loc. cit.* 17, pp. 144—145
25 *ibid.*, p. 150
26 Radcliffe-Brown, A. R. — "Some Problems of Bantu Sociology", *Bantu Studies*, October 1922, pp. 40—41, also quoted by Leslie A. White in *The Science of Culture*, Farrar, Straus and Company, New York, USA., 1949, p. 241
27 Roscoe, J. — "The Bakitara or Banyoro", Cambridge University Press, England, 1923, pp. 2—3
28 *ibid.*, p. 1
29 Baker, Samuel — "Albert Nyanza", 1913, p. 69; also quoted in loc. cit. 27, p. 1
30 *loc. cit.* 27, p. 199
31 *ibid.*, p. 9
32 *ibid.*, p. 305
33 *ibid.*, p. 10
34 *ibid.*, p. 199
35 *ibid.*, pp. 5—6
36 Roscoe, J. — "The Soul of Central Africa", Cassel and Company, London, 1922, p. 167
37 *loc. cit.* 27, p. 6
38 *ibid.*, p. 232
39 Johnston, H. H. — "A Comparative Study of the Bantu and Semi-Bantu Languages", Oxford at the Clarendon Press, 1922, Vol. II, p. 15
40 Speke, J. H. — "Journal of the Discovery of the Sources of the Nile", Everyman's Library, London, 1922, p. 201;
Johnston, H. H. — "The Uganda Protectorate", Hutchinson, London, 1902;
loc. cit. 27; etc., etc.
41 *loc. cit.* 27, p. 9
42 Roscoe, J. — "The Northern Bantu", Cambridge at the University Press, 1915, pp. 4—5
43 *loc. cit.* 27, p. 9
44 *ibid.*, p. 9
45 *ibid.*, p. 7
46 *ibid.*, pp. 9—10
47 *ibid.*, p. 10
48 *ibid.*, p. 52
49 *ibid.*, p. 52
50 *ibid.*, p. 121
51 *ibid.*, p. 123
52 *ibid.*, p. 8
53 *ibid.*, p. 59
54 *ibid.*, p. 52
55 *ibid.*, p. 59
56 *ibid.*, p. 52
57 *ibid.*, p. 52
58 *ibid.*, p. 54
59 *ibid.*, p. 176
60 *ibid.*, p. 55
61 *ibid.*, pp. 54—56, etc.
62 *ibid.*, p. 176
63 *ibid.*, pp. 177—178
64 *ibid.*, p. 11
65 *loc. cit.* 39, Vol. II, p. 18
66 *loc. cit.* 27, p. 200
67 *ibid.*, pp. 10—11
68 *ibid.*, pp. 2—3
69 *ibid.*, p. 200
70 *ibid.*, pp. 14—18
71 *ibid.*, p. 10
72 *ibid.*, p. 52
73 *ibid.*, p. 176
74 *ibid.*, p. 178
75 *ibid.*, p. 200
76 *ibid.*, p. 11
77 Thomas, H. B. — "Arabic Correspondence Captured in Southwest Bunyoro in 1895; with a Note on Arab Traders in Bunyoro", *The Uganda Journal*, Kampala, Uganda, Vol. 13, No. 1, March 1949
78 Oberg, K. — "The Kingdom of Ankole in Uganda", in *African Political Systems*, edited by M. Fortes and E. E. Evans-Pritchard, Oxford University Press, 1948, p. 122
79 *ibid.*, p. 123
80 *ibid.*, pp. 124—125
81 *loc. cit.* 27, p. 306
82 *ibid.*, p. 3
83 *loc. cit.* 39, Vol. II, pp. 17—18
84 Guthrie, M. — "The Classification of the Bantu Languages", Oxford University Press, 1948, p. 77
85 *loc. cit.* 39, Vol. II, p. 2
86 *loc. cit.* 84, p. 77
87 *loc. cit.* 78, p. 121
88 *ibid.*, p. 122
89 *ibid.*, p. 122
90 *ibid.*, p. 122
91 *ibid.*, p. 128
92 Roscoe, J. — "The Banyankole", Cambridge at the University Press, 1923, p. 2
93 *loc. cit.*, 78, p. 130
94 *ibid.*, p. 131
95 *ibid.*, p. 131
96 *ibid.*, p. 157
97 *ibid.*, pp. 157—158
98 *ibid.*, p. 136

99 *ibid.*, p. 137
100 *ibid.*, p. 138
101 *loc. cit.* 92, p. 12
102 *ibid.*, p. 12
103 *ibid.*, p. 12
104 *ibid.*, p. 13
105 *ibid.*, pp. 14—15
106 *ibid.*, p. 16
107 *ibid.*, p. 16
108 *ibid.*, p. 15
109 *ibid.*, p. 15
110 *ibid.*, pp. 1—2
111 *ibid.*, p. 78
112 *loc. cit.* 42, p. 110
113 *loc. cit.* 92, p. 2
114 *ibid.*, p. 13
115 *loc. cit.* 42, p. 102
116 *ibid.*, p. 103
117 *loc. cit.* 92, p. 1
118 *ibid.*, p. 151
119 *loc. cit.* 78, p. 133
120 *loc. cit.* 42, pp. 117—118
121 *loc. cit.* 92, p. 18
122 *loc. cit.* 78, p. 135
123 *ibid.*, pp. 130—131
124 *ibid.*, p. 135
125 *loc. cit.* 92, p. 2
126 *ibid.*, p. 17
127 Roscoe, J. — "Twenty-Five Years in East Africa", Cambridge at the University Press, 1921, p. 199
128 *loc. cit.* 36, p. 123
129 *loc. cit.* 127, pp. 199—200
130 *ibid.*, p. 204
131 *ibid.*, p. 200
132 *ibid.*, p. 200
133 *loc. cit.* 39, Vol. II, p. 2
134 *loc. cit.* 127, p. 200
135 *ibid.*, p. 200
136 *loc. cit.* 78, p. 135
137 *loc. cit.* 84, p. 77
138 Roscoe, J. — "The Bagesu and Other Tribes of the Uganda Protectorate", Cambridge at the University Press, 1924, p. 96
139 *ibid.*, p. 97
140 *ibid.*, p. 110
141 *ibid.*, p. 98
142 *ibid.*, p. 109
143 *ibid.*, p. 109
144 *loc. cit.* 78, p. 130
145 *loc. cit.* 138, p. 109
146 *ibid.*, p. 111
147 *ibid.*, pp. 110—111
148 *ibid.*, p. 111
149 *ibid.*, p. 114
150 *ibid.*, p. 115
151 *ibid.*, p. 115
152 *ibid.*, p. 97
153 *ibid.*, p. 99
154 *ibid.*, p. 97
155 *ibid.*, p. 99
156 *ibid.*, p. 100
157 *ibid.*, p. 100
158 *ibid.*, p. 100
159 *ibid.*, pp. 97—98
160 *ibid.*, p. 98
161 Roscoe, J. — "The Baganda: An Account of Their Native Customs and Beliefs", Macmillan and Company Limited, London, 1911, p. 234
162 *loc. cit.* 138, p. 98
163 *loc. cit.* 161, p. 234
164 *loc. cit.* 42, p. 199
165 *loc. cit.* 138, p. 98
166 *ibid.*, p. 98
167 *ibid.*, p. 98
168 (a) Westermann, D. — "The Shilluk People", 1912, p. 1
(b) Driberg, J. H. — "The Lango: A Nilotic Tribe of Uganda", T. Fisher Unwin Ltd., London, 1923, pp. 25—29
(c) Schweinfurth, G. — "Emin Pasha in Central Africa", 1888, p. 155
169 Unpublished notes of the author collected during his stay in Acholi country during 1949—1950
170 Crazzolara, J. P. — "A Study of the Acholi Language, Grammar and Vocabulary", International Institute of African Languages and Cultures, London, 1938, pp. ix—x
171 (a) *ibid.*, p. x
(b) Butt, A. — The Nilotes of the Anglo-Egyptian Sudan and Uganda", International African Institute, London, 1952, p. 82
172 loc. cit. 171(b), p. 82
173 Personal communication of the author with Dr. F. K. Girling in 1951
174 *loc. cit.* 171(b), p. 83
175 *ibid.*, p. 81
176 *ibid.*, p. 83
177 *loc. cit.* 170, p. x
178 Nalder, L. F. (Ed.) — "A Tribal Survey of Mongalla Province", 1937, p. 145
179 *loc. cit.* 171(b), p. 83
180 *ibid.*, p. 84
181 *ibid.*, p. 83
182 *ibid.*, p. 84
183 *ibid.*, p. 82
184 *loc. cit.* 168(b), p. 33
185 Gray, J. M. — "Rwot Ochama of Payera", *The Uganda Journal*, The Uganda Society, Kampala, Uganda, Vol. 12, No. 2, 1948, p. 121
186 *loc. cit.* 168(b), p. 71

187 *ibid.*, p. 71
188 *ibid.*, p. 71
189 *ibid.*, p. 70
190 *ibid.*, p. 170
191 *ibid.*, p. 208
192 *loc. cit.* 171(b), p. 93
193 *loc. cit.* 168(b), p. 106
194 *ibid.*, p. 107
195 *ibid.*, p. 107
196 *ibid.*, p. 107
197 *ibid.*, p. 111
198 *ibid.*, p. 204
199 *ibid.*, pp. 205—206
200 *ibid.*, p. 206
201 *loc. cit.* 171(b), p. 94
202 *loc. cit.* 168(b), p. 206
203 *loc. cit.* 171(b), p. 94
204 *loc. cit.* 168(b), p. 206
205 *ibid.*, p. 206
206 *ibid.*, p. 206
207 *ibid.*, p. 206
208 *ibid.*, p. 206
209 *ibid.*, p. 206
210 *ibid.*, pp. 206—207
211 *ibid.*, pp. 80—101
212 *ibid.*, p. 72
213 Gulliver, P. H. — "The Karamajong Cluster", *Africa*, The Journal of the International African Institute, London, Vol. XXII, No. 1, 1952, pp. 6—7
214 *ibid.*, p. 7
215 Huntingford, G. W. B. — "The Northern Nilo-Hamites", Ethnographic Survey of Africa — East Central Africa, Part VI, International African Institute, London, 1953, p. 9
216 *ibid.*, pp. 14—15
217 *loc. cit.* 213, p. 3—4
218 Bernardi, B. — "The Age System of the Nilo-Hamitic Peoples", *Africa*, The Journal of the International African Institute, London, Vol. XXII, No. 4, 1952, pp. 323—324
219 *ibid.*, p. 323
220 *ibid.*, p. 325
221 *ibid.*, p. 331
222 *ibid.*, p. 325
223 *loc. cit.* 213, p. 14
224 *ibid.*, p. 16
225 *ibid.*, p. 19
226 *ibid.*, p. 20
227 *ibid.*, p. 20
228 Lawrence, J. C. D. — "The Karamajong Cluster: A Note", *Africa*, The Journal of the International African Institute, Vol. XXIII, No. 3, 1953, p. 245
229 *ibid.*, p. 244
230 *ibid.*, p. 247
231 *ibid.*, p. 247
232 *ibid*, p. 248
233 *ibid.*, p. 248
234 *loc. cit.* 138, p. 91
235 *ibid.*, p. 162
236 *ibid.*, p. 163
237 *ibid.*, p. 163
238 *ibid.*, p. 163
239 *ibid.*, p. 172—173
240 *ibid.*, p. 173
241 *ibid.*, p. 163
242 *ibid.*, pp. 163—164
243 *ibid.*, p. 164
244 *ibid.*, p. 163
245 *ibid.*, p. 168
246 *ibid.*, p. 168
247 *ibid.*, p. 165
247 *ibid.*, p. 165
248 *ibid.*, p. 169
249 *ibid.*, p. 138
250 *ibid.*, p. 138
251 *ibid.*, p. 137
252 *ibid.*, pp. 137—138
253 *ibid.*, p. 138
254 *loc. cit.* 171(b), p. 175
255 *loc. cit.* 138, p. 147
256 *ibid.*, p. 148
257 *ibid.*, p. 147
258 *ibid.*, p. 147
259 *ibid.*, pp. 149—150
260 *ibid.*, p. 150
261 *ibid.*, pp. 1—2
262 *loc. cit.* 42, p. 168
263 *ibid.*, pp. 165—166
264 *ibid.*, p. 166
265 *loc. cit.* 138, p. 13
266 *ibid.*, p. 20
267 *ibid.*, pp. 20—21
268 *ibid.*, p. 3
269 *loc. cit.* 42, p. 169
270 *loc. cit.* 138, pp. 12—13
271 *loc. cit.* 42, p. 162
272 *ibid.*, p. 169
273 *loc. cit.* 138, p. 7
274 *loc. cit.* 42, p. 169
275 *loc. cit.* 138, p. 7
276 *ibid.*, p. 51
277 *ibid.* p. 51
278 *ibid.*, pp. 51—52
279 *ibid.*, p. 52
280 *ibid.*, p. 62
281 *ibid.*, p. 65
282 *ibid.*, p. 66
283 *ibid.*, p. 52
284 *ibid.*, p. 54
285 *ibid.*, p. 54
286 *ibid.*, pp. 54—55
287 *loc. cit.* 42, p. 145

288 *ibid.*, p. 146
289 *ibid.*, p. 153
290 *ibid.*, p. 153
291 *ibid.*, p. 153
292 *loc. cit.* 40, pp. 232—359
293 Lugard, F. D. — "The Rise of Our East African Empire", William Blackwood & Sons, Edinburgh and London, 1893, p. 1
294 *loc. cit.* 161, p. 186
295 *loc. cit.* 40, p. 206; *loc. cit.* 161, p. 231
296 *loc. cit.* 161, p. 137
297 Kagwa, Sir Apolo — "The Customs of the Baganda", Translated by Ernest B. Kalibala and Edited by May Mandelbaum Edel, Columbia University Press, New York, 1934, p. 18
298 *ibid.*, pp. 9—11
299 *ibid.*, p. 164
300 *ibid.*, p. 19
301 *ibid.*, pp. 9, 112, 161, etc.; *loc. cit.* 161
302 *loc. cit.* 297, pp. 10—11, etc.
303 *loc. cit.* 161, p. 215
304 *loc. cit.* 297, p. 20
305 *ibid.*, p. 97
306 *ibid.*, p. 16
307 *loc. cit.* 40, pp. 205—206
308 *loc. cit.* 27, 36, 42, 78, 92, 127, 138, 161, etc.
309 *loc. cit.* 161, p. 236
310 *ibid.*, pp. 233—234
311 *ibid.*, p. 239
312 *ibid.*, pp. 239—240
313 *ibid.*, p. 246
314 *ibid.*, p. 240
315 *ibid.*, pp. 240—241
316 *ibid.*, p. 241
317 *ibid.*, p. 238
318 *ibid.*, p. 238
319 *loc. cit.* 127, pp. 83—84
320 *loc. cit.* 161, pp. 268—269
321 *ibid.*, p. 244
322 *ibid.*, p. 245
323 *ibid.*, p. 246
324 *ibid.*, p. 245
325 *ibid.*, p. 246
326 *ibid.*, pp. 242—243
327 *ibid.*, p. 243
328 *ibid.*, p. 247
329 *ibid.*, p. 245
330 *ibid.*, p. 234
331 *ibid.*, p. 234
332 *loc. cit.* 127, pp. 76—78
333 *ibid.*, p. 79
334 *ibid*; pp. 79—80
335 *loc. cit.* 27, p. 1
336 *loc. cit.* 161, p. 13
337 *ibid.*, p. 269
338 *ibid.*, p. 269
339 *ibid.*, p. 415
340 *ibid.*, p. 268
341 *ibid.*, p. 246
342 *ibid.*, pp. 383—384
343 *ibid.*, pp. 384—385
344 *ibid.*, p. 410
345 *loc. cit.* 127, p. 79
346 *loc. cit.* 297, pp. 67—68
347 *loc. cit.* 161, p. 225
348 *loc. cit.* 297, p. 36
349 *loc. cit.* 161, p. 403
350 *ibid.*, p. 225
351 *ibid.*, p. 225
352 *loc. cit.* 297, p. 40
353 *loc. cit.* 127, p. 79
354 *loc. cit.* 161, pp. 413—414
355 *ibid.*, p. 414
356 *ibid.*, p. 414
357 *ibid.*, p. 226
358 *ibid.*, p. 227
359 *loc. cit.* 297, p. 79
360 *ibid.*, p. 98
361 *ibid.*, p. 23, etc.
362 *loc. cit.* 161, p. 227
363 *ibid.*, p. 227
364 *ibid.*, p. 2
365 *loc. cit.* 40, p. 206
366 *loc. cit.* 161, p. 228
367 *ibid.*, p. 229
368 *ibid.*, pp. 227—228
369 *ibid.*, p. 443
370 *loc. cit.* 297, p. 88, also 14, 43, 51, 102, etc.
371 *ibid.*, pp. 78, 87, 88, etc.
372 *ibid.*, p. 98
373 *ibid.*, p. 87
374 *ibid.*, p. 87
375 *loc. cit.* 161, p. 229
376 *ibid.*, pp. 228—229
377 *loc. cit.* 77, pp. 31—32
378 *loc. cit.* 127, pp. 97—98
379 *loc. cit.* 161, p. 14
380 *loc. cit.* 168(c), p. 123
381 *loc. cit.* 161, p. 229

CHAPTER IV.

UGANDA UNDER BRITISH RULE

With man-traps stole upon you those hunters
whose fierceness was keener than the fangs of your wolves,
whose pride was blinder than your lightless forests.
The savage greed of the civilised stripped naked its unashamed
inhumanity.
You wept and your cry was smothered,
your forest trails became muddy with tears and blood,
while the nailed boots of the robbers left their indelible prints
along the history of your indignity.
And all the time across the sea,
church bells were ringing in their towns and villages,
the children were lulled in mothers' arms,
and poets sang hymns to Beauty.

Today when on the western horizon
the sun-set sky is stifled with dust-storm,
when the beast, creeping out of its dark den,
proclaims the death of the day with ghastly howls,
come, you poet of the fatal hour,
stand at that ravished woman's door,
ask for her forgiveness,
and let that be the last great word
in the midst of the delirium of a diseased Continent.

[From the poem entitled "Africa" by Rabindra Nath Tagore, published in Bengali in 1937 and in English in *Poems*, Visva-Bharati, Calcutta, 1946, pp. 159—160.]

What happened to the Uganda societies after they came under British rule? Did they undergo any further development by the eradication of the evils in the societies and encouragement of their progressive features? Opinions do not differ that with the advent of the British Uganda entered upon a new course of history. But what was this new course? Was it the road to peace, progress and prosperity under a Christian Government of Western Civilisation in place of the primitive "tribal" life under the savage rule of heathen African despots and chiefs? Or, was it something else?

Chapter III has shown that while very few of the African peoples in Uganda remained in their tribal stage, the Uganda societies could be divided into two categories: (1) those in which the institution of the state and contending classes had definitely developed, and (2) those in which although distinct classes had not yet emerged, the tribal societies were in the process of disintegration and incipient tendencies were present towards *differentiation* in the societies and the consequential development of centralised authority — the prelude to the formation of class societies and the establishment of the institution of the state. The conclusion which would be obvious from this is that the Uganda societies did not need being *re-made* by the British, as acclaimed by Elspeth Huxley, for there was nothing peculiarly "primitive" or "obscurantist" in their course of development. Similar developments of societies, of course with local specialities, have been observed also in other parts of the world.

However, for further progress of the Uganda societies, many important developments were called for at the time when the British appeared in Uganda. Of these, changes in the political structure of the societies probably demanded first attention. What were mainly necessary in this respect were: — (1) where the institution of the state had developed, democratisation of the state-apparatus by the abolition of the powers of the kings and their appointed chiefs, and establishment of a government by popular representation, and (2) in other places, leading the societies to a higher level of democratic government, as suggested above, while freeing the people from chiefdom, where it had developed, and taking care that no other form of authority is *imposed from above* upon them or on the other peoples by appointing their rulers instead of allowing the peoples themselves to chose their own representative government.

Did the British undertake these responsibilities when they commenced "these Protectorates from motives of pure philanthropy"? This is the question which ought to be discussed following the previous account of the life of the Uganda Africans before the British appeared on their territory.

The evil effects of the autocratic powers of the kings and chiefs were known to the British from earliest times. From Speke to later comers, all of them have been very voluble on the subject of the despotic rule of the kings of Buganda and Kitara, and of the "barbaric" authority of the African chiefs. Indeed, the evil effects of such autocratic rule, finally exemplified by the participation of the ruling classes with the Arabs in hunting for slaves, could not have passed unnoticed. Thus, Speke, who thought that this was the main cause of stagnation of Africa, noted[1]:

> "Slavery begets slavery. To catch slaves is the first thought of every chief in the interior; hence fights and slavery impoverish the land and that is the reason both why Africa does not improve, and why we find men of all tribes and tongues on the coast. The ethnologist need only go to Zanzibar to become acquainted with all the different tribes to the centres of the continent on that side, or to Congo to find the other half south of the equator there."

Samuel Baker noted about the Acholi chief, Rwot Ochama of Payera, whom he called Rwot Jarma[2]:

> "He had determined to offer his allegiance and he and all adjacent countries will serve the government faithfully, in return for protection and justice."

Evidently, even though for some Acholi chiefs alliance with the Arab traders might have been profitable, for others and certainly for the general run of the people this was then the worst calamity. Therefore, they offered their allegiance to the British mainly for protection from the predatory slave traders and cattle hunters. And how the despotic rulers of Buganda and Kitara and their chiefs oppressed and robbed the weaker peoples has been particularly illustrated by the case of the Busoga who sought for British protection in order to get rid of the oppressive rule of the kingdom of Buganda. Thus, the writings of the first British explorers, missionaries and administrators in Uganda are full of horrible accounts of how the ruling classes in these states cynically dealt with the life, property and labour of the common people and still more oppressed those outside their kingdom to whom they had access through wars and raids and for whom they had no compunction whatsoever[3].

Such descriptions, of course, formed the basis for the British declaration that they had come to Uganda to bring peace to the country and to lead the Africans from "savagery to civilisation". Nearly a century after Speke, Elspeth Huxley visited Uganda and declared that the British came there as "protectors" to save the Africans from the autocratic rule of "the king and his henchmen". But did the British really have this aim in view? Or, is it true that this was only one of their pretexts to annex Uganda? And that by harping only on this feature of African life they have greatly helped in confusing the mind of the world public in regard to the real situation in Uganda and the life of the Uganda Africans before the Europeans came into their country?

Furthermore, after their rule was spread all over Uganda, did the British do away with these kings and their appointed chiefs, or, on the contrary, did they further stabilise the interests of these native oppressors and at the same time establish themselves over and above them as the supreme authority? And did they not simultaneously develop a similar machinery by importing and selecting chiefs even for those societies where democracy, although in the stages of tribal organisation or *military democracy*, was present before? And thus did they not spread the pre-British forms of oppression all over the Protectorate, that is, beyond the established kingdoms and chiefdoms, while in addition to it the people had to bear henceforth the burden of the British ruling class placed on top of all these societies? The following pages will examine this question, whether the peoples of Uganda faced a better situation after they came under British rule or whether since then they have been compelled to bear the double burden of feudal and colonial exploitation.

There is also the question as to whether the British changed qualitatively the state of productive forces in the Uganda societies, which with their avowed aim to free the peoples from the oppression of native rulers could have led the Africans on a course of peace, prosperity and progress. But did the British rulers undertake this task?

Early British accounts of the life of Uganda Africans are full of the "primitive" features of their economic and social life, and it is true that, even though there is nothing peculiar in it, the state of productive forces in pre-British Uganda was ill-developed and the standard of living was low. As noted before, although agriculture had become the main occupation for most of the peoples, human labour was practically the only motive power for production; plough cultivation having not developed yet. The productive techniques in agriculture were also poor, and the same is true for their industries. The upshot was that the living standards were low, and consequently the societies maintained many customs which could be labelled as "obscurantist" by a superficial European observer, for he would not care to find out the reason for the persistence of such customs in society, and, on the contrary, would be inclined to interpret them from the standpoint of his own social environment which had, no doubt, progressed farther than the African societies due to specific internal and external causes known to everyone today.

There is, of course, no doubt that many such customs will not fit in with modern conditions, and instead of glorifying them and proposing to revert back to the "idyllic life" of ancient Africa they must be changed. But should it not be questioned at the same time whether or not the *atavistic* presence of such customs even today is due to the fact that the normal course of development of these societies was cut short with the advent of the British? Was is not very probable that given the time for the employment of the primary accumulation of capital from trading with the Arabs into some industries and for importing *socially necessary goods*, the economy of the kingdoms of Buganda and Kitara would have been further improved and this constant improvement would in due course have permeated into other societies, eventually developing the state of productive forces of all the societies? And, in that case, would not the economic progress also have influenced, in turn, the social fabrics of the societies (just as it had began to be felt in Buganda during the reign of Mutesa), so that in their normal course of development these societies would have emerged as modern as can be found in other parts of the world?

Such however was not the course of history, for before all this could happen a British Protectorate was established over the territory now known as Uganda. Hence, there is all the more reason why with the appearance of the British in this area, with a superior culture and "philanthropic" motives, everybody would have expected that henceforth the Uganda Africans would be raised to ever-progressing levels of economic prosperity, which, in the end, could not but have eradicated their social evils and done away with their "obscurantist" customs. Therefore, besides democratisation of their political structures, this was the second most essential demand of the Uganda Africans on their new rulers.

A constant improvement of the state of productive forces in all societies, so that the peoples of Uganda could embark upon a life of peace, progress and prosperity under modern conditions while maintaining their sovereignty and independence, was the demand of the day. Sovereignty and independence the Uganda Africans lost to the British, as will be explained below; but, even so, did the British uplift the African societies from their low standards? Did they improve the state of productive forces in Uganda and develop the economic life of the people? Or, on the contrary did they put the Uganda Africans under bondage only in order to further the monopoly interests of the ruling class of their own country, while what superficially appeared as socio-economic progress of the Protectorate during British rule were mainly the bye-products of their colonial policy, which following the dictates of history could hardly be otherwise in their general characteristics? The following pages will endeavour to discuss this question.

However, to understand why the British behaved as they actually did in Uganda, from the beginning of their rule, it is first necessary to examine why they came there at all, and how they established themselves in this territory. Were they guided by "motives of pure philanthropy", as Johnston declared in 1902? Were they dictated by their "aims of idealism", as Elspeth Huxley asserted nearly half a century later? Was it due to what, in the words of Elspeth Huxley, the British were "impelled to make by inner forces, not by entreaties from without; something undertaken, unaware no doubt, to heal a conscience roughened by memories of slavery and exploitation, not in response to any African demand"? These are the questions which will be discussed first: whether it was the dictates of idealism of pure philanthropists or pangs of conscience of slave-hunters which brought the British to Uganda; or whether, irrespective of "any African demand", they came there with cool calculations with something else in view; and how then they went about giving vent to their "inner forces" on the peoples and societies of Uganda.

1. Why British Rule?

Even up to the third quarter of the last century Africa was only of secondary interest to the European capitalist powers. In East Africa they kept themselves confined only to some patches on the coast, and that also primarily as outposts of their lucrative possessions in the East. The Portuguese held an unstable position there from the fifteenth to eighteenth centuries, while it was the Arabs who were in a dominant position. "As early as the eight century A. D., Oman Arabs, borne southward in their open dhows by gentle monsoons, founded the towns of Kilwa, Mombasa, and Zanzibar" on the East African coast[4]. "In the tenth century, the Zenj Empire — a federation of Arab states which controlled the coast of what is now Kenya and Tanganyika — was founded."[5] The Portuguese, the first European power to arrive in East Africa, appeared on the scene on the 7th of April 1498, when Vasco da Gama on his way to India anchored off Mombasa. Thence he sailed to Malinde, and then to India, and stopped again in Malinde next year on his return voyage from India.

"The next few years were spent by the Portuguese in establishing their supremacy along the whole East African Coast. All the principal towns and Sultanates, such as Zanzibar, Kilwa, Pemba, Barawa and Sofala fell before the invader."[6] In 1505 Mombasa fell into the hands of the Portuguese, in 1506 the King of Lamu surrendered to the Portuguese, and "in 1509 the appointment was made from Lisbon of a Governor of the provinces of Aethiopia and Arabia to reside at Malinde"[7]. The Portuguese, however, never lived peacefully in this area; they were all along harrassed by the Arabs.

Meanwhile other European Powers followed the Portuguese. "The decline of Portugal after its annexation by Philip II of Spain in 1580 coincided with the invasion of the East by rival European peoples along the sea-way Portugal had opened; and in the course of the next century the Portuguese were robbed of all their eastern realm except one or two ports in the Indies and that fever-ridden stretch of African coastland between Cape Delgado and Delago Bay."[8]

Those were the days of the merchant capitalists of Europe, searching in every nook and corner of the earth for loot[9]. The looting expeditions were not, however, yet directed to the heart of Africa. The coast merely served for their "filling stations" on the voyage to the East.

"The first French ship rounded the Cape of Good Hope in 1529", but "French overseas enterprise was engaged for a time elsewhere", and so they did not pursue the Portuguese for a while[10]. The first English ship to round the Cape of Good Hope was in 1580, "homeward bound on Drake's world-voyage", and the first Dutch ship in 1595[11]. Then, "in 1591 Lancaster, with 'three tall ships', was sent to India by da Gama's East African route: and in 1600 the English East India Company was founded. Thereafter, both from England and from the Netherlands, armed trading-fleets were regularly sailing east and spreading out over the whole commercial sphere from the Red Sea to the Philippines"[12]. In 1664 the French East India Company was formed, and "the French entered the field mainly by the sea-route", having occupied "Bourbon, an island outpost in the Indian Ocean" in 1642, Mauritius in 1715, and repeating attempts to occupy Madagascar in the meanwhile[13]. In the end, the decisive mastery of Britain over India gave her the supreme position in this part of the world. "After 1815 the British were masters of its (Indian Ocean's) waters and able, when and where they chose, to dominate its coasts."[14]

During all this period, however, the European Powers hardly at all interested themselves in colonising Africa. Although Zanzibar, the coastal town in East Africa, presented itself to Lancaster in a favourable light, "the idea of developing any trade with Zanzibar did not occur to Lancaster —except for 'some thousand weight of a kind of grey and white gum', presumably gum-copal, Lancaster found nothing worth taking aboard"[15]. "The French, for their part, concentrated on the Mascerene islands and shunned the mainland."[16] It was only the Dutch who consistently struggled to oust the Portuguese from the coast, probably owing to the "lure of 'Sofala gold' "[17]. "The reason for this neglect of the coast was almost certainly its bad hygienic reputation. Nearly all the early travellers speak of the heat and unwholesomeness of Mozambique and its neighbourhood."[18] Therefore, when the hands of these merchant-capitalists were full with "commercial plans" for the East, and their attention was keenly directed to usurping, at the cost of others, the entire trade therein[19], East Africa remained for them only as coastal outposts.

Yet the capitalist Powers of Europe did not give a blank cheque to the petty Arab traders. Their interest in trading along the African coast became more and more felt in the era of Industrial Capital. It is interesting to note in this connection that even American companies were involved in it. "In 1830 the American firm John Bertram settled in Zanzibar, and found there a large market for their cotton goods and hardware, in exchange for which they exported ivory and rubber"[20]. "As early as 1833 the United States made a commercial treaty with the Sultan, and three years later they established a trading consulate at Zanzibar. Great Britain had close commercial relations with East Africa, on account of its proximity to India, and during a few months in 1824 had exercised a Protectorate over Mombasa, and established a Consulate. In 1844 the French followed, and were in great favour —especially under Sultan Said. In the years 1847—1849, under Captain Guillain, they took a great part in the exploration of East Africa. Shortly afterwards the Hamburg firm of O'Swald and Co. established themselves in Zanzibar. From the west coast, where they had been trading, they sent, in 1846, a sailing-ship to East Africa to get cowrie-shells, which were used as coinage on the west coast. The favourable commercial relations they found at Zanzibar encouraged the firm to stay there, and they acquired a site near the Customs House, which is one of the best in the whole town. In 1859 the Hanseatic States made a treaty with the Sultan (the Arab Lord of Zanzibar), and established a Consulate under the management of O'Swald representative. Some years later the Hamburg firm, Hansing and Co. opened a branch at Zanzibar."[21]

Those were the days characterised by the slogan of "Free Trade". Britain was then the "workshop of the world". "Raw materials were drawn from all over the world. The products of British machine industry dominated the markets of every country. British shipping, under the protection of the British Navy, dominated world trade. . . . In this era of Britain's nineteenth-century Industrial supremacy the unchallenged domination of British machine industry appeared able to break down every obstacle in all countries, not only in countries directly ruled by Britain, but also in foreign countries independent of Britain. This superior economic power, which found its expression in the doctrines of *laissez-faire* and free trade, seemed so invincible to the new ruling class representatives of the British manufacturers that conceptions began to gain currency during the middle nineteenth century which dismissed the whole colonial system as a superfluous extravagance and an obsolete relic."[22] The idea was shared also by other capitalist powers. For instance, Bismarck wrote to von Roon in 1868[23]:

> "All the advantages claimed for the Mother Country are for the most part illusions. England is abandoning her Colonial Policy; she finds it too costly."

In Britain, the picture of colonies was painted by Disraeli as "a millstone round our necks". Therefore, while holding tenaciously to the old colonies, like India, Britain with little reluctance gave up its control of Mombasa in 1827.

But only half a century later, like the other European capitalist Powers, Britain was bent on controlling the whole of East Africa, and not merely the port of Mombasa. This change in the attitude of Britain and her rival European Powers was directly linked up with the growth of monopoly capitalism in their countries.

The incipient change in the British attitude to East Africa was evident from the opening of the Suez Canal in 1869. There was then a strategic reason to get a firm hold over East Africa in order to safeguard the new route to India, the treasure-trove of the British capitalists. There was also the reason to protect the backdoor to Egypt where also they were becoming more and more interested. By 1876, British loans to Egypt for railways, canals, bridges, etc., which were built by British contractors, drew an annual interest of six million pounds sterling[24].

Typical were then the moves of Britain in spreading her interest in the east coast of Africa. Just as she had done so often, and almost always successfully in India[25] Britain began to play the game of umpiring the local contestants striving to gain a position. The Sultan of Zanzibar, "Seyid Said died at sea in October, 1856. His eldest surviving son

Seyid Thwain, became by his father's will Lord of Oman, whilst Seyid Majid, the next son, succeeded to the Government of Zanzibar and the East African coast. The former, dissatisfied with his heritage, threatened, in 1859, an attack on Zanzibar, but was prevented by English cruisers. The dispute was referred to Lord Canning, then Viceroy of India, by whose award Zanzibar was declared independent of Oman, and the Imam was paid an annual subsidy by India. This payment placed Zanzibar in the position of a subsidised dependency of India"[26]. Britain spread her influence in this favourable situation. Later, after the death of Seyid Majid in 1870 and the succession by his brother, Seyid Barghash, to the Sultanate, Britain further consolidated her position. "In 1872 the British India Steam Navigation Company started a regular service between Aden and Zanzibar. In 1873 Great Britain caused Seyid Barghash to sign a treaty for the abolition of slavery, and English influence grew so rapidly that the same Sultan offered to the Chairman of the British Shipping Company a lease of all his Custom duties and other royalties."[27] Henceforth. Britain began to control the waters of this region "by a force of British warships ostensibly present to prevent trading in slaves"[28].

The decisive change, however, came with the transformation of Britain's economy from that of industrial capitalism to monopoly capitalism. The new industrial states of the world, viz. Germany and the United States of America, were outstripping Britain in industrial production. "During the last quarter of the nineteenth century Britain lost industrial supremacy, first to the United States, and then to Germany. In 1880 British steel output stood at 1.3 million tons, American at 1.2 million and German at 700,000; by 1900 American steel output had reached 10.2 million tons, German 6.4 million and Britain 4.9 million; and by 1913 American steel output had reached 31.3 million tons, German 18.9 million, and British 7.7 million. Britain still maintained the first position in the export of manufactured goods, but with a lessening proportion. Between 1880—4 and 1900—4 British exports of manufactures increased 8 per cent., German 40 per cent. and American 230 per cent."[29] Britain could no more claim world supremacy in industrial production. The slogan of "Free Trade" had become obsolete.

The time was ripe for Britain to pass over from the saturation point reached in the progressive development of capitalism in her land to socialism. The people of Britain were striving for this progressive change in their life. "Socialist agitation arose anew in Britain from the eighteen-eighties, with the formation of the Social Democratic Federation, which later merged into the Communist Party. The modern labour movement derives from the work of the early socialist pioneers of the eighteen-eighties."[30] But the British bourgeoisie took another course to maintain their rule. While losing in the sphere of industrial production, "in the sphere of the export of capital and colonial expansion Britain led the way"[31]. Instead of moving towards socialism, Britain passed over to the era of finance-capitalism.

The key feature of the age was the "transformation of competition into monopoly" — "one of the most important — if not the most important — phenomena of modern capitalist economy"[32]. To quote further from this study on the "Highest Stage of Capitalism"[33]

> "Competition becomes transformed into monopoly. The result is immense progress in the socialization of production. In particular, the process of technical invention and improvement becomes socialized.
>
> This is something quite different from the old free competition between manufacturers, scattered and out of touch with one another, and producing for an unknown market. Concentration has reached the point at which it is possible to make an approximate estimate of all sources of raw materials (for example, the iron ore deposits) of a country and even, as we shall see, of several countries, or of the whole world. Not only are such estimates made, but these sources are captured by gigantic monopolist combines. An approximate

estimate of the capacity of markets is also made, and the combines 'divide' up amongst themselves by agreement. Skilled labour is monopolized, the best engineers are engaged; the means of transport are captured: railways in America, shipping companies in Europe and America. Capitalism in its imperialist stage leads right up to the most comprehensive socialization of production; it, so to speak, drags the capitalists, against their will and consciousness, into some sort of a new social order, a transitional one from complete free competition to complete socialization.

Production becomes social, but appropriation remains private. The social means of production remain the private property of a few. The general framework of formally recognized free competition remains, but the yoke of a few monopolists on the rest of the population becomes a hundred times heavier, more burdensome and intolerable."

This tendency, while most marked in the iron and steel industry and in textiles, became also apparent in other industries, such as chemical manufacturing or manufacture of soap and margarine, etc. "The Levers and Brunner Mond of the late nineteenth century held the seeds of the great Unilever and I. C. I. combines of today."[34]

The other notable feature of this age was the monopolist trend of the banking institution. "As banking develops and becomes concentrated in a small number of establishments, the banks grow from humble middlemen into powerful monopolies having at their command almost the whole of the money capital of all the capitalists and small businessmen and also the larger part of the means of production and of the sources of raw materials of the given country and in a number of countries. This transformation of numerous humble middlemen into a handful of monopolists represents one of the fundamental processes in the growth of capitalism into capitalist imperialism; . . ."[35]

Finally, the fusion of industrial and banking capital decides the new era of finance-capitalism, and supplies the motive power to this new and the highest stage of capitalist development. As it was noted[36]:

> "Scattered capitalists are transformed into a single collective capitalist. When carrying the current account of a few capitalists, a bank, as it were, transacts a purely technical and exclusively auxiliary operation. When, however, this operation grows to enormous dimensions we find that a handful of monopolists subordinate to their will all the operations, both commercial and industrial, of the whole capitalist society; for they obtain the opportunity — by means of their banking connections, their current account and other financial operations — first, to *ascertain exactly* the financial position of the various capitalists, then to control them, to influence them by restricting or enlarging, facilitating or hindering credits, and finally *entirely determine* their fate, determine their income, deprive them of capital, or permit them to increase their capital rapidly and to enormous dimensions, etc."

"Thus, the twentieth century marks the turning point from the old capitalism to the new, from the domination of capital in general to the domination of finance-capital."[37]

Transition to this new era began in the last quarter of the nineteenth century. Referring to this transitional period Dobb noted[38]:

> "What has become known as the Great Depression, which started in 1873 and, broken by bursts of recovery in 1880 and 1888, continued into the middle of '90' s, has come to be regarded as forming a watershed between two stages of Capitalism: the earlier vigorous, prosperous and flushed with adventurous optimism; the later more troubled, more hesitant and, some would say, already bearing the marks of senility and decay. This was the period of which Engels spoke his well-known phrase about 'the breakdown of . . . England's industrial monopoly', in which the English working class would 'lose its privileged position' and 'there (would) be Socialism again in England'."

Then again Dobb wrote[39]:

"If there may be some obscurity about the causation of the Great Depression, there is much less about its effects on British Capitalism. Having witnessed the drastic effect of competition in cutting prices and profit-margins, businessmen showed increasing fondness for measures whereby competition could be restricted, such as the protected or privileged market and the price and output agreement. This enhanced concern with the dangers of unrestrained competition came at a time when the growing concentration of production, especially in heavy industry, was laying the foundation for great centralization of ownership and of control of business policy."

Similar was the development in other capitalist countries. To quote Dobb once again[40]:

"In the newer industry of Germany and the United States this centralization was to be earlier on the scene than in Britain, where the structure of business, with its foundations firmly laid in the first part of the century, had developed according to a more individualist pattern, and the tradition attaching to this structure was more stubborn tn survival. In the structure of economics as of human organisms ageing bones are apt to grow rigid. In America the '70's saw the rise of the trusts, which had sufficiently grown in extent and structure to provoke the legislation against trust companies in the late '80's and the more sweeping Sherman Act of 1890 directed against 'combination in restraint of trade'. In Germany associations of producers in the iron industry and the coal industry were formed in the '70's, and over the next three decades multiplied in these and other industries, until in 1905 there were stated (by the Kartell-Commission of that year) to be something in the neighbourhood of 400 cartels: a development which, in the words of Liefmann, a well-known apologist for cartels, was 'a product . . . of the entire modern development of industry, with its increasing competition, the increasing risks of capital and the failing profit'."

The characteristic feature of the capitalists in this period was "a preoccupation with privileged spheres of foreign trade" and "an interest in privileged spheres of foreign investment"[41]. "New sources of supply of raw materials were urgently needed to feed the mighty industrial machine that concentration of capital had made possible. New and more profitable fields of investment than were available in Britain were needed to absorb the immense balances thad had been built up. But as capitalism in France and in Germany had after 1880 much the same needs as Britain, a feverish struggle was begun between the Great Powers for *exclusive* supplies of raw materials and *exclusive* rights of capital investment."[42] Giving a comprehensive account of this situation, Dobb wrote[43]:

"This concern with foreign investment was a distinctive mark of the new period, having no close likeness in its prototype. The difference marked the contrast between an age of undeveloped capital accumulation and the latter days of industrial Capitalism. Of this mature Capitalism, impelled by the need to find new extensions of the investment-field, export of capital and of capital goods constituted a leading feature. In the '80's there awakened a new-found sense of the economic value of colonies: an awakening which occurred with remarkable simultaneity among the three leading industrial Powers of Europe. During that decade, as Mr. Leonard Woolf has pointed out, 'five million square miles of African territory, containing a population of over 60 millions, were seized by and subjected to European States. In Asia during the same ten years Britain annexed Burma and subjected to her control the Malay Peninsula and Baluchistan; while France took the first steps towards subjecting or breaking up China by seizing Annam and Tonking. At the same time there took place a scramble for the islands of the Pacific between the three 'Great Powers.' Business interests in centres like Birmingham and Sheffield began to raise the demand that 'to make good the loss of the American market we ought to have the colonial market'; and Joseph Chamberlain was to call on the Government to give protection to markets at home while taking steps to 'create new markets' abroad, and to raise his glass in simultaneous toast of 'Commerce and Empire, because, gentlemen, the Empire, to parody a celebrated expression, is Commerce'. In similar vein, writers in Germany at the turn of the century were talking of the participation of Germany 'in the policy of expansion out of Europe, at first modestly, of late with growing decision', as being compelled by 'enormous increase of its industrial production and its trade', and of German activities in the Near East as 'doing what we are doing in other parts of the world — seeking new markets for our exports and new spheres of investment for our capital'. Mr. Rostow has summed up the effect on capitalists of their experience in the Great Depression as follows: they 'began to search for an escape (from narrower profit-margins) in the insured foreign markets of positive imperialism, in tariffs, monopolies, employers' associations'. The extension of the investment-field and the search for the stimuli of new markets to keep productive equipment working to capacity, the race to partition the undeveloped parts of the globe into exclusive territories and privileged markets, were quickly to become the orders of the day. Price agreements, it is true, were no new thing — they had been common among ironmasters quite early in the century — and export of capital was no sudden novelty. But this new preoccupation represented a very different focus of interest and yielded a very different design of economic strategy from that which had held the minds of the industrial pioneers of Ricardo's day."

It was this period in which Africa was opened up. It was at this time that the *Scramble for Africa* began, which was one of the most distinctive features of the race of the Colonial Powers for "the final partition of the globe"[44].

This was necessary not only for the export of capital from the metropolitan countries, but, as far as Africa was then directly concerned, to forestall rival powers in the prospective wealth of each and every region in the world. For[45]:

"Finance capital is interested not only in the already discovered sources of raw materials but also in potential sources, because present-day technical development is extremely rapid, and land which is useless today may be made fertile tomorrow if new methods are applied (to devise these new methods a big bank can equip a special expedition of engineers, agricultural experts, etc.), and if large amounts of capital are invested. This also applies to prospecting for minerals, to new methods of working up and utilizing raw materials, etc., etc. Hence, the inevitable striving of finance capital to enlarge its economic territory and even its territory in general. In the same way that the trusts capitalize their property at two or three times its value, taking into account its 'potential' (and not present) profits, and the further results of monopoly, so finance capital strives in general to seize the largest possible amount of land of all kinds in all places, and by every means, fearing to be left behind in the fierce struggle for the last scraps of undivided territory, or for the repartition of those that have been already divided."

Therefore, simultaneous with the export of capital went on land-grabbing by the states representing the interests of the imperialist bourgeoisie. In 1862, Great Britain had invested 3.6 billion francs of capital abroad; in 1872, the figure rose to 15 billions; in 1882, to 22 billions; in 1893, to 42 billions; in 1902, to 62 billions; and in 1914, from 75 to 100 billions. In 1869, France had invested 10 billion francs of capital abroad; in 1902, the figure was between 27 and 37 billions; and, in 1914, 60 billions. The newly rising power — Germany — was also coming up to the level. The capital invested abroad by Germany was 12.5 billions franc in 1902; in 1914, it rose to 44 billions[46].

At the same time, colonial possessions of these Powers went up by leaps and bounds. In million square miles Great Britain's colonial possession was 2.5 in 1860; in 1880, the figure rose to 7.7; and, in 1899, it was 9.3[47]. Thus, an increase of nearly four times was made in the colonial possession of Britain within a period of thirty-nine years; that is, at the average rate of adding territory nearly twice the size of the United Kingdom every year to the old colonial possession.

As regards Africa, in 1876, only 10.8 per cent of African territory was ruled by the European colonial powers; in 1900, only 9.6 per cent of the continent was not so held[48]. By the beginning of the present century, Africa fell under the jackboots of world imperialism.

In this onslaught Britain had the lion's share. Before the First World War, France came first in territorial responsibility, having acquired 4 million square miles of African land, excluding her control over Morocco. Great Britain was then second with 2.8 square miles of territory under direct Crown control and 473,000 square miles in South Africa. But having acquired the bulk of the German territory in Africa after the 1914—1919 war, which in total amounted to 1.13 million square miles, she further fattened herself. Belgium was next to Germany with over 900,000 square miles of the Free Congo State; and then was placed Portugal with 793,000 square miles of African territory. Italy increased her control from 200,000 square miles of African land to 600,000 square miles after her break with Turkey in 1911. And last came Spain with a total possession of less than 90,000 square miles, excluding her interests in Morocco[49].

Moreover, even if in territorial possession Britain could not out-rival France, in regard to the population under her control she topped the list. Under British control were placed 35 million individuals, of whom 29 millions were under Crown Government and the remaining 6 millions in the Union of South Africa —a British Dominion. France came next with her control over 25 million souls; and prior to the First World War, Germany exercised control over 12 million individuals, Belgium over 10 millions, Portugal over 8 millions, and the remaining Africans under bondage (a little more than 9 millions in number) were placed under Italy, Spain, etc.[50],

The biggest Colonial Power in the world thus became also the biggest occupying power in Africa.

2. Invasion of Uganda

One of the precious spoils of Britain in this loot of Africa was the possession of East Africa. Starting from Kenya, the British Power spread over Uganda, and after the First World War also annexed Tanganyika from the Germans. Apparently, the British had "a prescriptive right in East Africa". "The journeys of Livingstone were mostly to the south of the regions with which we are here concerned; but in 1857 Burton and Speke made their way from Bagamoyo to Tanganyika. Speke parted from Burton, and discovered Lake Victoria" in 1858[51]. It was also noted that Speke made a second expedition with Grant, "visited King Mtesa of Uganda and descended the Nile with Sir Samuel Baker in 1862"[52]. "In 1864 Baker discovered Lake Albert. In the seventies Stanley and Cameron crossed equatorial Africa: to the former was due the foundation of the Congo Free State, and the commencement of our interest in Uganda, which ended in the establishment of the Protectorate"[53]. Lugard declared in 1893[54]:

> "We have a prescriptive right in East Africa and its lakes. They were all discovered by British explorers. ... Our Missionaries first penetrated to Uganda in the footstep of our explorers."

In 1894, the "Protectorate over Uganda was proclaimed, and in 1895 . . . the East Africa Protectorate was constituted" over Kenya and Uganda[55]. By 1896, occupation of Uganda was complete; "the Protectorate was extended to most of the other regions which are now included within the present 'Uganda', and this term was thereafter applied to the whole territory"[56].

The invasion of Uganda followed "the classic African pattern —the use of armed force, the use of missionaries as pawns in the imperialist game, the encouragement of commercial enterprise in the form of a Chartered Company, and the final assumption of full power by the State"[57]. No doubt, some individuals came to dedicate their life in Africa as Christian missionaries to uplift the "benighted" Africans, and some also came for individual gain irrespective of any political interest. Thus Lugard gave a few instances of the true missionary spirit of the Christian Fathers in the early days of contact between black and white in Uganda[58]; and Johnston spoke of one Mr. Stokes, an Irish ex-missionary merchant, who wanted to remain as an adventurer without declaring his permanent allegiance to any European Power. But this Mr. Stokes was finally the victim of a "cold blooded and wholly indefensible murder"[59]. This showed how impossible it was, even at that period, to remain an "individual" for personal gain only. The upshot was that in the entire process of change which was set in motion after the initial contact between the black and white, the missionaries and the travellers and adventurers were either completely identified with the governing interests of the ruling classes of their countries, and thus served as the "spiritual" or the "pioneer" flank of the army of the official and non-official rulers, or they were some how eliminated from the society's life, preferably of course by removing them form the areas of occupation.

It may be recalled that Speke, the first European explorer to visit Uganda, was entrusted with finding out ways and means to open trade between Britain and the kingdoms of Buganda, Bunyoro, etc., for the purpose of which he wrote in his narrative that he carried a message form the Queen Victoria of England[60]. David Livingstone, the most publicised explorer of Africa, told Lord Palmerston that the best boon the latter could confer upon him was to "open the Portuguese ports in East Africa to free trade"[61]. The other renowned British explorer, H. M. Stanley, whose name is intimately connected with the first phase of spreading British influence in Uganda, was even more useful to the bourgeoisie of his country. He has been described by the British pioneers in East Africa as "a man with strong religious convictions and he understood the value of quotations from the Bible with a certain section of the public"[62]. He, who "was a popular hero" in England, did a lot to open up Uganda for the missionary activities of the Europeans. But was the spiritual uplift of the Black Africans his only, or even the primary, motive? "Stanley first inspired European missionaries with an interest in this country, and the Church Missionary Society sent a party of pioneers in 1877, who were soon followed by a Roman Catholic mission"[63], of the Order of the White Fathers of Algeria. At Stanley's request King Mutesa of Buganda agreed to the importation of Christian missionaries into his kingdom. But, as is evident from the following words of encouragement and wisdom emanating from Stanley to his compatriots, behind his religious zeal were the commercial possibilities of Uganda and of the neighbouring region. To quote[64]:

> "There are forty million people beyond the gateway to the Congo, and the cotton-spinners of Manchester are waiting to clothe them. Birmingham foundries are glowing with the red metal that will presently be made into ironwork for them and the trinkets that shall adorn those dusky bosoms, and the ministers of Christ are zealous to bring them, the poor benighted heathen, into the Christian fold."

More will be said later of Stanley's consistent endeavour to further the interests of the British capitalists in Uganda.

To continue, now, with the role of the Christian missionaries in the first phase of the European invasion of Uganda, it was "only after the commercial possibilities of Uganda were seen that more positive encouragement was given to the missions to carry out work in East Africa"[65]. The missionaries also did not fail to accept Stanley's advice, and follow the will of the ruling classes of their respective countries. Lugard mentions how frequently the missionaries became involved in factional intrigues as well as in national politics. The rival claims of the British and German Powers (the latter supported by the French) were ably upheld by these ministers of God in their day to day work among the "heathen" Africans. The Wa-Ingleza and Wa-Fransa factions were set up and their African followers were encouraged to fight, not for their own freedom or for the prosperity of their land, but in order to further the cause of respective foreign Powers. Some of these missionaries went to such a length that they quite openly refused to dissociate themselves from politics, declaring that it was part of their duty to instruct their African followers in secular matters also[66]! Eliot, who was the Commissioner for the East Africa Protectorate, thus commented on the situation[67]:

> "The early days of Christianity in Uganda were very stormy. The missions met with alternate encouragement and persecutions; the Protestants fought with Roman Catholics, and both with Mohammedans. The rivalry between the first two sects was particularly severe and practically political, since they represented the British and French parties."

General Gordon, an administrator of Uganda in those days, speaking of the Uganda Mission said, "as it is composed, it is more secular than spiritual"[68]. He, therefore, wrote to the missionaries indicating what political attitude they should take in this "Scramble for Africa".

The Christian missionaries were utilised by the colonising Powers in another more ignoble way — as "pawns in the imperialist game". "The French and especially the British enlisted the support of missionaries whose work could be at once 'protected' by armed force should the indigenous people object to conversion by attacking the missionaries. ... In 1885 a missionary with all his escort was killed by vassals of the Kabaka of Buganda. He was neither the first nor the last. In 1890 a British Protectorate was declared over Uganda to 'protect the mission.'"[69] Quite characteristically, Sir Charles Lucas in his lectures to the Royal Colonial Institute in London declared in 1921 that the missionaries were "in a marked degree pioneers of the Empire"[70].

Besides the missionaries, there were more direct pioneers of the Empire. The missionaries might have had reservations in regard to their role in colonial politics, and possibly they could not always be directly used by the colonising Powers to further their aims. But the role of the administrators — the front-rank fighters for colonial possession — was frank and direct. There was no inconsistency in the attitude of these invading administrators in regard to their mission. Thus, Lugard, who, as noted before, arrived in Uganda to establish the rule of the Imperial British East Africa Company, declared quite candidly[71]:

> "The 'Scramble for Africa' by the nations of Europe — an incident without parallel in the history of the world — was due to the growing commercial rivalry, which brought home to civilised nations the vital necessity of securing the only remaining fields for industrial enterprise and expansion. It is well, then, to realise that it is for our advantage — and not alone at the dictates of duty — that we have under-taken responsibilities in East Africa. It is in order to foster the growth of the trade of this country, and to find an outlet for our manufactures and our surplus energy, that our far-seeing statesmen and our commercial men advocate colonial expansion.
>
> I do not believe that in these days our national policy is based on motives of philanthropy only. ... There are some who say we have no *right* in Africa at all, that 'it belongs to the natives'. I hold that our right is the necessity that is upon us to provide for our

evergrowing population — either by opening new fields for emigration, or by providing work and employment which the development of over-sea extension entails — and to stimulate trade by finding new markets, since we know what misery trade depression brings at home."

Evidently, the invading administrator Lugard was more truthful than the later administrator Johnston who announced nine years later that Britain had "commenced these Protectorates from motives of pure philanthropy".

Of course, these pioneers — explorers, missionaries and administrators — could not by themselves decide about the occupation of Uganda. But there cannot be any doubt that the glowing tributes paid by these pioneers to the riches of Uganda obviously influenced this decision eventually taken by the ruling forces in Britain. Lugard spoke very highly of Uganda[72]. Of Stanley's concern something has already been said. Of his consistent efforts to bring Uganda under British domination it is further worthy of note that in 1892 he informed the British capitalists of the possibilities of production of cotton in Uganda for which the latter were then paying forty million pounds yearly to America and were still confronted with hostile tariffs and fluctuations in supply[73]. (As will be seen later, Uganda's colonial economy finally blossomed into the production of raw cotton by African capital and labour.) Besides these two stalwarts of British colonial system, Sir Gerald Portal, who came to Uganda after Lugard, and Sir H. H. Johnston, who later became the High Commissioner for Uganda, were even more lavish in their praises[74]. These concerted efforts were not in vain; their recommendations bore fruit. Lugard has noted[75]:

"The Chamber of Commerce of the United Kingdom have unanimously urged the retention of East Africa on the grounds of commercial advantage. The Presidents of the London and Liverpool Chambers attended a deputation (on 20th October 1892) to Her Majesty's Minister for Foreign Affairs to urge 'the absolute necessity, for the prosperity of this country, that new avenues for commerce such as that in East Equatorial Africa should be opened up, in view of the hostile tariffs with which British manufactures are being everywhere confronted'. Manchester followed with a similar declaration; Glasgow, Birmingham, Edinburgh, and other commercial centres gave it as their opinion that 'there is practically no middle course for this country, between a reversal of the free-trade policy to which it is pledged, on the one hand, and a prudent but continuous territorial extension for the creation of new markets, on the other hand."

The Government was also no less willing to serve the cause of these "commercial men" of Britain. The campaign got into full swing. On 6th February, 1892, the then Foreign Secretary and Prime Minister, Lord Salisbury, "spoke strongly in this sense" at Liverpool. The next Foreign Secretary, Mr. Chamberlain, went even further and "pointed out at Birmingham how directly to the advantage of the workingmen this policy of prudent but continuous extension is"[76]. The *Times* of London, in its issue of the 2nd of June 1892, reported his speech as follows:

" I should like to say to you in passing that this question of Uganda, and all questions which affect the extension of the Empire, have a very pressing interest for workingmen. Those people who want you to have a little Empire must make up their mind that with a little Empire will go a little trade. . . . Your hope of continuous employment depends upon our foreign commerce, and now that other nations are closing their ports to us, and everywhere we see that they are endeavouring to create a monopoly for their own benefit — I say that the future of working classes of this country depends upon our success in maintaining the Empire as it at present stands, and in taking every wise and legitimate opportunity of extending it."

The British ruling class did not remain content just with the knowledge of their true mission in Uganda, and with explaining the necessity of such a mission to the British people for their "benefit". As cool businessmen, they also calculated the debit and credit sides of the venture. The London Chamber of Commerce noted in their Annual Report of 20th April 1893 in "Section Uganda":

> "New colonial developments cannot be expected to pay at once, but almost invariably they do so in the long run. The uniform experience of this country from 1568 down to the present regime is that colonies amply repay the first expenditure in blood and money, and that they pay both by extension of trade and shipping, and in the growth of national power and status. In regard to Uganda, it is not easy — nor is it even material — to estimate when the country, if annexed, is likely to pay. It should be sufficient for us to know that investments of this class are invariably good in the long run, and that a few years more or less are immaterial from a national point of view."

Here is colonial book-keeping in its simplest terms; "blood and money" on the debit side, and "extension of trade and shipping" and the "growth of national power and status" on the credit side. The *businessmen* had made up their mind.

Having thus decided upon the soundness of the scheme, Britain's next move followed logically. Pilot work in trade and evangelical missions had already been started. A Chartered Company, by the name of the Imperial British East Africa Company, had also been formed in 1888 with a capital of 500,000 pounds sterling. Now was the time to annex the territory completely. This was not a difficult task.

In order to launch large-scale trading with Uganda, the Imperial British East Africa Company had begun to build a road from Mombasa to Lake Victoria. The Directors of the Company also intended to build a railway, for till then the only method of transporting goods from Uganda was by African porters, which was not sufficiently quick and efficient. But the Company did not have enough finance to launch such a project. Under the circumstances, "the Company found the occupation of Uganda a heavier drain on their resources than they had anticipated, and were anxious that Government should relieve them of their burden. . . . Sir Gerald Portal, then Agent and Consul-General at Zanzibar, was sent on an expedition to Uganda to report 'on the best means of dealing with the country, whether through Zanzibar or otherwise'. In his report he deprecated the abandonment of Uganda, for both religious and commercial reasons, and advocated the establishment of an official administration and the construction of railway. This was at the end of 1893, and the rights over the various East African territories in the British sphere soon passed form the Company to the Government"[77]. Uganda became a British Protectorate.

3. Rivalry over Uganda

Whatever difficulty arose while occupying Uganda, very little of it came form the simple and good-natured Africans. But from the rival powers of Germany, France and Belgium there was fierce opposition. In the following pages therefore a brief account is first given of the fight between the colonising powers to occupy Uganda.

Of the three powers rival to Britain, France received little attention in East Africa, for a timely warning from the British administrators to Her Majesty's Government at the time when France was about to enter the field of conquest of Uganda maintained the balance in favour of the British[78]. With Belgium however the situation was not so easy, and with Germany it was extremely difficult.

Germany was then looking for colonies of which so far she had practically none. She was the first to embark openly upon a policy of colonisation of East Africa. One Dr. Peters, who was to have an eventful career in East Africa as a representative of the German Power, sailed to East Africa in 1884. "In the early eighties German merchants began to be active in the coast towns, and made several treaties with the chiefs of the hinterland. In 1885 the German Government, who were anxious to inaugurate a colonial policy, gave the Society of German Colonisation, which had acquired these treaty rights, a charter of protection. From this date begins the partition of East Africa, and its administration by Great Britain and Germany"[79]. Dr. Peters was the man who received this Charter for the Gesellschaft für deutsche Kolonisation. In the same year "a German Protectorate over the independent State of Witu was declared"[80].

Britain stirred herself up at the prospect of being outmanoeuvred by Germany. In 1885 "a joint British, French, and German commission was appointed to decide the exact limits of the sovereignty of Zanzibar, the British commissioner being the future Lord Kitchener. The Commissioners reported in 1886. In 1886 Great Britain and Germany agreed on respective spheres of influence in East Africa, and the Sultan, acting throughout very reluctantly on the advice and pressure of the British Government, accepted the arrangement. In 1887 the Sultan gave a concession to a company formed by Sir William Mackinnon and his friends, which, in 1888, received a Royal Charter as the Imperial British East Africa Company*. In 1888 he ceded administration in the German sphere to a German Company, and in 1899 a German Protectorate over the German sphere was formally proclaimed. Finally, in 1890, the great Anglo-German treaty was concluded, which settled the main boundaries between Great Britain and Germany in all the parts of Africa where the two nations were side by side, and under which, as against the cession of Heligoland to Germany, Great Britain secured German recognition of an exclusive British Protectorate over what still remained under the rule of the Sultan of Zanzibar"[81].

The treaty of 1890, however, did not solve the rivalry between the British and the German Powers, just as the previous agreements were also not scrupulously observed. "In 1890 the German explorer Peters made a journey to the interior, during which he made many treaties with chiefs in Uganda and elsewhere, which would probably have created much trouble had they not fallen through in consequence of the Anglo-German agreement made in the same year. In virtue of that agreement the British East Africa Company proceeded to occupy Uganda, and sent Lugard thither. He restored order in a manner which has won high praise; . . ."[82]

The German side, however, did not quieten down. "The London agreement of July 1, 1890, was a great disappointment to all those who had an interest in German colonisation. Enthusiasts dreamt already of a huge German colonial empire, and Dr. Peters and his friends especially found their hopes sadly chilled when they learnt that the treaties he had made during his famous Emin Pasha Expedition were annuled by the fact that the country over which he had acquired a possible Protectorate fell under the English sphere of influence."[83] Skirmishes went on at the Anglo-German frontiers. "In the last years of the

* It may be of interest to note that the Mackinnon and Mackenzie Company is one of the oldest British firms in India. During British rule it ramified all over the sub-continent and began to control a good deal of India's economy. It secured cheap raw materials for the British manufacturers and sold British merchandise to the Indian's who, having lost their previous industrial production due to British oppression in the first phase of ruling India (see, for instance, *loc. cit.* 19, Chapters 4, 5 and 6), had to depend on importing manufactured goods and exporting agricultural and industrial raw materials, such as cotton, jute, tea, manganese ore, mica, etc. Also, as agents, the Mackinnon and Mackenzie Company controlled important shipping concerns like the *P & O* and the British India Steam Navigation Company. Sir William Mackinnon and "his friends" had thus considerable experience in how to organise a systematic exploitation of East Africa by means of the Imperial British East Africa Company.

nineteenth century German East Africa had become a terribly pushful neighbour. In 1900, for example, a German post was established near Kakuto in Buddu . . . which was the international boundary established by the Anglo-German Agreement of 1st July, 1890. There was, therefore, a real need for an early boundary settlement."[84] Another boundary commission between the British and Germans was called for during 1902—04, and the rivalry between Britain, Germany and Belgium was finally settled on 14th May, 1910, when "three agreements were signed at Brussels by the delegates of the three Governments whereby their respective boundaries were, in general terms, decided upon"[85].

In this fight between Britain and Germany over East Africa, France played the second fiddle. It has been noted earlier that in 1885 France joined the commission with Britain and Germany "to decide the exact limits of the sovereignty of Zanzibar". "The *locus standi* of the French was that, in 1862, they had joined with the British in guaranteeing the independence of Zanzibar. France had not, and has not, any territorial interest in the eastern side of the African continent south of the Red Sea, but has always had interests and claims on the island of Madagascar. At the end of 1885 a French Protectorate over Madagascar was proclaimed, and in 1890 both Great Britain and Germany recognized this Protectorate, while France recognized the British Protectorate over the island of Zanzibar and Pemba, and the German Protectorate in East Africa."[86]

Although it is thus asserted that France had no ambitions in East Africa, the action of the French in Uganda does not always bear this out. In Uganda they were represented by the Roman Catholic Missions, of which Cardinal Lavigerie "was placing all the local influence of the Roman Catholic Missions on the side of German trade, in consideration of the support which Germany was prepared to give his missions by the exclusion of British influence"[87]. Lourdel, the "First Catholic in Uganda", while asking for the "protection" of the British in a letter to Sir Frederick Jackson (the representative of the Imperial British East Africa Company)[88], came to an agreement with Dr. Peters (the representative of the Gesellschaft für deutsche Kolonisation) "to prevent the British party from obtaining any influence in Uganda"[89]. Such an attitude of the French missionaries shortly after the Franco-Prussian War cannot but lead one to think that by taking advantage of the two contestants over Uganda, the French Power aimed to serve its own ends. Such a view is also suggested not only by the use France made of her missionaries in order to further her policy elsewhere, but also by the direct part these missionaries took in Uganda in inciting their African followers against the British even after Germany had relinquished her rights over Uganda by the treaty of 1890. This finally led to serious trouble for Britain in her design to annex Uganda. Jackson wrote on this occasion[90]:

> "The political situation had become acute. The Roman Catholic part, encouraged by rumours of the withdrawal of the Company, were preparing for war. Mwanga [then ruler of the kingdom of Buganda] thought his personal safety lay with them. Lugard offered the French priests the protection of the Fort of Kampala, but they feared to be separated from their followers and eventually joined Mwanga, who had fled to Bulinguge [an island on the lake]. From here they tried to enter Budu, which province adjoined the German sphere, and constant information reached Lugard of negotiations with the Germans."

There is likely to be a good deal of truth in the supposition put forward to account for such mutual understanding between France and Germany against Britain, which the action of the French missionaries in Uganda so clearly revealed. Lucas, for instance, has noted[91]:

> "While Germany took good care of herself in the parcelling out of Africa, German statecraft, in the earlier stages of the scramble, was by no means hostile to French claims and aspirations in Africa. They tended to distract French attention from their lost provinces

> at home, and to result, as they did result, in renewed rivalry between France and Great Britain, and in a diversion of French bitterness from Germany to Great Britain."

Thus, if not directly for the French occupation of Uganda, the path which the French Power laid down for its missionaries in Uganda, and which the latter so admirably played, served its interest in other parts of Africa and at home. A detailed discussion of this issue is not within the scope of this book, and it should also be noted that the French were not a decisive obstruction to Britain in colonising Uganda.

With Belgium, however, the situation was different. With a cleverer tactic than that of Britain, Belgium outmanoeuvred her and took away the Congo from her prospective domination. Moreover, Belgium all along tried to push herself into Uganda from the north-western region of the present boundary of the Protectorate.

Loss of the Congo was a great disappointment to the British. The course of the Congo river was verified by Stanley, who visited this part of African territory several times, and had made "treaties" with the African chiefs dwelling in that region. He had obtained "concessions" from them, and had founded posts. By 1884 the Congo was regarded as a British sphere of influence with Sir Francis de Winton, who was deputising there in absence of Stanley, as the Governor of the area.[92] Meanwhile, in 1876, "an International African Association for exploring and civilizing Africa and putting down the slave trade" had been formed at Brussels with Leopold, King of the Belgians, as the President of the Association.[93] This "international" association gave birth, in 1878, to a "special committee, styled the Comité des Études du Haut Congo", which later took the title of the International Association of the Congo.[94] In 1885, all the trimmings of "internationalism" were dropped off, and "the formation of the Independent State of the Congo with King Leopold for its sovereign was announced"[95]. "In 1889, King Leopold made a will bequeathing his sovereign rights to Belgium.. In 1890 Belgium agreed to make a loan to the State for ten years on condition of being allowed to annex it at the end of that time."[96] "In November 1908 the State was at last transferred to Belgium, and became a Belgian dependency, the Belgian Congo"[97].

A smooth piece of work for Belgium, and Britain had to swallow it. For this was also the time when Britain was facing formidable opposition from the Germans in East Africa. Therefore, accepting the bitter pill of a "Belgian" Congo, she felt it wiser to abide by the saying: "half a loaf is better than no bread".

Belgium, however, did not stop at the Congo. While ousting Britain from this part of Africa, she began to make incursions into what is today the Western Province of the Uganda Protectorate. Thomas and Spencer have noted[98]:

> "Between the years 1907 and 1912 the Western Province was the scene of a succession of important survey operations for the most part arising out of the settlement of the western frontiers of the Protectorate."

The Uganda-Congo Boundary Commission of 1907—08, however, did not finally resolve the tension between the two Powers. "During 1909 there was a good deal of local tension between the British and Belgian authorities in the Mufumbiro region, and a British force, known as the Kivu Mission, endeavoured, by occupying various posts, to establish claims to territory extending to Lake Kivu, which had been admitted by Germany in an agreement signed at Berlin on 19th May, 1909. A stand-still arrangement was come to with the Belgians lest incidents should occur, . . ."[99]

Thus, the claims of even the biggest Colonial Power in the world could not be settled amicably with the rival Powers. Indeed, as the above lines show, such was their lust for occupation that sometimes they had even taken up arms to establish their claims. And all this happened in Africa where they declared that the "savages" were fighting against one another, and so they had come to bring "peace" and "civilisation" to the country.

Finally, in 1910, the rival colonising powers agreed to accept Britain as the undisputed ruler of Uganda.

4. Occupation of Uganda

All these hagglings between the colonising powers to annex Uganda were, no doubt, conducted over the heads of the African peoples whose native soil they had come to occupy. It was possible to dupe the Africans so easily, firstly, because they were completely unaware of the colonising designs of the representatives of a superior civilisation, and, secondly, because many of them at that time were striving for a change. This was so because just then most of the societies in present-day Uganda Protectorate were at the cross-roads, and reflected the contradiction between the apparent difference in the economic and political life of the different African societies and the underlying uniformity in the ways of living of the common people. This situation in Uganda at that time needs a deeper understanding than it appears superficially; otherwise, one may come to extreme conclusions: either that the Uganda societies were smoothly developing to a higher life of economic prosperity and social progress or that it is mere wilful imagination to say that there were qualitative changes incipient in them (as explained earlier), which given the time for emergence would have solved this contradiction.

Thus, although Emin Pasha was then impressed by the spectacular growth in southern parts of Uganda because of the contact between the Africans and the Arab slave-traders, and, as noted before, came to the conclusion that the people were in consequence much more developed in the south than in the northern regions, in fact the life of the common people in all areas was still not very different. They lived on a subsistence economy, growing their food crops, rearing their cattle, and hunting for meat and skins. In agriculture, the primitive method of hoe-cultivation remained the only technique of production in all areas. The state of productive forces was similar, and almost equally primitive, in both the northern and southern regions. Human labour was the primary, if not the only, item to reckon with in producing for society. Prevalence of domestic slavery in some of these societies had some effect, but within the limited scope of a subsistence economy it could not appreciably affect the social relations of production. In short, the demand of the day was for further changes in the state of productive forces and relations of production, leading to a qualitative transformation of the societies.

Such a demand, however, had not yet been fulfilled. Some areas were, of course, opening up and external trade had begun to develop which could have had a significant effect in breaking up the parochial existence of these societies and in their eventual development in all aspects. But, firstly, it was still in its early stages and had not yet affected the general run of the people, except perhaps round about the capital of the kingdom of Buganda, and, secondly, the external trade took the form of slave-hunting in the main, thereby complicating the situation. Uganda was then at a critical stage, with all the pains and expectations of a new birth. The pain had started, but it was prolonged by the changes occuring mainly in the political superstructure of society because of the peculiar development of slave-trade carried on by the ruling class with the Arabs.

Local despotism coupled with the appearance of the Arab traders on their territories had weakened the bonds within the societies and also disturbed the stability between them. The common people everywhere were at the mercy of the chiefs and kings who would catch them on any pretext or at an unguarded movement and pass them over to the Arab traders. For such "acquisition of slaves" the kings and chiefs were also intriguing all the time for greater power and annexation of larger territories which would bring them more slaves for the market[100]. Predatory raids into others' territories by the stronger powers, and internecine struggles between them, ultimately for the slave-market, had become the order of the day. But the profit which came to the ruling classes of the "Big Powers" from such business transactions with the Arab traders was not yet invested to develop the state of productive forces of the country. It is true that some changes were effected in society, since the merchandise imported by the Arabs was in the long run also partly available to the common people. But even in the kingdom of Buganda where such changes were most marked, they did not yet significantly develop the state of productive forces or improve the mode of production. And in other kingdoms, for instance, in the kingdom of Kitara, even such changes were still less visible.

This point is important to note, for it appears that when the Europeans first came to this country the societies of Uganda, or at any rate the most important ones of them, were at a turning-point. Changes were being effected mainly in the political sphere, giving rise to African despots. But such changes in the political superstructure of the societies without a basic foundation in a developing economy were unstable and fissiparous. The upshot was, as noted before, internecine struggles between the kingdoms and the emergence of ever-more autocratic rulers who cynically decided the fate of the people. This has been profusely mentioned by those British explorers, missionaries and administrators who first came to Uganda[101]. Thus, while the economic life of the people remained stagnant, futile political intrigues and slave-trade were devitalising the common people in all societies as well as entire communities of the weaker peoples. Consequently, these peoples were intent on breaking through the vicious bondage of despotic rule. What was then needed in Uganda was progressive government and progressive economy.

It is, of course, somewhat conjectural where the tension in the life of the Uganda Africans would have led them if they had been left to themselves. Nevertheless, one would imagine that, at however slow a tempo, ultimately this ever-increasing tension within and between the societies would have burst asunder the despotic rule and initiated progress in the communities. But this normal course of development was checked by the advent of the British in the last quarter of the nineteenth century. And they appeared at the most opportune moment, for on account of this weakness in the social structure of the African societies they could so easily play off one against another and ultimately overpower them all, and establish themselves on top of all these societies. Thus, while some of the weaker communities, like the Basoga or the Acholi chief Ochama, offered their allegiance to the British to get rid of the menace of despotic rule and slave-hunting, some chiefs of the established kingdoms also joined in league with them to undermine the influence of some of their compatriots and colleagues. A typical example of the second situation is Kasagama of Toro. Undoubtedly, the situation for the British could not have been better. They became the champion of the oppressed and selfless supporters of their allies, and thus very ably utilised the situation for their own ends.

Indeed, it would be no exaggaration to say that the contact between the British and Africans in Uganda in its earliest stage was characterised by an open campaign by the immigrants to grab the native's territory. This they did either by offering them peace and security or by putting before them wonderful trade offers. Occasionally, of course, they had to brow-beat some Africans if they would not listen to their suggestions and were strong enough not to be eliminated easily. In such circumstances, if the Africans, individually or in groups, showed reluctance to accept such "offers", they were cowed by means of superior military strength; and if they showed any effective opposition, such resistance was ruthlessly smashed.

A "treaty-making competition" with the African chieftains to establish political rights for the rival colonising powers was started from the beginning. In fact, this was first started by the Germans, and the British then ridiculed and resented it. Jackson, on his arrival in East Africa, found that the "Germans were inaugurating their scheme for Colonial possession by a well-thought out and unscrupulous gang of bogus treaty-makers"[102]. The British, however, did not take long to learn this trick of bogus treaty-making, and soon outstripped the Germans in the game. Thomas noted[103]:

> "Colonel Colvile, on his appointment as Commissioner in August 1893, had indeed already been directed to make a new series of treaties on the western confines of the British sphere. This had now become a matter of moment and urgency since there was uneasiness regarding the proceedings of the Belgians from the Congo. ... This sense of urgency had its share in deciding Colonel Colvile to launch the expedition against Bunyoro in December 1893. ... Immediately after the occupation of Hoima he despatched Major Owen ... to Wadelai, where, on 4th February 1894, he made a treaty with Chief Ali and arranged to subsidize him to maintain a garrison of irregulars. On 24th March, Capt. Thurston made a similar treaty at Mahagi Saghir on the 'west shore of Lake Albert' with chief Maswa. Later, ... Dr. R. U. Moffat, ... on the 18th April made a like treaty at 'Injarri on the Albert Lake in Kavalli's country' with the chief Kavalli. ... Thurston himself made one more treaty to complete in Colvile's words 'a chain of friendly chieftains from the south-eastern extremity of Lake Albert to Wadelai'. This was on 24th April 1894 with 'Tukenda chief of the country between Gutta and Mahagi Saghir on the Albert Nyanza and the hill country behind it'. ... By this time, however, the treaty-making competition in this region had lost its importance for on 12th May 1894 an Agreement was made between Great Britain and the King of the Belgians, which settled for the time being the boundary between the British sphere and the Congo Free State. ... A result of the 1894 Agreement was that it was no longer essential to provide support for the chief of Wadelai since the western bank of the Nile now fell to the Belgians, and his garrison was soon disbanded."

Such treaties were also made to pit one African chief against another, or to line up a number of chieftainships against a powerful African ruler, such as Kabarega, the king of Kitara. Thus, a "treaty was made on 15th February 1896 ... with Abura, a substantial Acholi chief. ... Again the treaty had no international significance. But Abura was a potential ally against Kabarega, who was still at large; and he would be a useful look-out for any southward movement of the Mahdists"[104].

And what was the "modus operandi" of this "amiable farce" of treaty-making between the "primitive tribal" peoples and the bearers of the superior western civilisation? Major A. F. Thurston, who as stated above was one of these treaty-makers gives some account[105]:

> "I had a bundle of printed treaties which I was to make as many people sign as possible. This signing is an amiable farce, which is supposed to impose on foreign governments, and to be equivalent of an occupation.
>
> The *modus operandi* is somewhat as follows. A ragged, untidy European lands at a native village, the people run away; he shouts after them to come back, holding out a shilling's worth of beads. Someone, braver than the rest, at last comes up; he is given a string of beads, and is told that if the chief comes he will get a great many more. Cupidity is, in the end, stronger than fear; the chief comes up and receives his present, the so-called interpreter pretends to explain the treaty. The chief does not understand a word of it, but he looks pleased, as he receives another present of beads; a mark is made on the printed treaty by the chief, and another by the interpreter, the vagrant who profess to be the

representative of a great empire, signs his name. The chief takes the paper, but with some hesitation, as he regards the whole performance as a new and therefore dangerous piece of witchcraft. The boat sails away, and the new ally and protégé of England or France immediately throws the treaty into the fire."

It is important to note that even such a political "ceremony" was performed only in the case of an established chieftainship; otherwise, the "sites" were simply "appropriated". Thus, Thomas and Spencer have noted[106]:

> "In the nineties of last century, under the Chartered Company's regime and in the early years of the British Protectorate, there was little time for study of the ethical aspects of native land tenure. ... The early administrators adapted themselves to facts as they encountered them. A native potentate who could impress his personality on an area was regarded as competent to grant such land as was required for forts or stations; but where only primitive tribal organizations were visible, the agents of British interests frequently acted, as doubtless the tribes had acted before them, and appropriated sites from unoccupied land without permission from anyone. If at all possible, however, the 'amiable farce' of treaty-making was, on account of its political significance, gone through."

While those parts of the present territory of Uganda which were not within the kingdoms of Buganda and Bunyoro could thus be occupied, the negotiations with Mwanga (son of Mutesa and King of Buganda) and Kabarega (son of Kamrasi and King of Bunyoro-Kitara) were not so simple. Although decisively in their favour, the British had yet to come to a definite "agreement" with the Kingdom of Buganda, and from Kabarega they could get Bunyoro only after a number of full-scale battles[107].

With the fathers of both these kings, Speke had arranged for trading in British merchandise in return for ivory, etc. Trading operations did not succeed so well in Bunyoro, for the king was not taken in by the British and went on supporting the Arab traders in the main, from whom he received guns, ammunitions, etc., in return for slaves and ivory. But in Buganda the situation was different. Here Mutesa started the game of playing off the British and the Arabs one against the other, hoping to get the most out of the tussle. Roscoe noted[108]:

> "Until Europeans made their appearance in Uganda, both Arab and Swahili traders had striven to impress Mutesa with the greatness of his own kingdom and power, in the hope of getting presents from him by this flattery and of obtaining better terms in their bargains. Mutesa soon realised the hostility between the Arabs and the missionaries, and began to play off one against the other; ... He soon discovered that the English were far more clever than the Arabs, and he also found that he could extort large sums from them in the form of cotton goods on the most trivial grounds."

But this game of playing with the Europeans, which Mutesa's son, Mwanga, started in a larger scale by allowing also the Germans to come on the scene after the British and Arabs had already been present there for a longer period, doomed the future of the kingdom of Buganda. African diplomacy in these matters was much too crude compared with the profound experience of the greatest colonial power of the world.

Mutesa's permission to trade in Buganda was the thin end of the wedge with which the British had entered. Within a generation the simple traders were demanding political concessions. The situation was appropriate for such a demand. After Mutesa's death in 1884 the country was in a flux. The Arabs were making a "vigorous attempt to obtain ascendence over the English. They worked upon the feelings of the old people who were devoted to their Lubare worship, and these in turn worked upon the weakness of the young king", namely Mwanga, the son of Mutesa[109]. Mwanga acquiesced to their demand, putting serious obstacles in the way of missionary activities in his kingdom. But he was also unwilling to identify himself completely with the Islam party. In the end, he stood alone, without a following amongst either the Christians or the Mohammedans. At this opportune moment the Arab faction raised a rebellion, deposed Mwanga, who "was expelled and fled to the lake, where he went by canoe to one of the Roman Catholic stations in Bukoba"[110]. The Arab faction first put up one of Mwanga's brother, Kiwewa, as the king, but when he also refused to become a full-fledged Mohammedan, his brother Kalema became the king and Kiwewa was burnt to death on the occasion, so that he might not aspire for the throne any longer. "Kalema, the new king, redistributed the land among his followers, appointing new chiefs to rule the districts. The people who remained in the country were now under a million in number: for a few months a Mohammedan king was the nominal sovereign, while the country was really ruled by Arabs and by Swahili traders."[111] The state of affairs was not at all congenial to the British, but here also lay their chance to gain an effective control over Uganda. Mwanga was then wavering between the British and the Germans, giving his allegiance at one moment to Dr. Peters and at the next to the representative of the British East Africa Company. At last he made up his mind and wrote to the British Company, offering it the "monopoly of commerce" as well as "provisioning" the Company's staff and accepting its flag in return for British support in regaining his throne[112]. The Company did not miss this opportunity. Lugard, who as stated before was despatched by the Company to establish its influence in Uganda, set about fulfilling his mission. In Uganda he "found himself in a hornet's nest of unrest. 'To guarantee peace' was easy on paper. Nothing but a common fear of the Mohammedan menace kept the two Christian factions from open war"[113]. Eventually a united front of all Christians, headed by the Company, was established, and the Company offered to help Mwanga to regain his throne, provided he signed a treaty with it. "With great patience Lugard explained the details of his proposed treaty to each party [of the Christians] in turn: King Mwanga was most unwilling to sign *any treaty*. One of the important clauses prohibited slave-trading, another banned the import of guns and ammunition."[114] It is interesting to note that when Speke first concluded friendship with the King of Buganda (Mutesa) he did it by offering a substantial quantity of arms and ammunition. But when the chance came to occupy Uganda, his successor wanted to disarm their vassal king. "Finally an agreement was reached and Mwanga signed a treaty, valid for two years."[115]

To reinstate Mwanga on the throne was not a difficult task for the British with their superior military strength. "The Islam party were attacked on the lake shore some distance from Mengo, the capital, and after several battles they were driven back, and Mwanga was enabled to land and to return to his old capital."[116] Mwanga "drove the Mohammedans to the north-west towards the Bunyoro confines, where they settled"[117], and the British were rid of the Arab menace to the kingdom.

Mwanga, however, did not want to settle down under British influence. Secretly the shrewd king began to lean towards the Germans to gain his own ends, of course. In this he was ably supported by the Roman Catholic priests who also had their own axes to grind. Following the dictates of the French ruling class they "thought that the country was leaning too much to the English, and [they] therefore urged King Mwanga and their converts to inaugurate a new policy, which would give many advantages to Roman Catholics"[118]. The situation again became tense. The "French party were determined to prosecute their plans and secure their end by force of arms, and war broke out again, this time between the two Christian parties"[119]. Roscoe commented on the situation[120]:

> "These stirring times of civil wars had not been without their excitements and dangers; one of the worst conflicts between the two parties, French and English, had revealed to

us how bitter people could become in their desire to be the dominant party; clans and even brothers were divided and took opposite sides. Such a state of clan animosity would have been inconceivable before the people began to take political sides; the old ties of primitive customs had now broken down, and there was nothing to bind the families as before."

In the name of religion, the invaders belonging to a superior civilisation thus broke up the whole society, and rent the sublime bond of a family.

On the British side there was an advantage. Organised military force was brought to bear upon the duped Africans to acquiesce them to British interest. "The officers of the I. B. E. A. Co. [Imperial British East Africa Company] considered the English party to be in the right and joined forces with them, and after a sharp battle the French party was driven from the capital; Mwanga fled by lake to southern Budu where he was joined by his army."[121] But the trump-card was in the hands of the British; so the later events came out fully in their favour. "Lugard threatened to make Mbogo [chief of the Mohammedan party] Kabaka, and Mwanga, at last realising how desperate his position was, escaped from his Roman Catholic followers and arrived in Kampala at the end of March. Lugard was now able to treat with the different parties separately and at last a treaty was signed, much to the same effect as the first one which he had drafted. The Company's flag was hoisted over the Kabaka's *lubriri* (enclosure)"[122]. Lugard noted on the occasion[123]:

> "Mwanga's sole object in dealing with Europeans appeared to consist in a hope that, by playing them off against each other, he would be able to secure presents of arms from both factions, and temporary assistance against his enemies, while he refused from giving any equivalent."

The scheme was of course to end in failure; the crude diplomacy of the king of a "barbaric civilisation" was much too obvious to the astute British administrator Lugard. The British became the *de facto* rulers of the kingdom.

Mwanga made one more attempt to be free from foreign rulers. When, shortly after his second return to the capital, the Sudanese troops stationed in Uganda rose against the British army, he "fled and joined the rebels in the district of Budu in southern Uganda; here the rebels consisted mainly of Baganda who wished to throw off the restraints of the British Government and join the Sudanese. For several months this part of the country had been known to be unsettled, before the Sudanese rose. But this disturbance could not be dealt with efficiently owing to the greater rebellion, and measures had to be delayed until the Sudanese rising had been suppressed. At length the rebel troops were overcome, and the Baganda forces were free to deal with their own countrymen. Under the leadership of two or three British officers, the natives then began to attack Mwanga and his followers, who had by this time joined Kabarega, the rebel king of Bunyoro in the Nile region."[124]

From the beginning Kabarega had maintained a consistent anti-British attitude. Before the British had established their rule in Bunyoro, "it had been impossible to enter the country owing to the fierce and determined opposition of Kabarega, the former king, who strongly condemned the king of Uganda [Buganda] and his subjects for having invited the British into Uganda. To make his strong disapproval more acutely felt, king Kabarega began a series of petty invasions into the states of Uganda bordering on Bunyoro. These invasions had to be resisted by Baganda troops, who endeavoured to police the frontier and stave off marauding armies; but at length the raids became so persistent, that the British took up the cause and organised an expedition, in which the king was driven from his capital and became a fugitive"[125]. "Yet for three years, after he had been driven out of his country, and was a fugitive, he kept together a sufficiently large force to put up several stiff fights, and to the very last was a nasty thorn in our [British] side."[126] And now, "when

Mwanga, the king of Buganda, had rebelled and joined him, a formidable expedition under British officers swept the country"[127]. Mwanga and Kabarega were hounded out from one place to another. "At length the rebel kings were brought to a stand, and were no longer able to escape, their place of hiding having been made known by some of their followers who turned traitors. In a battle which followed the king of Bunyoro fought bravely, until one of his arms was shot through and so shattered that he could no longer hold his gun; afterwards the arm had to be amputated. Both the kings were captured and taken to Mengo, and from thence were exiled to the Seychelles."[128]

Thus, even though Mwanga in the end joined his traditional enemy, the king of Bunyoro, and even though the latter all along maintained his courageous stand and bravely fought the British until he was incapacitated, both of them failed in driving the British from their soil. Kabarega made an attempt to take his life in the prison, but the kind British saved him, and exiled them both to one of the most unhealthy places of the continent. The last attempt of these two kings to free Uganda from foreign domination ended with the decisive consolidation of British rule over the whole area which became a British "Protectorate". The measuring rod of traders finally emerged as the sceptre of rulers, just as it had done a century and a half before in India.[129]

In this way, by making bogus treaties or just by appropriating the "sites" of the Africans, splitting the African society and even families into political factions, and lastly driving out the independent kings by means of a superior military force, Britain brought about the "bloodless revolution . . . without breathing a word of it to the outside world", of which Elspeth Huxley spoke so glibly half a century later.

5. Consolidation of British Power

The next stage in British policy after occupying Uganda was directed to the consolidation of their position in this area. For this they found it most suitable not to do away with the oppressive African kings and chiefs, but to absorb them into the machinery of colonial administration. This, they know, would lead to several points of advantage to them. Firstly, in this way they could for a time hood-wink the peoples into believing that no qualitative change had taken place in Uganda since the country came under British control; secondly, by absorbing the native rulers into the machinery of colonial rule (with the assurance of considerable economic gain while taking away their political and military rights and reducing them to the position of low-grade vassals or of petty administrative officers of a foreign government), they could build up a wall of defence between themselves and the peoples of Uganda.

This *modus operandi* they knew from their rich experience in colonial rule to be of use when because of ruthless exploitation the peoples would eventually rise against foreign rulers. British rule in India since 1757 had given them this insight. There during the reign of the East India Company they had played the same game with considerable success. Vassalage over the Nawabs of Bengal, Carnatic and Oudh (and indirect domination over the Nizam of Hyderabad) in the later part of the eighteenth century was the thin end of the wedge which successfully increased British power in India, eventually leading them to be the masters of the whole subcontinent[130]. Later, during the office of Lord Dalhousie as the Governor-General of India during 1848—56, an attempt was made to oust the native rulers of India by bringing their "kingdoms" under direct British administration; but soon they realised their mistake when this was found to be one of the factors leading to the first large-scale Indian uprising against British rule which went by the name of Sepoy Mutiny of

1857[131]. And there also they found that where "native rule" was maintained, the uprising was not so severe. Consequently, as their future policy they preferred to keep a large part of India under the rule of Indian Maharajas and princes. Up to the time of British withdrawal from India, these "native states" comprised two-fifths of the total area of the subcontinent[132]. Referring to this policy of the British, Lord Canning, the Governor-General of India, quite candidly remarked in his Minute of 30th April, 1860[133]:

> "The safety of our rule is increased, not diminished, by the maintenance of native chiefs well affected to us. Should the day come when India shall be threatened by an external enemy, or when the interests of England elsewhere may require that her Eastern Empire shall incur more than ordinary risk, one of our best mainstays will be found in these Native States. But to make them so we must treat their chiefs and influential families with consideration and generosity, teaching them that, in spite of all suspicion to the contrary, their independence is safe, that we are not waiting for plausible opportunities to convert their country into British territory."

Basing itself upon such profound experience, the British Colonial Government in Uganda stopped the excesses of the African despots (which apparently had a favourable reaction upon the peoples in regard to over-all British influence on their territories), but the foreign masters did not do away with the rule of these African kings and chiefs over their people. The kings could no longer, at their whim, send their subjects to the slaughter place or hand them over to the slave-traders; predatory raids into others' territory and internecine wars were also stopped. But the forms of the pre-British political system were rigorously maintained to the extent that the hierarchy of chiefs and kings, while playing second fiddle to the supreme power of the British, kept the people subjected to themselves.

Indeed, this form of indirect government was found to be so useful that a similar political apparatus of ruling the people by placing them under loyal chiefs was also extended to those societies in the Protectorate in which chieftainship as a political authority had not yet developed to a significant extent. In India the British were the greatest king-makers: in Uganda they became chief-makers. And this method of "preservation" or promotion of African oppressors of the people to safeguard the interests of the foreign rulers went by the names of Indirect Rule and Native Administration. This has been the essence of colonial administration of Uganda since the beginning of British rule.

However, before introducing the institutions of Indirect Rule and Native Administration, the first move of the Colonial Power was to weaken the powerful kingdoms of Buganda and Bunyoro-Kitara, while preserving the skeleton of their political systems. The weaker cummunities of Uganda Africans living under different chieftainships or otherwise were not a serious concern to the British. But it was necessary to curb the power of the kingdom of Buganda, reduce the rebel kingdom of Bunyoro to complete submission, and institute full control over them, so that the British could be sure of no further recrudescence of opposition from the rulers of these states.

Therefore, as a first step, Busoga, the most important tributary state of the kingdom of Buganda, was taken out of its orbit. Thereby while playing the policy of *divide et impera*, some apparent relief was given to the oppressed people like Basoga. What a mirage this relief was will be discussed later.

Similarly, to weaken the kingdom of Bunyoro, Toro was legally established as an independent kingdom with Kasagama, who had rendered the British Company "valuable assistance" during the phase of occupying Uganda, "as its first ruler". Moreover, "in due course the authority of Kasagama was further extended over the country of Busongora [first given to the kingdom of Ankole] and over the Bakonjo hill tribes"[134]. Finally, in

1900, when an Agreement "was made with Kasagama and the five chiefs of the Confederacy, Kasagama was recognised under the title of Kabaka, as supreme over these chiefs, and also over certain others who had previously been semi-independent"[135]. Such was the reward given to Kasagama for his loyalty to the invading power, and such was the punishment meted out to the independent Banyoro for their desire to live in freedom.

Furthermore, while creating the kingdom of Toro out of the kingdom of Bunyoro-Kitara, some other parts of this rebellious kingdom, such as Bugangadzi, Buyaga, Buruli, northern Singo and northern Bugerere, were taken away from her sphere of influence and placed with the more reliable kingdom of Buganda,[136] where the British had then firmly established themselves and had the fullest control. Also, the loyal Native Government of Buganda, formed by the British, had deserved this reward for their help in fighting against Kabarega and in bringing Bunyoro under complete subjugation.

Concurrently, to counterpoise both these two developed kingdoms of Buganda and Bunyoro, the weakest link in the chain of feudatory states in this region, namely the Kingdom of Ankole, which had all along exhibited little ambition and capabilities to extend its power and could therefore be more effectively controlled than the others, was re-constituted "much bigger than when it was an independent state"[137]. Just after their occupation of Uganda, "the British Government has added to it Mpororo, Egara, Bwezu and Busongora, small states which were originally separate kingdoms, entirely distinct and even antagonistic. The Mugabe of Ankole is now, under the British Government, ruler of all five states"[138].

The fierce resistance of the ruling class of Bunyoro was also successfully broken down in this period. Jackson, who in 1899 was in "charge of the kingdom, in residence at Kampala", found the mother of Kabarega "an irreconcilable old spitfire". She had been captured in 1895 and was under detention, but she would not come to see Jackson, and so he had to go to her residence. However, shortly afterwards, she came down to see him in his office. This time "she had no complaints to make, and had very clearly only come to beg, and as a few rupees were handy she soon made a move to go"[139].

Meanwhile, Mwanga's and Kabarega's sons were installed on the thrones of the kingdoms of Buganda and Bunyoro, respectively. In Buganda, Daudi Cwa, "the prince chosen by the British Government as king", was placed under an English tutor for proper education. In Bunyoro "a grown-up son of Kabarega was appointed to succeed his father, but he soon showed that he was not a fit person to rule; he was therefore deposed, and another son was placed on the throne"[140]. Why he was not *fit* to rule may be a point for discussion, but leaving that aside there could be no doubt that the British were now the king-makers of Uganda.

The British also choose other native rulers. The young king of Buganda was placed under three Regents during his minority. All of them were faithful allies of the British. One of the Regents, who was also the Prime Minister of the kingdom, was Apolo Kagwa. He had consistently served the cause of the British all through their campaign in Uganda. Later, he was adorned with the title of "Sir" for his faithful service. Those of the other chiefs, who like Kagwa had remained faithful to the British, were retained in power; others were deposed and from the loyal allies of the British new chiefs were created. Most of them were Christians owing their allegiance to the British Uganda Mission[141]. In other areas also the situation was not different. Moreover, to consolidate British rule all over Uganda, loyal Baganda Christians were sent as chiefs to distant places, in the lands of Bateso, Basoga, Baghisu, etc., of which more will be said later.

While thus readjusting the territories of different kingdoms and putting the right men in the right place, the time became ripe for the imposition of the systems of Indirect Rule and Native Administration on the people. The system of Indirect Rule was established in the kingdoms of Buganda, Bunyoro, Ankole and Toro, of which that in the kingdom of Buganda has been regarded to have developed best of all. The other communities of African peoples in Uganda who were under different chiefdoms were brought within the system of Native Administration, and this system was extended to those societies where chiefdoms had not existed before.

Instead of being relieved of their oppressive native rulers, henceforth the people had to bear the additional burden of colonial rule over and above that of the African rulers. And both these burdens were imposed on the previously "free" societies of Uganda where there was not yet any centralised authority. To understand the political problems of Uganda today, it is therefore necessary to know first how this form of colonial administration was set up.

6. Indirect Rule

Of the two institutions, Indirect Rule and Native Administration, the former is fairly well known internationally, for the spokesmen of British Government have often declared this to be a remarkably progressive feature of British Colonial Administration. This, according to them, is *democracy put into practice* for peoples "not yet able to stand by themselves under the strenuous conditions of the modern world". Some British social scientists of repute have also spoken in the same vein. Thus Lord Hailey, who is regarded as an authority on African problems, defined the term Indirect Rule "to denote the system by which the tutelary power recognizes existing African societies and assists them to adapt themselves to the functions of local government"[142]. Lucy Mair, a pupil of the famous anthropologist Malinowski and herself an anthropologist of repute, explained the function of Indirect Rule as "the progressive adaptation of native institutions to modern conditions"[143]. She, who had spent some time among the Baganda as an anthropologist, further explained the broader implications of Indirect Rule as follows[144]:

> "The real meaning of Indirect Rule cannot be summed up in the phrase, 'Find the Chief'. It consists in an understanding of the structure of native society and the inter-relation of its parts, which precludee the possibility of assuming that it can be suddenly modernized from outside, and at the same time reveals the points at which changed circumstances call for readjustment, and the bases on which necessary innovations can be firmly established."

Elspeth Huxley, an ardent supporter of Indirect Rule in Uganda, noted in 1948[145]:

> "Official policy is ... to offload more and more control on to 'the people'."

The role of Indirect Rule in Uganda should be examined against the background of such viewpoints.

The famous Uganda Agreement of 1900 which left the administration of the kingdom of Buganda to the native king, his three Regents to act on his behalf until he reached his majority, and to the Council of native chiefs (Lukiko) —all re-constituted within the ancient form of political structure of Buganda —led to the establishment of Indirect Rule among the Baganda. Since, of all other kingdoms and native areas in British Africa, the system of Indirect Rule has worked out "so successfully in Buganda", the situation in this kingdom after the introduction of Indirect Rule will be treated in some detail in the following pages. Similar Agreements were also made with the kingdoms of Ankole and Toro, but while resembling the Uganda Agreement in all essentials they were in diminutive forms. Moreover, these two kingdoms were "subject to such general laws and regulations as are generally in force throughout the Uganda Protectorate"[146]. With the kingdom of Bunyoro-Kitara no such Agreement was signed at the beginning, probably because of its consistent anti-British policy in the past, but "it has been administered in much the same way as the other states"[147], and in 1933 an Agreement was made "but on a smaller scale, and with less extensive powers"[148].

To discuss first the kingdom of Buganda in order to explain the situation in these states after the introduction of Indirect Rule, the following extracts from the Uganda Agreement of 1900 (which, for the reason stated earlier, meant only for Buganda) should be noted.

Clause 2. The Kabaka and chiefs of Uganda hereby agree henceforth to renounce in favour of Her Majesty the Queen any claims to tribute they may have had on the adjoining provinces of the Uganda Protectorate.

Clause 3. The Kingdom of Uganda in the administration of the Uganda Protectorate shall rank as a province of equal rank with any other provinces into which the Protectorate may be divided.

Clause 4. The revenue of the Kingdom of Uganda, collected by the Uganda Administration, will be merged in the general revenue of the Uganda Protectorate, as will that of the other provinces of this Protectorate.

Clause 5. The laws made for the general governance of the Uganda Protectorate by Her Majesty's Government will be equally applicable to the Kingdom of Uganda, except in so far as they may in any particular conflict with the terms of this agreement, in which case the terms of this agreement will constitute a special exception in regard to the Kingdom of Uganda.

Clause 6. So long as the Kabaka, chiefs, and people of Uganda shall conform to the laws and regulations instituted for their governance by Her Majesty's Government, and shall co-operate loyally with Her Majesty's Government in the organisation and administration of the said Kingdom of Uganda, Her Majesty's Government agrees to recognise the Kabaka of Uganda as the native ruler of the province of Uganda under Her Majesty's protection and over-rule. . . .

The name of the person chosen by the native council must be submitted to Her Majesty's Government for approval, and no person shall be recognised as Kabaka of Uganda whose election has not received the approval of Her Majesty's Government. . . .

The Kabaka of Uganda shall be guaranteed by Her Majesty's Government from out of the total revenue of the Uganda Protectorate a minimum yearly allowance of £ 1,500 a year. . . .

Clause 7. The Namasole, or mother of the present Kabaka (Chua), shall be paid during her lifetime an allowance at the rate of £ 50 a year. This allowance shall not necessarily be continued to the mothers of other Kabakas.

Clause 9. . . . At the head of each county shall be placed a chief who shall be selected by the Kabaka's Government, but whose name shall be submitted for approval to Her Majesty's representative. This chief, when approved by Her Majesty's representative, shall be guaranteed from out of the revenue of Uganda a salary at the rate of £ 200 a year. To the chief of a county will be entrusted by Her Majesty's Government, and by the Kabaka, the task of administering justice amongst the natives dwelling in his country, the assessment and collection of taxes, the up-keep of the main roads, and the general supervision of native affairs. . . . For the assessment and payment of taxes, the chief of a county shall be immediately responsible to Her Majesty's representative, and should he fail in this duties in this respect, Her Majesty's representative shall have the right to call upon the Kabaka to dismiss him from his duties and to appoint another chief in his stead. In each county an estate, not exceeding an area of eight square miles, shall be attributed to the chieftainship of a county, and its usufruct shall be enjoyed by the person occupying, for the time being, the position of chief of the county.

Clause 10. To assist the Kabaka of Uganda in the government of his people he shall be allowed to appoint three native officers of state, with the sanction and approval of Her Majesty's representative in Uganda (without whose sanction such appointments shall not be valid): —

A Prime Minister, otherwise known as Katikiro; a Chief Justice; and a Treasurer or Controller of the Kabaka's revenues. These officials shall be paid at the rate of £ 300 a year. Their salaries shall be guaranteed them by Her Majesty's Government from out of the funds of the Uganda Protectorate. During the minority of the Kabaka these three officials shall be constituted the Regents, and when acting in that capacity shall receive salary at the rate of £ 400 a year....

Clause 11. The Lukiko, or native council, shall be constituted as follows: — In addition to the three native ministers, who shall be *ex officio* senior members of the council, each chief of a county (twenty in all) shall be *ex officio* a member of the council. Also each chief of a county shall be permitted to appoint a person to act as his lieutenant in this respect to attend the meetings of the council during his absence, and to speak and vote in his name. The chief of a county, however, and his lieutenant may not both appear simultaneously, at the council. In addition, the Kabaka shall select from each county three notables, whom he shall appoint during his pleasure, to be members of the Lukiko or native council. The Kabaka may also, in addition to the foregoing, appoint six other persons of importance in the country to be members of the native council. The Kabaka may at any time deprive any individual of the right to sit on the native council, but in such a case shall intimate his attention to Her Majesty's representative in Uganda, and receive his assent thereto before dismissing the member. The functions of the council will be to discuss all matters concerning the native administration of Uganda, and to forward to the Kabaka resolutions which may be voted by a majority regarding measures to be adopted by the said administration. The Kabaka shall further consult with Her Majesty's representative in Uganda before giving effect to any such resolutions voted by the native council, and shall, in this matter, explicitly follow the advice of Her Majesty's representative. ...

Clause 12. In order to contribute to a reasonable extent towards the general cost of the maintenance of the Uganda Protectorate, there shall be established the following taxation for Imperial purposes, that is to say, the proceeds of the collection of these taxes shall be handed over intact to Her Majesty's representative in Uganda as the contribution of the Uganda province towards the general revenue of the Protectorate.

The taxes agreed upon at present shall be the following: —

(a) A hut tax of three rupees, or 4s. per annum, on any house, hut, or habitation, used as a dwelling place.

(b) A gun tax of three rupees, or 4s. per annum, to be paid by any person who possesses or uses a gun, rifle, or pistol.

The Kingdom of Uganda shall be subject to the same Customs Regulations, Porter Regulations, and so forth, which may, with the approval of Her Majesty, be instituted for the Uganda Protectorate generally, which may be described in a sense as exterior taxation, but no further interior taxation, other than the hut tax, shall be imposed on the natives of the province of Uganda without the agreement of the Kabaka, who in this matter shall be guided by the majority of votes in his Native Council. This arrangement, however, will not affect the question of township rates, lighting rates, water rates, market dues, and so forth, which may be treated apart as matters affecting municipalities or townships; nor will it absolve natives from obligations as regards military service, or the up-keep of main roads passing through the lands on which they dwell. ...

The possession of any cannon or machine gun is hereby forbidden to any native of Uganda. ...

The Kabaka will be credited with fifty gun licenses free, by which he may arm as many as fifty of his household. The Queen Mother will, in like manner, be granted ten free licenses annually, by which she may arm as many as ten persons of her household; each of the three native ministers (Katikiro, Native Chief Justice, and Treasurer of the Kabaka's revenue) shall be granted twenty free gun licenses annually, by which they may severally arm twenty persons of their household. Chiefs of counties will be similarly granted ten annual free gun licenses; all other members of the Lukiko or native council, not chiefs of counties, three gun licenses, and all landed proprietors in the country, with estates exceeding 500 acres in extent, one free annual gun license.

Clause 13. Nothing in this agreement shall be held to invalidate the pre-existing right of the Kabaka of Uganda to call upon every able-bodied male among his subjects for military service in defence of the country; but the Kabaka henceforth will only exercise this right of conscription, or of levying native troops, under the advice of Her Majesty's principal representative in the Protectorate. ...

Clause 14. The chief of a county shall have right to call upon each native town, village, or commune, to furnish labourers in the proportion of one to every three huts or houses, to assist in keeping the established roads in repair, provided that no labourers shall be called upon to work on the roads for more than one month in each year. ...

And so on, with *Clause 15,* dealing with rights over lands in the kingdom, of which more will be said later; *Clause 16,* dealing with forestry regulations which had yet to be formulated; *Clause 17,* dealing with mineral rights, vested on private estates in the owners subject to a 10 per cent. *ad valorem* duty to the British administration "when the minerals are worked", and belonging directly to the British Administration outside the private estates; *Clause 18,* conferring a sum of £ 500 to the Kabaka and to the three Regents a total of £ 600 (of which, £ 300 to the Katikiro) "in return for the cession to Her Majesty's Government of the right of control over 10,550 square miles of waste, cultivated, uncultivated, or forest lands"; *Clause 19,* agreeing "to pay to the Muhammedan Uganda chief, Mbogo, a pension for life of £ 250 a year, on the understanding that all rights which he may claim (except such as are guaranteed in the foregoing clauses) are ceded to Her Majesty's Government"; *Clause 20,* declaring that the Agreement would become null and void if the Kindgom of Buganda fails in the first two years to provide the British Administration with at least half of the total native taxation as calculated "in proportion to the number of inhabitants", or if "the Kabaka, chiefs, or people of Uganda, pursue, at any time, a policy which is distinctly disloyal to the British Protectorate", and, on the other hand, promising the Kabaka, his ministers and chiefs "the right to appeal to Her Majesty's Government for an increase" in their subsidy and stipends, provided they can run the kingdom so efficiently that "the revenue derived from the hut and gun tax exceed two years running a total value of £ 45,000 a year"; *Clause 21,* clarifying some terms used in the Agreement; and *Clause 22,* finally stating that "in the interpretation of this agreement the English text shall be the version which is binding on both parties", although the Agreement was prepared both in English and in Luganda (the language of the Baganda people) and obviously those Baganda who could read English in those days could be counted on one's fingers. Evidently, in the minds of the foreign rulers, the Agreement, which sealed the possibilities of independent development of Buganda, was of no concern to the people.

Moreover, it is only too evident from the above extracts of the Uganda Agreement of 1900 that behind the facade of respecting the traditional institutions of the African people, the British reduced the feudal king to the status of a low-grade vassal who could not only be disposed off at the pleasure of the British Government (and, in fact, his son was so treated in 1953), but could not also have any effective political, military, or economic power in his kingdom. His kingdom was henceforth merely a province of the entire colony in which again he could not make any law or take any move without the previous permission of the British administrator placed over him. His grand-father, Mutesa, kept an impressive army and navy, but henceforth he could build a "native army" only when permitted, or actually ordered to do so, by the British Administration. Mutesa toyed with firearms[149], but now his grandson could keep, at the pleasure of the foreign rulers, a maximum number of thirty guns in his possession, and of course no cannon or machine-gun which, in order to uphold

British rule, were henceforth tabooed to the Baganda. From now on the king would not even have the right to obtain taxes from his subjects, for which his chiefs would be directly responsible to the British Administration. Eventually, therefore, the Kabaka came to be regarded as occupying a position "which is tantamount to that of an ordinary Paramount Chief of one of the second-rate native tribes of Africa"[150].

And if such became the position of the king, that of the chiefs — the new ones exalted to this position or the previous docile ones — turned out even worse. They became merely a stratum of the "native civil service" in the colony. Even the mild criticism made by Buell of the administration of the kingdom of Buganda by the British is of interest to note here[151].

> "Nevertheless, the visitor who has witnessed tribal institutions in other parts of Africa has certain misgivings at the ease with which the British Government shifts a chief from one county to another, just as if he were a British official. In fact, the British Government aims to make out of these chiefs a regular native civil service, composed of educated young men who receive their apprenticeship in a British office. It seems that the British authorities exercise a high degree of control over the appointment and designation of these chiefs, and for this reason, there is a danger that the Buganda Government will lose the autonomy which the presence of hereditary rulers would insure.
>
> In carrying out this aim of establishing a native civil service, it appears that British officials have dealt directly with many native chiefs, instead of first going to the Kabaka."

In short, eventually the chiefs, trained previously as apprentices "in a British office", were turned out as "glorified clerks" of the foreign administration, as happened in all other parts of the Protectorate.

The upshot was that the Kabaka "has now come to be looked upon by these Chiefs as merely the headman or superior Chief of the Natives of Buganda on more or less the same level and receiving an annual salary from the Protectorate Government in the same way as themselves"; and the new chiefs "are now beginning to flout the authority of the Native Government of Buganda, and to consider themselves more as Native Officers of the British Government to whom they are responsible for all their administrative duties in their respective areas"[152].

In return for this docile attitude of the king and chiefs, "by a series of agreements modifying the original treaty of 1900", the salaries of the Kabaka, the ministers, and the chiefs were steadily increased. In the estimates of 1938, £ 30,880 was provided for this purpose[153]; in 1947 the figure came to £ 133,058, or 48.4 per cent of the total expenditure of the Buganda Native Government[154]. And of the total amount of £ 274,766 as the estimated revenue and expenditure of Buganda in 1947, £ 225,093, or 82 per cent of the total amount, was obtained as taxation of the African people living in Buganda[155].

It should further be noted that while 48.4 per cent of the revenue was expended as emoluments and allowances of the native rulers, and 6.6 per cent on the police and prisons maintained by the local government, 1.4 per cent of the total amount was spent on "agriculture" (which is the main and virtually the only source of support of the African people), 2.3 per cent on their medical benefit (even though horrible diseases, including leprosy, are rampant in Uganda), and 4.8 per cent on their education (in spite of the fact that very few Africans are acquainted with letters).[156]

Thus, instead of instituting a progressive government in the Buganda territory, the oppressive feudal rulers of pre-British days were given a long lease of life and were maintained on the labour of the people, while their position was reduced to the position of low-grade civil servants of the British Colonial Administration.

In the other kingdoms, viz. Bunyoro-Kitara, Ankole and Toro, which now belong to the Western Province of the Uganda Protectorate, the same state of affairs prevailed. But here the position of the kings and chiefs was even worse, although as in Buganda the facade of the previous political structure was rigorously maintained in the name of preserving the traditional institutions. Hailey noted regarding local administration in the Western Province[157]:

> "Viewed from the standpoint of native administration, the Province presents a feature of special interest, owing to the existence of the Agreements made at different times with the Ruling Houses in Bunyoro, Ankole and Toro, though as will subsequently be seen, the Agreements did not create in those three areas a constitutional position similar to that existing in Buganda."

In these states, while as before the kings with their subsidiary County, District and Parish chiefs dominated over the peoples, they were henceforth treated more openly as Paramount Chiefs under British Colonial Administration. And, of course, like in Buganda, these feudatory oppressors were shorn off their previous political, military and economic power, and this was done more openly than in the kingdom of Buganda.

Thus the agreement made between the kingdom of Bunyoro-Kitara and the foreign rulers "acknowledged the Mukama as 'the Native Ruler of Bunyoro-Kitara', but laid down that in all matters of administration he should follow the advice of the Protectorate Government, communicated through its local officers. All the legislation of the Protectorate was to extend to the Bunyoro district. The Mukama was to appoint and dismiss his ministers and the senior grades of chiefs, but his action in this respect was to be subject to the approval of the Governor, though no such approval was to be necessary in the case of the junior grades of chiefs."[158]

Similarly, the agreement made with Toro "provided that the Toro District was to be subjected to the same laws and regulations as were in force throughout the Protectorate, but for the rest, one of its chief objects seem to have been to satisfy the claims of the Ruling House and chiefs in regard to land"[159]. In other words, here also the colonial administration did not do away with the previous form of political domination; on the contrary, the position of the feudal oppressors was stabilised and at the same time while taking away their effective power they were pacified by land-grants. Of this more will be said later.

Likewise, in Ankole the "Agreement closely resembles that concluded with Toro, and, as in that case, many of its articles have now been superseded in subsequent usage, so that they no longer define the actual character of the relations of the Protectorate Government with the Ankole Native Administration. It recognized the Kabaka (who is now known under the traditional title of Mugabe) as supreme over the ten chiefs who were described in it as holding sub-divisions. These chiefs were to be allowed to nominate their successors, but it may be noted that the successors thus nominated have (with one exception) since died or been retired, and apart from the one exception the chiefs now in charge of counties are subsequent appointments which carry no hereditary status"[160].

To sum up then, in the words of Hailey who is a staunch supporter of the system of Indirect Rule, the retention in these kingdoms of the post of Ruling Chief "gives him an influence which can be of great value to the Protectorate Government"[161]. And with regard to the lesser chiefs, although they are formally appointed by the Ruling Chiefs, it is "in most cases usual to consult the District Commissioner both as to their appointments and their transfer"[162]. Moreover, these chiefs have no traditional recognition in the societies. As Hailey has described, they are selected in accordance of their meritorious service in the Native Governments, Police Departments, etc.[163]. In short, the entire hierarchy of pre-British rulers of these territories, with the addition of those who had no position in the society before but have now become chiefs because of their loyal service to the colonial government, now serve as low-grade civil servants of the British Colonial Administration in the Uganda Protectorate.

Thus established, these native rulers, like their colleagues in Buganda, live on the labour of the people. At the beginning there was a slight difference between the native bureaucracies in these states and those of Buganda. This was that the former received a percentage of the poll-tax collected from the people instead of receiving regular salaries as in Buganda. The Paramount Chiefs of Ankole and Toro, the Mugabe and the Kabaka or Mukama, were usually paid twenty per cent of the tax collected in these two kingdoms; while the Paramount Chief of Bunyoro, "because of the poverty of the kingdom", was given thirty per cent. The ordinary chiefs received ten per cent of the tax collected, of which seventy per cent was reserved for the County Chiefs and the remaining thirty per cent was for the District Chiefs. There are reasons to believe that in these states, which at the beginning were not so efficiently organised as Buganda, payment of a percentage of the poll-tax instead of a fixed salary was meant as an incentive to the native bureaucracies to extort as much as they could from the people*. Of this poll-tax more will be said later, but it is interesting to note that after a few decades these native officials also were turned into salaried underlings of the British colonial administration. And how they have a similar existence as the native rulers of Buganda is evident form the following table prepared from the data supplied by Hailey[164].

Table IV.1

Kingdom	Finance of Native Governments in 1946	
	Percentage of taxation to total income	Percentage of personal emoluments of native rulers to total expenditure
(1)	(2)	(3)
Bunyoro	74	38
Toro	75	49
Ankole	79	47
Buganda 1947	82	48

Such then is the true meaning of the system of Indirect Rule, which, as referred to before, in the words of Hailey, was to give the Ruling Chief "an influence which can be of great value to the Protectorate Government".

Furthermore, another purpose for establishing Indirect Rule was to minimise expenses on colonial administration, since otherwise many British officers at a much higher expense would have been needed to run the apparatus as efficiently as could be done by fewer native officials at a comparatively much lower cost. Thus Buell noted while describing the native administration of the kingdom of Buganda[165]:

> "By means of this native machinery, the administrative officials are relieved of an immense amount of work assumed by officials in the French and Belgian colonies, and in such places as Kenya."

In addition, the policy of the colonial administrators could be executed more successfully by these native officials than by the foreigners, for the former had an intimate knowledge of their people and so knew how to go about spreading British influence all over the country.

---

* It is of interest to note that in India, after the East India Company undertook the civil administration of Bengal, the Tax Collectors "were encouraged to replenish the exchequer by the grant of a percentage on their collections, in addition to their salaries". (cf. *loc. cit.* 133, p. 199).

And thus, while ensnaring the peoples under the bondage of British rule and making necessary changes in society to suit to the needs of the British ruling class, the entire process could be executed behind a fairly peaceful atmosphere in the colony. For, at the same time, the people could be hood-winked in a certain measure with the show of respecting their traditional institutions but making lambs out of lions, or good Christians out of the previous autocratic rulers, under the benevolent protection of the British.

The system of Indirect Rule had still another purpose; a more sinister one. This was to build a wall of defence in the interest of the foreign rulers between them and the people. It was best executed in the kingdom of Buganda by Clause 15 of the 1900 Agreement, dealing with the land system in Buganda. No doubt, particularly in Buganda, to examine the effect of Indirect Rule on the life of the people, attention should primarily be drawn to how it affected the relation of the people to land. For, living almost entirely on an agrarian economy, the possession of land, their primary means of production, would vitally affect the well-being of the people.

Prior to British occupation, this was not a problem at all, for land was held communally by the clans and the people were characterised by their clan areas. This was noted by Roscoe in his monograph on the Baganda[166]. Moreover, Lucy Mair, who made a scientific investigation on the life of the Baganda people, noted in 1934[167]:

> "Right up to the time of the Uganda Agreement in 1900 the head of each clan lived on the land which was supposed to have been settled by the first ancestor, and one way of describing a member of the clan, especially on ceremonious occasions was to say that he came from that place. ... Each clan was further divided into the descendants of sons of the original founder, who formed a *siga*, and again into those of his grandsons or great-grandsons, who formed a *mutuba*, but even these sub-divisions were still very large."

Under this arrangement the king and chiefs could exercise only political rights over the people; they had no right over land. This point has been noted before. The following extract from Mair further clarifies the situation[168].

> "Theoretically, the whole country and all that was in it was not only subject to the king's authority but actually belonged to him; hence his power of life and death, his right to the spoils of war, to tax his subjects, and make use of their labour. He made his will effective throughout his kingdom by the appointment of chiefs, of whom it may with equal truth be said that each controlled a certain area of land or that he controlled the whole population within that area. ... [But] The obligations of the peasant to his chief in pre-Europeandays ... were not analogous to a payment of rent. There was no conception of the landas a possession of the chief from which he derived profit by letting other people work it. His right to admit and evict peasants was part and parcel of his general position as the political authority over the area in question; as the natives themselves put it, *Tafuga taka, afuga abantu*, 'He has rights not over land but over people'."

But now the British policy of creating a new subservient native bureaucracy under the system of Indirect Rule made a drastic change in the relations of people to the land. To quote Mair again[169]:

> "As far as land is concerned (the Uganda Agreement of 1900) created rights which were unknown before. For it not only alloted large estates in freehold tenure to the king and his relatives and to the three chiefs who were appointed to act as regents during the king's minority, but asigned 'the estates of which they are already in possession and which are computed at an average of 8 square miles per individual' to 'one thousand chiefs and private landowners'. The allotment was made by the native council of chiefs, Lukiko, which the agreement set up. In a good many cases this council disregarded the rights of the bataka (clan heads), who were not represented there.

> The famous Sir Apolo Kagwa, K.C.M.G., with whom, as Regent, the agreement was concluded, profited largely from this disregard."

Not only the chief bureaucrat, but the entire newly-made bureaucracy profited enormously by this agreement. The total area of Buganda, excluding open water, is 17,310 square miles. Out of this total, nearly 1,000 square miles (or 5 per cent) went to the Royal Family; 8,003 square miles (or 46 per cent) were shared among the new bureaucrats; and the remaining 49 per cent of Buganda land passed to the British Government for direct colonial enterprises (for the purpose of which a part of the Crown land was first alienated to establish various firms and for prospecting for extraction of the mineral wealth of the country) and with a view to White Settlement (which was then favoured by the British administrators like Lugard and Johnston)[170].

Land no longer belonged to the people. As noted by Thomas and Spencer[171]:

> "Thus, in regard to land tenure, the 1900 Agreement had brought about the almost complete suppression of custom by law."

And this was done by the British Government at the same time as they declared that Indirect Rule respects the traditional system of social organisation.

Evidently, because of this immense economic gain to the native bureaucracy, practically no instance occured in the kingdom of Buganda expressing the resistance of the native rulers to the newly-formed colonial administration. Although even in this developed kingdom the African rulers lost the political, military, and economic powers enjoyed by their forerunners, they were amply recompensed by steady incomes from the Treasury and from the lands given to them. And, in return for their strict obedience to the instructions issued by the Colonial Government, they were assured of "all their rights and liberties" and "all their lands, possessions and ancient privileges" by the then Under-Secretary of State for the Colonies, Mr. Winston Churchill (now Sir Winston), who visited Uganda in 1907. He, further, described the Uganda Agreement as a "human document", and assured the chiefs who had gathered to listen to his address that they "have no fear that it will be encroached upon or melted away so long as they themselves and the people of Uganda faithfully adhere to their portion of the contract"[172].

Thus, basing itself upon the pre-British feudal structure of society there developed a well-satisfied native bureaucracy in Buganda, deeply concerned in the peaceful continuation of British rule in this territory, on which depended their existence and prosperity. How they were obliged by the colonial government in this respect would be further evident from the following statement made by Hailey as late as in 1950[173]:

> "The field of legislation left to the Native Government has related primarily to purposes arising out of the conditions of land tenure in Buganda, such as the Mailo Law of 1908, the Land Survey and Boundary Laws of 1909, 1911 and 1933, the Buganda Land Succession Law of 1912, the Busulu and Envujo Law of 1927, the Buganda Township Law of 1932, the Land Tax Law of 1933, and the Land (Agreements) and Land (Sale and Purchase) Laws of 1939, and the important measure enacted in 1945, the Law for Selecting Unofficial Representatives for the Buganda Councils."

The newly-created bureaucracy could thus thrive as landed aristocracy for which alone they got authority from the colonial administration as evident from the foregoing. It is also worthy of note that a system of *Bulungi bwa nsi* was instituted by which the plot-holders were obliged to contribute labour for "local benefit", and this "may in time become the basis of a Local Rate to be administered by miruka or gombolola Councils (Parish and District authorities), if they acquire powers as local government bodies"[174]. To whose benefit such taxes go will be discussed later.

In the other kingdoms of the Uganda Protectorate also, which according to Thomas and Spencer had reached the stage of "savage regime"[175], but not that of Lugard's "barbaric civilisation[176]" as in Buganda, the imposition of Indirect Rule brought about the same form of change in society as that observed among the Baganda. Here also land was given to the sets of native bureaucracy, loyal to the prolongation of British rule in their territories, in order to stabilise further their interest in the welfare of the foreign regime.

This "land distribution" to the native bureaucracies in the kingdoms of Bunyoro-Kitara, Ankole and Toro, however, took place at a later date. It is not unlikely that at the beginning land was not a serious concern with them, for they led mainly a pastoral life. But the immense "prosperity" enjoyed by their colleagues in Buganda soon roused their cupidity and therefore they also pressed their masters for land-rights. This was so, for, as will be explained later, the introduction and encouragement of cotton production as the pivot of the colonial economy, soon brought home to these people the importance of possessing large tracts of land. From the onset of British rule in Uganda active preparations were set on foot to reduce the whole territory into an area for the production of raw materials as needed by the capitalists in the metropolitan country. In this scheme, the cultivation of cotton played ultimately the most conspicuous role. By 1911 cotton became the most important cash crop of the Africans all over Uganda, and thus, for the benefit of the foreign capitalists, as will be explained later, the whole Protectorate was brought under this form of economic production. It may not therefore be quite fortuitous that in 1911 a committee was appointed "to consider the question of Native Land Settlement in Ankole, Bunyoro, Busoga and Toro"[177].

As will be noticed, in the above mentioned list Busoga was also included, probably because it was a developed tributary country of the kingdom of Buganda and had from the beginning offered its allegiance to the British. Buell noted[178]:

> "While it has no Paramount Chief, Busoga is the most advanced native government, next to Buganda. The President of the Lukiko draws a salary of five hundred and fifty pounds a year."

There might have been also another reason for giving so much attention to Busoga a few years after the establishment of British rule in this territory. For, although in the pre-British era the Basoga had offered their allegiance to the British to get rid of the Baganda oppressors, a few years of British rule found them in a concerted opposition to the British and the imported Baganda chiefs in their territory (about which more will be said later). Probably as a countermove it was therefore considered necessary to create quickly a contented native bureaucracy in Busoga in about the same form as applied to the developed African kingdoms in the Protectorate.

Whatever the reason may be, the native rulers of Busoga benefited most by the recommendations made by the committee appointed in 1911. The comittee's report came out in 1912, and, in the words of Thomas and Spencer, who had made a close study of Uganda lands, "made a disappointing display". To quote these two authorities[179]:

> "Satisfied that 'according to native custom no land was owned by anyone except the chiefs and that the peasants were merely occupiers' the Committee could devise no better solution than an allotment of estates to two or three hundred chiefs in each district, which estates were to be subject to much the same conditions as applied to Buganda mailo land [cf. Uganda Agreement of 1900]. In Ankole about 16%, in Bunyoro 11%, in Busoga 27% and in Toro 15% of the land area of the district was thus to be disposed of, for the most part as the private property of chiefs, the remainder in each district being available for alienation by the Crown. Peasant occupiers were to become rent paying tenants and some security from arbitrary eviction by private landlords was to be ensured by legislation."

Landlordism was thus implanted also in these territories of Uganda to secure firmly the alliance between the native bureaucracy and the foreign rulers.

It will be recalled that, as Hailey noted, the successive Agreements made by the Colonial Government with these kingdoms dealt mainly with "the claims of the Ruling House and chiefs in regard to land". This is the power the previous oppressive feudal rulers of these territories received from the colonial administration, while if it was the intention of the British to institute a progressive from of government in these territories, these creatures should have been removed from the political field and replaced by those who would be democratically elected by the people. But the oldest and wisest Colonial Power in the world knew what is good for itself. In India the British were not only king-makers; in large areas, like in the Subah of Bengal, where they apprehended uprisings of the people against their oppressive rule, they introduced landlordism in 1793 by means of the so-called Permanent Zamindari Settlement of land-tenure, in order to build a wall between themselves and the oppressed people. Lord William Bentinck, who was the Governor-General of India from 1828 to 1835, declared quite candidly in an official speech during his rule in India[180]:

> "If security was wanting against extensive popular tumult or revolution, I should say that the Permanent Settlement, though a failure in many other respects and in most important essentials, has this great advantage at least, of having created a vast body of rich landed proprietors deeply interested in the continuance of the British Dominion and having complete command over the mass of the people."

A little more than a century later, the same policy was applied in another colony, and in those regions of it where the stage of social development was conducive to it. As in major parts of India before they came under British rule, private ownership of land was unknown in Uganda. Here the tribal democracy of a long past maintained its last hold on the clan-possession of landholdings. But here also the British not only became chief-makers, but also the creator of landlordism. This finally stabilised the position of the previous oppressive rulers in the societies, and over all of them ruled the supreme power of Britain through the Colonial Administration.

Such was the outcome of the introduction of the system of Indirect Rule in the African kingdoms of Uganda.

7. Native Administration

The establishment of the system of Native Administration in the remaining parts of the Uganda Protectorate had also a similar aim, that is, to create allies for the foreign administration. This system was found more suitable to those societies the political structure of which was much simpler than in the African kingdoms. As noted before, in some of these societies chiefdoms had developed in the pre-British days. Now therefore the attempt was made to maintain the facade of respecting the traditional institutions of the people by apparently retaining the authority of the chiefs, while making them tools in the hands of the British administrators.

Moreover, with the same purpose in view, in those remaining societies where chieftainship as a political institution had not yet developed, this was now imported without any pretence of respecting the traditional institutions of the people. Indeed, although the system of Native Administration was declared to be based on the "old tribal organisation and authority"[181], how barefaced was this claim would be at once realised if those societies are first discussed in which chiefdoms had not developed earlier. It has been noted earlier that in pre-British Uganda chiefdom was not present among the Nilo-Hamites, and some other peoples, like the Bagishu or the Bakiga, who were either still in the tribal stage or their tribal organisation was in the process of disintegration. Among these people now chieftainship was imposed from above.

Thus, among the Jie "there were no chiefs or formal leaders", but since the introduction of British rule, "as one county in the Karamoja district of Uganda, the Jie have an administrative system of the usual Uganda hierarchical type"[182].

Similarly, the Karamajong also had no institutionalised authority of chiefs. But after the British came, the situation became different. Hailey noted[183]:

> "In the Karamoja District there are no chiefs commanding a traditional position, and in the earlier days of rule the Protectorate Government relied largely on the introduction of Agents from Acholi."

It will be remembered that among the Acholi chiefdom was a well-developed institution in the pre-British days, and that some big Acholi chiefs, like the Ochama of Payera, had sought for British protection at the time the latter were intent on colonising Uganda. Now they were rewarded by being made into "chiefs" of the Karamajong, although, as is quite clear from Hailey's account, such an imposition of foreign "chiefs" or even their creation from the local elements was quite contrary to the "old tribal organisation and authority" vested in the Karamajong society. To quote Hailey[184]:

> "The real centre of local authority appears to be in the manyatta or kraal heads, and this, combined with other circumstances, has made it difficult to form suitable administrative units on the model followed elsewhere in the Protectorate, or to find authorities competent to take charge of them."

Nevertheless, the system of "Native Administration" was introduced in this society, using the Acholi "chiefs" at the beginning and eventually manufacturing Karamajong "chiefs" out of the loyal elements in society.

In this way, in the Karamoja District, inhabited by the Karamajong in one region and by other peoples in the adjoining territories, in whichever form democracy existed previously in the political structure of the societies, it was now totally suppressed; and, in order to keep these peoples securely under British rule, allies of the foreign government, imported or locally manufactured, were imposed on them as their "chiefs". To quote Hailey again[185]:

> "Six counties have, however, now been created, three in areas predominantly Karamajong, and the rest embracing other tribes, though they do not correspond strictly with tribal boundaries. The county unit is an innovation and there is no local name for it, but the county chiefs have the title of ekapalon or leader. Below the county chiefs are divisional chiefs (jakait), but the divisions do not seem to have well defined geographical boundaries. Below the jakait are numbers of group headmen."

Thus, the "usual Uganda hierarchical type" was well-imposed by the foreign rulers upon the Karamajong and neighbouring Nilo-Hamites among whom there were no institutionalised chiefdoms before. Gulliver noted[186]:

> "Today the Karamajong are divided into three administrative counties, each with its own chief, sub-chiefs and courts."

Of these "chiefs", one is still an imported Acholi, but the others have in the meanwhile been manufactured from the local society[187].

The same was the process among other Nilo-Hamites. For instance, among the Iteso, although some changes were taking place towards the emergence of chiefdoms just before the advent of the British, local authority still rested "in the leaders of family groups". Hence, Hailey admitted[188]:

"The institution of a system based on the division of the country into territorial units was therefore an innovation, but its justification lay in the fact that the boundaries of the areas in which clan leaders exercised authority were ill-defined, and the extent of the authority exercised by them was limited, being mainly concerned with questions relating to the land, or arbitration in judicial issues."

And it followed from the above argument[189]:

"In Teso District the county chiefs do not now include any clan or other traditional leaders and represent in each case the result of promotion from the gombolola chiefdoms."

In this way, the entire hierarchy of chieftainships was created among the Iteso. Lawrence noted more candidly[190]:

"There is no indigenous political system based on clans and clan leaders, nor are there indigenous chiefs. The political system now prevailing, with its hierarchy of chiefs, was imported by the Baganda and perpetuated by the British. . . . In recent years an attempt was made to base the political system on clan leaders; but these clan leaders are in effect henchmen of the Government, their duties are similar to those of a policeman or, at most, of a justice of the peace."

Nothing could be more explicit. More will be said later on how the traditional institution of the Iteso was drastically altered by the foreign rulers with the export of Baganda "chiefs" to this and other regions of the Protectorate.

The situation was similar among all other Uganda Africans who in the pre-British period of their history still retained some degree of tribal democracy in their political life and had not yet come decisively under the domination of a hierarchy of chiefs.

Thus, taking for instance, the Bakiga, among whom, as noted before, institutionalised form of chieftainship had not developed earlier, it was now imposed first by exporting Baganda as "chiefs" to the Kigezi country and eventually manufacturing the "chiefs" from the local people. Hailey noted[191]:

"Kigezi District was originally ruled through Baganda Agents, and the system of chiefdom follows the Baganda model, no clan or other traditional heads being recognized."

About these Baganda Agents more will be said later; here the above extract shows how the traditional institutions of the Bakiga people were respected by the colonial administration.

Similarly, in the West Nile District of the Northern Province, "the county chiefs, other than the two hereditary chiefs of the Alur and Jo-nam . . ., have no traditional position and are chosen for personal qualifications"[192]. In other words, they were manufactured, as the following extract from Hailey makes quite clear[193]:

"In the West Nile District the Belgians had left a number of chiefs (sultans), some of whom had served as soldiers in Emin Pasha's army. The British at the outset recognized the two hereditary chiefs (opi) of the Alur and the Jo-nam; but elsewhere they relied on sultans chosen on grounds of service or local position."

The above extract also exemplifies the qualification to be a "chief".

And so it was in the Eastern Province, of which the Teso District has already been discussed. Referring to the Bukedi Native Administration in the Mbale District of this province, Hailey has noted[194]:

"The chiefs are only rarely clan or group leaders, and the possession of traditional authority of this nature is not regarded as a condition of their appointment; there seems indeed to be in Bukedi few personalities clearly recognized as possessing traditional or hereditary authority."

Nevertheless, the system of "Native Administration" was extended to the people living in this territory, and "chiefs" were eventually manufactured from the societies.

It was the same among the Bagishu who, as noted before, had not developed chiefdoms in the pre-British days. Hailey has noted[195]:

> "In most material respects the formation of the Bugishu Native Administration of the Mbale District is on the same lines as that of Bukedi. The county is a more or less artificial unit; the gombolola has some relation to the clan area but does not necessarily correspond to it.... The appointment of the chiefs depend on personality and prestige rather than on a traditional position, and even though it may have had some influence in making an initial appointment, it ceases to carry weight in the process of promotion or transfer."

This is Native Administration put into practice in those societies of Uganda where chiefdoms had not developed earlier. What Hailey noted for the Eastern Province of the Protectorate can be generalised for all such areas where democracy, whether in the tribal stage or in that of *military democracy*, was retained, to some extent at least, in the political life of the people. To quote Hailey[196]:

> "The system adopted, therefore, differed materially from that followed in some other territories, which sets out to adopt the tribal or other indigenous institutions as a unit of rule, giving them an authority prescribed by statute law, under the control of the Administrative head of the District. The principle which it followed may best be expressed as the creation in each District of a District Native Administration of which the District Commissioner was the controller and chief executive, operating through a hierarchy of chiefs and subordinate chiefs in charge of territorial units. Though given statutory powers as Native Authorities for the discharge of the duties imposed on them, these chiefs were not Native Authorities in the same sense as in some other territories, for though an effort was often made to select as chiefs men who were tribal or clan leaders, this was not the universal rule, and the basis of their authority lay not in their traditional status, but in the place they occupied as part of an official or semi-official organisation, the District Native Administration."

In short, these "chiefs" were nothing else than the lowest grades of civil servants of the Uganda Colonial Administration, and their existence depended on the pleasure of the foreign rulers. The "Native Administration" was thus only a mask behind which the foreign rulers stabilised their interests in these societies and perpetuated their rule.

Were British policy and its *modus operandi* in any way different in those other societies in Uganda where chieftainship as a political institution had developed in the pre-British period, as the above extract tends to suggest? The facts indicate that here while the facade of "old tribal organisation and authority" was maintained, its content was completely changed. Firstly, in these societies only those big chiefs were retained in office who had either already signed "treaties" with the British or gave their assurance to serve loyally the present regime. In the same way, only those smaller chiefs were allowed to remain in office who were willing to serve faithfully under British rule. And the discontented ones were replaced either by locally-found reliable individuals or by importing some Baganda (mostly loyal "Christians") as "chiefs" in these areas. With such reliable "leaders" of the people, the system of Native Administration was brought into force in these societies.

Secondly, here also a traditional position among the people was not to be the main criterion for being a chief. In running the Colonial Administration, the "chiefs" were selected according to their reliability and suitability in maintaining colonial law and order. And if such qualities could be found among those who had a traditional recognition in the societies as chiefs, so much the better. Otherwise, it was not considered as an indispensable qualification.

Thus, in the Busoga District, where the people were regarded as the best developed in the Uganda Protectorate after the Baganda and where chieftainships had earlier developed, out of the eight County chiefs:

> "two of the county chiefs are at present the heads of local clans, though working outside their own clan areas, and all represent promotions from the ranks of gombolola chiefs."[197]

More will be said later as to who these "chiefs" were, how they are transferred and promoted by the foreign administrators, and how the Baganda "chiefs" lorded over these people at the beginning of British rule under the system of Native Administration.

To take next the Lango, among whom as noted before despotic leadership in the forms of Twon Lwak, Rwot and Jago had made their appearance in the pre-British period, chiefs were appointed without any traditional and hereditary claims, and obviously because they could further the interests of the British. To quote Hailey[198]:

> "In the Lango District the relative weakness of the clan organization no doubt accounts for the fact that in the first instance the British Administration utilized as its agents "appointed" chiefs, who for the most part had no traditional position. There was, however, a considerable undercurrent of feeling in favour of the recognition of traditional or hereditary claims. . . . In its present form, introduced in 1946, the organization of the Native Administration has a close analogy to that of Acholi District."

But among the Acholi also the situation is not different. As noted before, in the pre-British period chiefdoms had developed in the Acholi society, and with regard to present times Hailey has noted that the "Acholi District is now the most advanced of the Districts in respect of the development of native administration"[199]. But this advancement is not due to the maintenance of "old tribal organisation and authority" among the Acholi. On the contrary, it is due to the infusion of a completely new content into the traditional form, or in the words of Hailey, "as a result of a series of changes in system"[200]. What are these changes? To quote Hailey again[201]:

> "The District came under regular administration between the years 1912 and 1914, and in the first instance local affairs were controlled through chiefs, some of whom were clan heads, while others were appointed on the ground that they commanded some measure of local authority.
>
> In the years 1935 and 1936 an effort was made to rely more definitely on the agency of the hereditary clan heads, but experience showed that this system had inherent defects, for the clan heads were so numerous that their salaries made an unreasonable charge on the funds available, and an effort to amalgamate the clans still left the District with some 50 divisions. The senior clan chiefs were formed into Courts of Appeal, but in other respects they failed to command any authority over other chiefs in their area. Moreover very few of the clan heads showed any aptitude as agents of local administration, and were in most cases so far dominated by the members of their clans that they proved useless for the maintenance of law and order."

Here one finds the real reason for removing the traditional chiefs from the political arena of these societies. For, as will be seen later, the expenditure on the "appointed" chiefs is quite substantial and the amount comes from the people; so that the question of expenses on the traditional chiefs cannot be the reason, and surely not the main reason, for replacing them by the "appointed" chiefs. On the other hand, it is admitted that the "hereditary clan heads", who have traditionally represented the Acholi society, are "dominated" by the people and, evidently, not so much by the foreign rulers as the latter might desire. These chiefs are therefore found to lack in "aptitude as agents of local administration", and are "proved useless for the maintenance of law and order". Hence to continue with Hailey[202]:

> "In 1943 therefore an experiment was made in the introduction of a system by which chiefs were selected on merit, though at the same time some consideration was also given to their traditional claims, and the authority of those selected was reinforced by associating with them an ascending chain of local councils."

Indeed, as will be shown later, very few of these chiefs have a traditional recognition in society.

This is how the system of Native Administration works in the Uganda Protectorate — over the heads of the natives, and according to the wishes of the foreign rulers. The following extracts from the Uganda Government publication entitled *Native Administration* which is meant for "Office Use Only" is revealing in this context, although it speaks most specifically for the Acholi society.[203]

> "There are, in any case, uncommonly few 'primitive people' left in Uganda, and they are rapidly becoming fewer. . . . (But) in spite of the diminution of family and clan solidarity, especially economic solidarity, resulting from the introduction of economic crops and currency, . . . in some cases, as in Acholi, a reconstruction has already been carried out with the object of making the old tribal organisation and authority the basis of the present Native Administration."

Not only the present Native Administration in these societies is based on the skeleton of the "old tribal organisation and authority", but it has been so from the beginning of British rule in the Uganda Protectorate as the above discussion has indicated. This will be more clearly revealed while discussing the role of the exported Baganda "chiefs" in these societies.

Who then are these "new" chiefs, either in those societies where chiefdom was present before or where it was created under the auspices of Native Administration ? As Hailey's account of them clearly reveals, barring a few who had a traditional position in society in earlier times, they were local policemen, or interpreters of some British officials, or clerks in the administrative departments, or at least had already given some other form of valuable service to the colonial regime. These individuals could thus be judged by the foreign administrators, and it was not difficult to instal them into office.

Moreover, thus appointed as "chiefs", they became no better than the lowest grades of civil servants in the colonial administration of Uganda. They are at the beck and call of the foreign administrators and rule the people over their heads — irrespective of the fact whether the people like to be governed by them or not. They are transferred from one place to another without any regard to what the people desire. And they are promoted or demoted — not as wished by the people in token of their appreciation of the services rendered to them by these "chiefs", but as the foreign rulers think best in relation to whether these individuals are most suitable to uphold the colonial interests of the British. These features of "Native Administration", of which something has already been noted in the above extracts, will be more clearly seen from the following pages.

Thus, in the Karamoja District, "the county (ekapalon) chiefs have been selected from local men who have served in the police or as interpreters, except for one who is an Acholi, in charge of a county for which it is not possible to find locally a suitable county head. The second class (jakait), are selected from local men, many of whom have been in the Native Administration police, and they are not treated as transferable. The headmen (wakungu) are mostly local men who have been members of the police or other services"[204].

Among the Acholi the author found in 1950 that out of the 6 and 22 county and sub-county chiefs respectively, 3 and 20 belonged to other clan or lineage areas. And all of them had either previously worked for the colonial administration or were near relatives (such as son or brother) of those who were also "appointed" chiefs.

Similarly, in the Lango District, "the county chiefs have for the most part been promoted from the second class (jagi) chiefs, the latter having been selected on personal grounds, but a number are clan headmen, not serving in their own areas. The jan jegi (sub-county) chiefs are local men, but liable to transfer"[205].

Also, in the West Nile District of the Northern Province where all chiefs are "appointed" except two hereditary ones, in 1946, "following on representations made in the District Native Council, a beginning was made in the process of transfer"[206]. It has also been noted before on what grounds these "chiefs" were selected.

Even in the district of Busoga, promotion to the class of county chiefs, "as in other classes of chiefs, is subject to passing an efficiency bar"[207]. And they are of course liable to transfer.

Then in the "Bukedi Native Administration of the Mbale District the chiefs are subject to the same conditions as to promotion, transfer, pensions and house accommodation as in Busoga"[208].

About the Kigezi and the Teso Districts, and the Bugishu Native Administration it has already been noted how the "chiefs" were manufactured and how they are liable to transfer, promotion, and such conditions and facilities applicable to other low-graded civil servants.

Indeed, this has been the general characteristic of Native Administration all over the Protectorate. Therefore to quote finally from Hailey's description of the system of Native Administration adopted in Uganda:[209]

> "... apart from the representatives of certain Ruling Houses, the personnel of its executive organization, unlike the personnel employed in systems which make use of traditional Native Authorities, are not now the customary representatives of a tribal or clan community and cannot command its instinctive support. Nor has the system made in the past any definite provision for the association of competent non-officials or representatives of popular opinion in the conduct of local affairs."

There lies the peculiar characteristic of the system of Native Administration in Uganda. It can neither command "instinctive support" from the traditional representatives of these societies nor is it based on a popular representation of the people. Its creation was not intended to fit either to traditional systems of these societies or to lead the people to a progressive form of government. It was just a measure suitable to the colonial regime in order to keep the people under control; a measure needed by the ruling class of Britain, as will be explained later, in order to exploit the resources and labour of these areas.

Indeed, how this system differs from the traditional institution can still be observed in those areas where chiefdoms were present before British rule and where still now there are a few traditional chiefs. In these regions, in the *bush*, it is found that an old chief would be respectful to an ordinary official from the Protectorate Administration, provided due regard is paid to his office. But a new chief, more often than not, is always anxious to serve as a "glorified clerk" of the present regime.

And that this native bureaucracy hardly at all depends on popular will needs no further comment.

To sum-up then, like the system of Indirect Rule, the introduction of Native Administration, whether or not political authority of chiefs existed in the areas concerned, did not lead to a progressive development of these societies in their political sphere. The point is not whether the traditional institutions were preserved or not. As noted at the beginning, when the British appeared in Uganda the central demand of the Africans in their political life was progressive and democratic governments. But not only was this task not undertaken by the British, but what they provided instead was even worse than what had gone before. In the name of preserving the traditional institutions they made an open mockery of them, and this was true for all societies in Uganda, whether they were at the feudal stage of development, or were in the stage of military democracy, or even in their tribal organisation. Nowhere was popular representation in the picture. On the contrary, whatever democracy these people enjoyed under military democracy or tribal organisation was now taken away from them. And in its place were placed firstly a host of native officials answerable to a foreign regime and not the peoples, and on top of all stood the supreme authority of the colonial power. Thus, for all peoples of Uganda the political changes under British rule were a mark of retrogression and not of progress.

How the native bureaucrats in the areas under Indirect Rule enjoy their exalted position at the expense of the people has been discussed earlier. In the areas under Native Administration also they were assured of a comfortable living by the foreign rulers, the source of such living being, as in the other areas, the labour of the common people. But in these societies the native bureaucracy was not immediately developed into a landed aristocracy, with the exception of the "chiefs" in Busoga. The special position of Busoga in this respect has been discussed before. Apart from Busoga, in a sense this policy of creating landlords was not particularly necessary for these "simpler" societies, for they were not so developed as the "kingdoms". Hence, such a thorough arrangement to stabilise the interest of the native rulers in the continuation of British rule was not called for. There was also the argument, probably not less important, that here the social organisations were not so autocratic as in the developed kingdoms. Consequently, the economic relations within the societies could not be so easily altered as the latter, and at the same time the facade of preserving the traditional system could be maintained.

Thus, while it is true that the common people in all areas of Uganda had similar ways of living and everywhere land was held communally by the clans or lineages, because of the presence of "servient peasantry" under the chiefs in the developed kingdoms, land-relations in the latter areas could be comparatively easily changed without the effect of such a change being at once felt by the people. Needless to say, the difference in the systems of land tenure in the "kingdoms" and in other societies of Uganda should not be exaggarated, as is often done by some authorities; but, however small, such differences should be taken into account, for they had some bearing upon the development of native bureaucracies in these two areas in somewhat different ways.

Referring to these two systems of land tenure in Uganda before the advent of the British. Thomas and Spencer noted[210]:

> "It is a convenient generalisation to regard land in Uganda as falling, prior to the British occupation, under one or other of two land tenure systems corresponding to two social systems — the one aristocratic, the other democratic. Under the former a dominant race or ruling family is associated with a servient peasantry, conditions which obtained wherever Bahima influence had established itself. Under the latter, characteristically among Nilotic tribes, the individual occupier was essentially a free cultivator."

However, as noted before, in both areas land belonged to the people living in their clan or lineage areas or in their family-groups. But because class differentiation among the peoples like the Nilotes was less marked than, say, among the Baganda, they could easily be distinguished as "essentially a free cultivator".

Because of this difference in the stages of development, the native bureaucracies of these "simpler" societies could not profit so much as those in the developed kingdoms. They could not at once develop into a landed aristocracy after British occupation, though as it will be noted later such tendencies are not altogether absent today when colonial exploitation through cotton cultivation has made deep in-roads into the Uganda society. As a result, for the top stratum of some of these societies also it has become a lucratcive source of income to become big land-holders and eventually turn into landlords. Naturally it is the native bureaucracy who could best utilise the situation in their favour, but in the beginning they neither had any such scope nor was there the need for it so long as cotton cultivation had not permeated all over the Protectorate and spread into the remote corners of the colony.

However, that does not mean that from the beginning these native bureaucrats had no incentive to align themselves with the foreign rulers. They had always full access to the other important emolument enjoyed by all the native rulers under colonial administration, namely a portion of the surplus labour of the people transformed into the colossal burden of government taxes. How heavy is this tax burden on the people and how they are liable to other forms of exactions will be discussed later. But here it may be of interest to note what proporation of the incomes of the native governments this direct taxation accounts for, and how much of the incomes go to these bureaucrats as their personal emoluments. The following table in this connection has been prepared from the "latest available figures" supplied by Hailey, which in most cases refer to the year 1946[211].

Table IV.2

District	Percentage of taxation to total income	Percentage of personal emoluments of native rulers to total expenditure
(1)	(2)	(3)
Kigezi	85	42
Busoga	82	44
Teso	76	42
Bukedi	71	63
Bugishu	70	53
West Nile	76	37
Acholi	55	32
Lango	49	39
Karamajong	48	41

The table shows that even in the most barren district of Karamajong direct taxation of the people accounts for nearly half of the total income of the native government, and in the fertile districts of Kigezi or Busoga more than four-fifths of the total income of the native governments are so accounted for. The above figures of course do not give a complete picture of the tax burden of the people, because besides filling the coffers of the native governments' treasuries they have to provide for the Protectorate Government run directly by the foreign rulers, and then there are many other forms of direct and indirect taxation which will be discussed later. But as far as the native governments are concerned, the table shows how their treasuries are furnished. And it will also be seen from the table that a great bulk of this income of the native governments is spent as personal emoluments of the native bureaucracies, leaving very little for other items of expenditure.

Of the other items of expenditure, police and prisons (to maintain colonial law and order) call for substantial amounts. But without going into that discussion here, it will be noticed from the table that from all these districts one-third or more of the total expenditure is ear-marked to meet the salaries, pensions, etc., of the native bureaucrats, and in Bukedi it is as high as 63 per cent. Such figures, of course, do not take account of other privileges enjoyed by these "chiefs" (officially acknowledged or not), such as use of labour for their personal needs, maintenance of their houses, etc.; yet they give an idea of the social character of the existence of these local rulers. For this reason, these bureaucrats, like their counterparts in the areas under Indirect Rule, remain loyal to the British regime with which their prosperity and future became inseparably linked up.

In this way, in the "simpler" societies also a wall of defence was built up by the colonial administration, which indeed was the real objective behind the introduction of the system of Native Administration in Uganda.

8. Export of Baganda "Chiefs"

Reference has been made to the fact that while introducing the system of Native Administration to the "simpler" societies of Uganda, reliable Baganda dressed-up as "chiefs" were sent to many of them. What was the purpose behind this move which made an open mockery of Native Administration as preserving the "old tribal organisation and authority"? An examination of it will make it further clear that behind the declaration of upholding the traditional rights and usages of the people there was the shrewd policy of the foreign rulers to consolidate their empire in Uganda and to prepare the country for colonial exploitation. The second part of the programme will be discussed later. But here it may be of interest to explain what role these Baganda played in the political life of these people, until the time was ripe for the creation of "chiefs" from within these societies, when the Baganda Agents were either unobstrusively withdrawn from the scene or if they protested vigorously they were appeased with a suitable reward.

For several reasons, in the less developed areas in Uganda, the British found the Baganda most suitable for the purpose of teaching the local chiefs the intricacies of colonial administration or to impose the system of chief's rule in places where there was no precedent. Firstly, they could be relied upon and properly selected because of their long association with the British. It should be borne in mind that the Christian missionaries had already worked in Buganda for a quarter of a century duiing which time they dominated Baganda life to the extent that even family-bonds were torn between the politico-religious factions of Wa-Ingleza and Wa-Fransa, and finally the Wa-Ingleza faction had won the battle. Thereby the influence of the British was decisively asserted over the Baganda. Secondly, it was in Buganda that, out of the previous rulers and loyal supporters of the British, the most powerful native bureaucracy in the Protectorate was created. Therefore, these Baganda could now serve as the most faithful advanced guard of the foreign power to bring the whole territory under complete subjugation. Hence Baganda "chiefs" were exported to other regions to establish the *pax Britanica.*

They were sent to developed areas, like Busoga and Ankole, in order to teach the local rulers the arts of Native Administration; and they were also sent to "simpler" societies to impose chief's rule. Thus, in the Mbale District of the Eastern Province, where live the Bagishu, Bakedi, Bagwere, Banyuli, Jopadhola, Bagwe, Basamia, etc., "the British and their Baganda Agents (who continued to be employed in decreased numbers until the early thirties) introduced the system of Buganda administration"[212]. Then they were in Kigezi, Teso, and some other districts of Uganda as mentioned earlier. The result of their imposition was obvious. To take the Iteso, for instance, at the beginning of the century they "were administered by Baganda agents, and the indigenous clan and community system was heavily overlaid with a Baganda-like political structure"[213]. With this importation of Baganda "chiefs", the age-class organisation of the Teso society was suppressed for good "by Kakunguru's conquering Baganda at the turn of the century though they are still practised in Usuku country"[214].

What was the role of these Baganda? According to Buell, who in general was not opposed to British Colonial Administration of Uganda, the role was to train the "local chiefs to meet the responsibilities connected with tax collection and the administration of justice". In other words, to pave the way for systematic colonial rule. To quote Buell more fully[215]:

> "It has been the policy of the British Government to introduce the system of native administration worked out so successfully in Buganda, into the other parts of the Protectorate where it did not originally exist. Thus it has established Native Courts in some fifteen different districts, and recognized and developed the power of local chiefs. These chiefs are assisted by native advisers or agents, who have usually been members of the ruling class in Buganda. These advisers differ from the German *akidas* in Tanganyika in that they are supposedly advisers, and not chiefs. Many of them have been tempted, however, to assume responsibility rather than merely control. While opinion is not unanimous, it appears that these agents under close and continuous supervision by district officers, have been more successful than European advisers would have been in training local chiefs to meet the responsibilities connected with tax collection and the administration of justice. At the present time, the Buganda advisers have been withdrawn except in certain backward areas where "county agents" still advise the saza chiefs and Lukikos. In Teso, all except two such advisers have been withdrawn, and the native chiefs are standing on their own feet. About six thousand pounds is still expended annually upon these agents and followers."

Thus, even after nearly three decades of British occupation of Uganda (when the above lines were written), in order to further the cause of the British, and since the Baganda Agents can do this better than the "European advisers", the Protectorate Government maintained, at a colossal expense, suitable Baganda as "chiefs" outside their jurisdiction.

There was another reason for exporting Baganda "chiefs". This was to keep the peoples of these societies under control. Indeed, from the beginning the foreign rulers felt the need for sending the Baganda "chiefs" to those areas where whatever might have been the attitude of the people before the British began their final onslaught in Uganda, their role in Buganda and Bunyoro had opened the eyes of the neighbouring people. Thus, although Roscoe justified the role of the British in Uganda, he had to note[216]:

> "In past years before the British occupation of the Nile Province it was unsafe to travel in districts inhabited by Bageshu without an armed escort. The tribes in those parts could only think of the white men as a foe, and it was a laudable action to attack and, if possible, to kill him. Reports had gone forth, spread chiefly by the Banyoro, that white men only sought to conquer countries for purposes of plunder and profit; hence tribes, shut off from the old trade routes, and only able to gain information from neighbours nearer Uganda (meaning Buganda — RKM), formed wrong impressions about Europeans, and tried to prevent them from entering their country."

But now under British occupation the situation was changed, and in this process the selected export of Baganda "chiefs" to these and such other societies which were not yet brought under full British control was of great help.

One of these important areas to which these Baganda "chiefs" were sent was the kingdom of Ankole. It has been noted earlier that this kingdom was weaker than the other two kingdoms in this region, so that it suffered at the hands of the Baganda and the Banyoro; and that it was considerably enlarged after it came under British domination, for the British could rely upon the passivity of the Banyankole. In this submission of the Banyankole the Baganda "chiefs" had an important role to play. Just after the British had occupied this area, these people had shown some serious resistance to the foreign regime. Buell had noted[217]:

> "In 1905, the British Government terminated the agreement with Ankole, following the murder of a district commissioner. Half a dozen natives were convicted by the High Court at Entebbe for this murder, but the case was appealed, through the good offices of the Church Missionary Society, to the Court of Appeal at Zanzibar which acquitted

> these natives. Upon their return to Ankole, the government was in a delicate position; the murder has been committed and the guilty person had not been apprehended. It finally decided to terminate the agreement, to deport the natives concerned, and to impose a collective fine of twenty thousand cattle upon the tribe."

A good example of colonial law and order. Even though the suspects were acquitted by the Court of Appeal they were deported, and the fine which was imposed on the people was according to Banyankole currency equivalent to some ten thousand male slaves[218].

No doubt the situation was tense. There was therefore an important reason to send the Baganda "chiefs" there to help in quietening the people. This was of course done successfully. Later, "some of the Ankole prisoners have been allowed to return, while the agreement was restored in 1912"[219]. Roscoe noted in 1921 that during his visit to this kingdom he found that "the king and almost all the chiefs have become Christians, and the country is being developed along lines laid down by British officers"[220]. *Pax Britanica* was thus established with the help of Baganda henchmen.

In Teso too the Baganda "chiefs" played a similar role. Roscoe noted in 1921 about this part of Uganda[221]:

> "The peaceable state in the country has been brought about through British occupation; a wise and tactful Muganda chief, Kakungulu (also known as Kakunguru —RKM), was placed over the country, and without resorting to arms in a single instance he has been able to bring every chief into happy relations with the Government. Roads were made into every part of the district from one chief's residence to another, and Baganda chiefs were established in each district to rule, collect taxes, and organise cotton and rubber plantations. Most of the chiefs appointed were Christian Baganda, who built their little mud and wattle churches, and began a system of daily services, to which numbers of the Bateso (also known as Iteso — RKM) youths went."

So much about the subordination of the Iteso and the religious zeal of the Baganda missionaries, and about the first inkling of the other important mission of the Baganda "chiefs", that is, to initiate the people to the production of rubber and cotton (of which more will be said later while discussing the nature of colonial exploitation in Uganda).

In Busoga, however, the situation was a little complicated. The Basoga had previously volunteered themselves to British rule in order to be relieved of the oppression of the Baganda king and his overlords. But after British occupation the eyes of these people also began to open. A serious famine ravaged the country which, in the words of Roscoe, "was not due to climatic causes, such as lack of rain, but was owing to the discontent among the women and their rebellious spirit, which caused them to refrain from cultivating their plantations because of the forced labour question"[222], which was initiated by the British (for reasons to be explained later) and kept the men away from their homes for long periods. Moreover, under British administration also Baganda "chiefs" were placed over them as in the old days. Naturally the Basoga strongly resented this, and at last the British realised their mistake. Thereafter they withdrew the Baganda "chiefs", and the "native administration" of Busoga was conducted by the hierarchy of Basoga chiefs. Interesting to recall here that besides in the kingdoms of Buganda, Bunyoro, Ankole and Toro, the native bureaucracy in Busoga also was made into a landed aristocracy. By the report of the committee set up to decide the land question, 27 per cent of the total land of Busoga was disposed of "for the most part as the private property of chiefs", as Thomas and Spencer had noted. Thus, by force or by bribery, the British had their way.

Indeed, force and bribery remained the two main courses for the foreign rulers in consolidating their empire in Uganda. Thus, although the Baganda "chiefs" were a great help to the British, occasionally they could create trouble when while looking after the interests of the colonial administration they also wanted to assert their own rights to exploit permanently the peoples placed under them. For instance, Buell has noted[223]:

> "An annual allowance of three hundred pounds is made to Kakunguru, a Bukedi chief. Originally he was a Buganda chief who assisted the British many years ago in the conquest of Bunyoro. Since he was more or less of a rival of the Katikiro (Apolo Kagwa, the devoted follower of the British — RKM), the British took him and four thousand followers to Mount Mbale, and told him that he could govern this country as his own. Kakunguru proceeded to put in the Buganda system of administration, with 'foreign' saza and gombolola chiefs. When in about 1901 the British extended their administration to this country, they encountered the opposition of Kakunguru who insisted that the British had promised that he could always govern it. But the British denied that they had made any such promise. They had come to realize that they could not permanently impose an alien Paramount Chief upon these people. When Kakunguru made trouble for them, they took away his power, but in return made him president of the Busoga Lukiko. Upon his failure here, he was made a saza chief, in which capacity he again made trouble by preaching the Malaki religion (an obscurantist variation of Christianity, supported by thousands of Baganda at that time, probably as a veiled opposition to the colonial rulers who professed "true" Christianity — RKM). His attitude led to his deposition (and putting into jail and eventual release after a hunger-strike — RKM), but he still draws an annual subsidy of three hundred pounds a year."

The above extract gives one a good glimpse of the role of the Baganda "chiefs" and of the consistent policy of the foreign rulers. Kakunguru had assisted the British in reducing the free kingdom of Bunyoro into a district of the colony, so he had to be rewarded and used for future services. But he was a rival of another important ally of the invaders, namely, the then Prime Minister of Buganda. Therefore he was removed from Buganda and made the ruler in the territory which for the time being the British could not bring under direct control. Here the despot ruled autocratically, while indirectly fulfilling the British mission by uprooting the traditional institutions of the local societies and imposing chief's rule which could later blossom forth under the aegis of Native Administration. So long as the British were busy in consolidating their rule in the other important parts of Uganda, such as Buganda, Bunyoro, etc., they did not concern themselves with bringing "civilisation" to this region. But in due course of time they appeared here to remove the "native" despot and to impose their rule instead. All their previous promises then went overboard, and they removed him. But when he was found to be in a position to start a serious opposition with the local malcontents who were restless under British rule, the shrewd rulers isolated him and silenced him with a substantial reward of "three hundred pounds a year."

It is true that not all the Baganda "chiefs" were given high promises and made into Paramount Chiefs even for a time. Kakunguru's was a special case, arising out of his special position in the Buganda society and his large number of followers. Otherwise, as a rule, the Baganda "chiefs", while sent out to those areas which were still "unsafe" for the foreign rulers even though they had been brought under British domination, were asked to bring the people under control without any promise of paramountcy. Therefore, such exigencies as with Kakunguru did not occur very frequently. But it is equally true that having the control of the entire territory, it was not difficult for the foreign rulers to solve such exigencies by additional force and sometimes with additional bribes.

In this way, the British utilised the loyal Baganda as harbingers of their colonial rule in the Uganda Protectorate, to create their allies and thus finally to stabilise their position over the whole area. In return, the devoted servants got their reward — in bribes or otherwise.

9. Colonial Revenue

And who supplied the bribe-money? Who fed and fattened the colonial bureaucracy and provided for running the colonial administration? It was the peoples of Uganda to keep whom under control the very apparatus was established.

How is the colonial revenue raised? Various forms of direct and indirect taxation were imposed on these peoples from the beginning of British rule. It has been noted earlier that before the advent of the British the peoples in the developed kingdoms had to pay regular tributes to the monarchs and chiefs, and in the established chiefdoms also they were under customary obligation to pay tributes. Such payments in a non-cash economy was not properly organised; but now not only the system was well arranged in the developed kingdoms and chiefdoms, but it was also spread to those parts where compulsory payment of tributes or taxation in a class society had not yet come into force.

To begin with, according to the 1900 Agreement discussed on earlier, a hut-tax of three rupees or four shillings on each hut or dwelling place was levied in the kingdom of Buganda. Besides filling the coffers of the Colonial Treasury, this had the purpose of forcing the introduction of a money economy among the Uganda Africans, which is so necessary for the flowering of a colonial economy. Therefore, the hut-tax was soon spread over other areas. Brode noted in 1911 that "in Uganda as much as Rs. 5 is levied in the more developed provinces"[224]. The administrators, however, were concerned that in the early stage of their rule this "new impost" should be imposed with caution, so that the people would not rebel against the burden. Therefore:

> "To facilitate the collection, it was also provided that the tax could be paid in kind, or that, in lieu of paying money, one could work for the Government during a certain time. In British East Africa a labour period of one month is regarded as equal to a tax of Rs. 3."[225]

This commutation of tax into labour had a significant bearing upon "developing" Uganda by building roads, railways, and trade-centres as preparatory measures to the introduction of a colonial economy. Of this more will be said later.

The hut-tax, however, soon fell into disfavour with the administrators. Brode noted[226]:

> "The hut-tax was in the earliest time thought to be the best system of taxation, because the number of the existing huts gave the surest base of control; but, on the other hand, it had its drawbacks, because it charged the domiciled people, whilst a great number of workmen, especially most of the porters, escaped free. A disinclination to create new huts was soon remarked, and it was often found that several families shared one hut in order to avoid the taxation.
>
> So, with the extension of administrative influence, the somewhat unjust hut-tax system has been augmented by a subsidiary poll-tax."

This "just" system of poll-tax could embrace all the people; not only those who stayed at home to produce crops, but also those who were drafted into forced labour to pave the way for British colonial exploitation. Moreover, the poll-tax being of a much higher amount than the hut-tax it could provide the Treasury with a big revenue. The government therefore altogether abandoned the hut-tax system in favour of the poll-tax.

The introduction of poll-tax had another advantage. A proper counting of the people could be henceforth made in order to organise a more efficient supply of forced-labour from the people. Thus Brode further noted[227]:

> "Apart from the good effect which a regular census has in accustoming the natives to work, it is a favourable source of revenue to the Governments."

The full implication of this "census" will be realised later when the economic "development" of Uganda under British rule will be dealt with.

To discuss now how the poll-tax became "a favourable source of revenue to the Governments", when this system was first introduced in Buganda by the Agreement of 1910, it was at the rate of 10 shillings. Later, it was "again raised to a maximum of 15 s. in an agreement of 1920"[228]. This rate was spread "throughout Ankole, Bunyoro, Buganda, and the greater part of the Eastern Province", and elsewhere it ranged "from six to ten shillings, according to the relative wealth of the district concerned"[229]. Later, this amount was further increased in some areas; for instance, since the last war the figure for the Acholi and Lango districts in the Northern Province is 15 shillings. Evidently, it has become a profitable source of income to the efficient administration. In 1926, out of the total Protectorate revenue of £ 1,306,761, the poll-tax accounted for £ 401,500, or 30.7 per cent[230]; in 1936, out of £ 1,712,940, it provided for £ 591,000, or 34.5 per cent[231]. In the post-war years also, even though the Protectorate revenue had very markedly increased from the items of indirect taxation (such as "customs and excise", "export taxes" on cotton, etc.), only the "Native Poll Tax", that is, excluding others who have also to pay a poll-tax, accounted for 15.1 per cent of the revenue in 1946[232].

What is of greater importance to note is that while the rate of taxation of the Uganda Africans was increased in course of time, the income-tax payable by the non-natives in Uganda was much less than the rates in the metropolitan country. This can been seen from the following table which has been deduced from the table on page 31 in the Annual Colonial Report of the British Government on Uganda for the year 1951 and the Notice on Income Tax Assessment in the United Kingdom for 1951—52.

Table IV.3

Income Tax Payable by 5 Classes of Tax Payers in Uganda and U.K.								
Class of Tax Payers	Annual Salary in pounds sterling							
	400		800		1200		1600	
	Uganda. £ — s	U.K. £ — s	Uganda. £ — s	U.K. £ — s	Uganda. £ — s	U.K. £ — s	Uganda. £ — s	U.K. £ — s
(1)	(2)	(3)	(4)	(5)	(6)	(7)	(8)	(9)
Single man	18—15	51—10	72—13	195—10	150— 0	347—10	240—0	499—10
Married man	3—15	29—10	35— 3	157—10	87—13	309—10	160—3	461—10
Married man with one child	nil	10— 5	24—15	124— 5	69—16	276— 5	136—6	428— 5
Married man with two children	nil	nil	20— 5	91— 0	61—11	243— 0	125—1	395— 0
Married man with three children	nil	nil	15—15	59—15	53—15	209—15	114—5	361—15

The table is revealing.

Because of such discrimination, even though the incomes of the foreigners in Uganda are very many times higher than that of the Africans, of the total amount realised as direct taxes in 1934, that from the Africans accounted for 95 per cent[233]. This shows how the colonial treasury used to be furnished until with the rapid growth of Uganda's colonial economy (which will be described later) indirect taxation of the peoples of Uganda became the chief source of colonial revenue.

Indeed, discrimination in direct taxation is not the only, or even the primary, feature of Uganda's colonial revenue today. This is much more marked in the scheme of indirect taxation in the Protectorate from which the colonial administration draws the bulk of its revenue. To give only a few notable examples, one of the forms of indirect taxation is the export-tax on African-grown cotton. This tax was imposed in 1918—19 for the first time, and is continued ever since. In 1929 the cotton-tax provided the colonial treasury with the sum of £ 231,000; and although due to a temporary fall in the world price of cotton it came down to £ 136,000 in 1930 and £ 129,000 in 1936, yet it accounted for an appreciable proportion of the Protectorate revenue, such as in 1936 for 7.5 per cent of the total revenue of £ 1,712,940 of the Protectorate[234]. In 1946, the "Export Tax" accounted for 14.6 per cent of the total revenue; in 1951, the proportion rose to 51.5 per cent[235].

Like African-grown cotton, while the African-grown coffee is also fully subject to export tax, "under the Coffee Export Duty Ordinance 1953, non-native growers will pay export duty but will get a refund of half the duty if they pay income tax"[236]. The non-native coffee-growers in the Protectorate are essentially the British who own about 93 per cent of Uganda's "non-native acreage under coffee". Since all of them earn a pretty good sum in the colony, they pay income tax. But not only their rate of income tax is much lower in the colony than in the mother country, but while paying income tax at a lower rate they further obtain a rebate on the export tax on coffee. The poor African growers, however, are not entitled to such benefits.

Another source of indirect taxation, which again affects the Africans adversely, is the differential rates of freightage in the State-owned Kenya-Uganda Railway. Hailey noted[237]:

> "The (Kenya-Uganda) railway is primarily a goods, not a passenger line, and rates have been adjusted to enable the principal crops of cotton from Uganda, and maize and coffee from Kenya, to be exported as economically as possible; it is therefore the policy to counter-balance low export rates, particularly on maize from Kenya and cotton from Uganda, by high import rates, . . ."

Export of cotton, as will be explained later, is the responsibility of the non-natives of Uganda. So the Uganda Africans do not benefit by the low freightage rate for export. But the bulk of the imported consumer goods — cotton piece-goods being the most important item of import in Uganda[238] — is sold to them. Thus they are directly affected by the higher rates of import freightage.

More revealing is the differential tariff in the British-owned territories in East Africa. Leys noted[239]:

> "The Kenya-Uganda tariff heavily taxes, not only luxuries, but all the articles necessary to civilised life, clothing, furniture and hardware, as well as flour and every sort of food-stuff. But its most remarkable feature is its free list. That list includes goods valued at 40 per cent of the total imports. In any ordinary country it would consist of the goods that the poorest must or ought to be encouraged to consume — tinned milk and calico, for instance. But none of such things is exempt from duty. Books and drugs are the only items on the free list that conceivably might, to some extent, reach African hands. Instead, the free list is almost entirely composed of articles that only Europeans buy, especially those of them who have land — such as lorries, machinery and fencing materials. . . . . All East African life in epitome lies in the facts that the bicycle to be used by the peasant in Uganda who grows cotton is taxed, while the European landowner's tractor is admitted duty free. . . . Hardly any East Africans can afford artificial light. The few use candles. The candles they use pay a higher duty than those intended for use in mining gold."

One should add to the above the information supplied by the Government Treasurer of the Uganda Protectorate, viz: "The articles in the tariff most productive of Uganda revenue are mainly those which concern native clothing"[240], and note further that imported cotton piece-goods in the Protectorate accounted for 32 % $\pm$ 1.5 % of the total import value in each year during the post-war period of 1944—47[241]; and then the picture is complete.

Thanks to this form of indirect taxation, "customs and excise" and "export taxes" accounted for 36.5 per cent of Uganda revenue in 1936*; in 1946, the figure was of 51.7 per cent; and in 1951 it rose to 75.1 per cent[242]. Simultaneously, direct taxation, in the form of Native Poll Tax, Non-native Poll Tax, Income Tax, and Estate Duty accounted for 34.5 per cent of the Protectorate revenue in 1936*, 24.5 per cent in 1946, and 7.9 per cent in 1951[243].

These figures should be compared against the composition of Great Britain's revenue from direct and indirect taxation, which for the year 1938 were 386 and 337 million pounds sterling, for 1944 — 1,869 and 1,088 million pounds sterling, and for 1951 — 1,973 and 1,348 million pounds sterling, respectively. Although not always for the same financial years, a comparison of the above figures for nearly the same periods would be revealing, and is therefore given below.

Table IV.4

Composition of Government Revenue from	Percentage of total revenue					
	Pre-war Period		Just after war		Present — 1951	
	Great Britain (1938)	Uganda (1936)	Great Britain (1944)	Uganda (1946)	Great Britain	Uganda
(1)	(2)	(3)	(4)	(5)	(6)	(7)
Direct Taxes	44	35	58	25	51	8
Indirect Taxes	39	37	33	52	35	75

It will be noticed from the table that while in the metropolitan country the indirect taxes, which affect the general run of the people, have been a little lessened in course of years, in the colony they have been doubled. On the other hand, while in regard to direct taxation the total share of non-natives has become greater than that of the Africans since 1949, its contribution to Uganda's total revenue has become negligible. In 1951, the share of non-natives in direct taxes was 60 per cent, but the total amount realised as direct taxes accounted for only 8 per cent as noted above. An essential characteristic of colonial revenue is thus revealed by the above table, particularly if it is borne in mind that the increasing burden of indirect taxation in Uganda lies primarily on the shoulders of the African peoples only.

It should further be noted that the burdens enumerated above are not the only burdens the Uganda Africans have to bear under British rule. As discussed before, in addition, the peoples of Uganda have to pay for the maintenance of the local bureaucracies. For this purpose, since the British occupation of Uganda, several other forms of taxes have been levied on the people, of which the following are most notable.

In Buganda a land-tax was imposed when the land system within the kingdom was drastically changed in order to develop a landed aristocracy. Hailey noted[244]:

> "Though there is no tax on plural wives, there are other demands on the native which materially raise the incidence of taxation. The Buganda landlords have in virtue of their quasi-freehold tenure, the right to take from tenants a rent fixed by law at 10s.; a land tax is levied by the native government on owners of more than 10 acres of land. The rate

* The figures for 1936 are probably underestimates, for the available data do not take account of taxes other than on cotton, and income tax and estate duties. But since the underestimation applies both to direct and indirect taxation, the figures for 1936 may not be regarded as totally incomparable with the other figures.

is 5s. for any landlords with fewer than five tenants; for others, 15s. or 25s. according to the district. In addition, 1s. 6d. out of every rent payment is payable to the native treasury, and is in fact collected by their agents from the tenants, who themselves deduct it from the rent payable to the landlord."

Thus, all taxation comes from the common people. In 1935, 41 per cent of the revenue of the Buganda Native Government was accounted for by this land-tax, fees and fines from the judicial courts, market dues and licences[245]. In 1947, out of the total estimated revenue of the Buganda Government of £ 274,766, £ 176,843, or 64.4 per cent, was from this land-tax and cash-commutation of slave-labour, *luwalo,* which was increased from 10 shillings to 14 shillings in 1947[246].

More will be said later on the slave-labour, *luwalo*, and on the pernicious effect of the land-tax and the British-introduced land-system in Buganda. But here it should be noted that except in Buganda this form of land-tax was not introduced in other parts of the Protectorate. However, that did not absolve the colonial administrators and the native bureaucracies from having access to various other forms of taxes and tributes. For instance, Hailey noted[247]:

"Outside Buganda occupiers of land do not pay rent, but tribute ranging from 1s. to 7s., known as *busulu,* is collected for the purpose of the native administration; the minimum rate payable in combined poll-tax and *busulu* is 21s. in Busoga."

This, however, is just one of the ways of increasing the tax-burden on the people. Thus, in Busoga, in addition to the 21 shilling of Native Poll Tax per adult male, "there is a Native Administration tax of 7s., imposed under the Native Administration Tax Ordinance 16 of 1938. In Mbale and Teso the corresponding figures are 18s., 4½ s., and 7s."[248]. Similarly, "The Native Administration Tax was 13s. in Bunyoro and 6s. elsewhere" in the Western Province, that is, in the districts of Toro, Ankole and Kigezi[249]. Then, for the Northern Province Hailey has noted[250]:

"In Acholi and Lango Districts the Protectorate Poll Tax is 15s., of which the Native Treasury receives 3s., the Native Administration Tax being 7s. In West Nile the Protectorate Poll Tax is 8s., of which the Native Treasury receives 20 per cent.; the Native Administration Tax is 8s. In Karamoja, the Protectorate Poll Tax is 9s., of which the Native Treasury share is 5s.; a Native Administration Tax of 3s. was imposed for the first time in 1947."

This impost of additional taxation on the Uganda Africans, in order to feed the local allies of the foreign government is nearly equal in amount to the poll-tax, of which the bulk goes to the Protectorate Government and a share to the Native Administration. It can be worked out from the figures supplied by Hailey and the Annual Colonial Reports of the British Government on Uganda that while in 1946 and 1947 the Native Poll Tax amounted to £ 613,563 and £ 662,002, respectively, the Native Administration Tax per year was of the order of £ 409,011[251]. Details are given in the table below.

Table IV.5

Provinces of Uganda Protectorate	Native Administration Tax (in £)	Remarks
(1)	(2)	(3)
Buganda	176,843	Estimate for 1947,
Eastern	116,775	"Latest figures"
Western	63,812	Estimate for 1946
Northern	51,581	Actual for 1946
Uganda Protectorate	409,011	[cf. *loc. cit.* 134, pp. 17, 36, 55, 66]

This is how the foreign rulers and their local henchmen live on the people.

Furthermore, how heavy this tax burden is on the peoples of Uganda may not be fully realised solely from the above enumeration of different forms of taxation in the Protectorate. For this reason it is of greater importance to note what the Government Treasurer noted in 1936 in his report on revenue and taxation in Uganda. According to him, 85 per cent of Uganda's revenue used to be derived from taxation in pre-war days. In 1951 also the proportion remained practically the same[252], while the relative importance of indirect and direct taxes had become markedly aggravated.

And how does taxation affect the Africans? To quote from the Treasurer's report[253]:

> "In Uganda the bulk of the taxation is paid by large numbers in rural areas and the amount paid by each individual represents a very large proportion of his money income; in many cases the proportion approaches 100 per cent .... In Uganda taxation is the principal incentive to labour and production."

Comments would be superfluous.

10. Colonial Expenditure

How is this revenue, extracted from the African peoples, spent in Uganda? How much of it goes back to the people in order to improve their socio-economic life? This is a question which would naturally follow from the previous account of heavy taxation of the Uganda Africans. In the following passages it will be shown that here also one finds the essential colonial characteristics of public expenditure in the Protectorate. The bulk of the expenses is in the running of the colonial administration or for maintaining law and order (in other words, to keep the people down under foreign rulers), and the remaining small sum is spent on social or economic developments. But here again the colonial officers and their local agents, who organise and control such socio-economic services, claim the greater part of the expenses on this account. Consequently, the net benefit accruing to the common people of Uganda from "public expenditure" is very little indeed.

To give some instances, in 1926, out of the total expenditure for the Protectorate, 53 per cent was spent on "General Administration", 6.7 per cent on "Police and Prison", and 5.8 per cent on "Military"[254]. These figures relate only to what was described as "ordinary expenditures excluding of railways". The railways, of course, accounted for a reasonable share of colonial expenditure, and, as will be shown in the following, besides affecting the Africans adversely by the differential rates of freightage, it is primarily meant to serve the colonial economy, that is, the interests of the foreign capitalists. However, one may exclude notice of the expenditure on railways; but one cannot fail to note how little was spent on the economic and cultural development of the people. And in that respect, in the year 1926, which was in no way different from any other year at that time, while 12.5 per cent of the total expenditure was on police, prisons and military (that is, to maintain colonial law and order), the combined expenses for economic and cultural developments (under the headings of "agriculture, veterinary and forest" and "education") was only 10.5 per cent; the former claiming 6.95 per cent and the latter 3.55 per cent of the total budget[255]. And this is so in a country where from the beginning of colonisation the professed aim has been to educate the "savages" and to uplift them from their "primitive" socio-economic life.

Even a decade later the situation was not very different. In 1937, out of the total budgeted expenditure, 74 per cent was ear-marked for "recurrent" expenditures like above[256]. Here again, from a partial analysis of the expenditure on the basis of the estimates of 1936—37, it is seen that the proportionate expenditure on "economic development" was lower than the combined expenditure on "debt charges other than railways", "defence" and "non-effective pensions, etc.". If the debt charges for railways could be taken into account, the comparison would have been too striking. Even without it, the former figure is 14.7 per cent, and the latter 17.3 per cent of the total estimated expenditure[257]. Furthermore, it should be borne in mind that very high salaries paid to the British officers placed in charge of such "economic developments" (and other advantageous conditions enjoyed by them, such as special allowances, etc.), tremendously inflate the real value which accrue to the African people from the expenditure on such developments. From the data available it is not possible to extricate such charges, but it can safely be said that in terms of money, the benefit going to the Africans is much less than the printed figure of 14.7 per cent of the colonial budget. The expenditure of 26.2 per cent of the total budget on "social services" in this year should also be accepted with the same proviso. And, in any case, it is worthy of note that while in that year 41.8 per cent of the total budget was expended on "administration", less than that amount, or 40.9 per cent, was spent on "economic development" and "social services". Furthermore, it should be borne in mind that in such "economic developments" and "social services" priority is given to the urban areas, whereby principally it is the non-natives inhabiting these places who benefit from the amenieties of civilisation introduced into Uganda by means of such schemes. The Africans continue with their "traditional" ways of living.

In recent times also the administration runs on the same lines. Even as late as in 1951 Uganda remained purely an agricultural country. Hailey noted in 1950[258]:

> "Agricultural production is therefore the predominant asset of the country and this, with the exception of the sugar and tea estates and a number of coffee estates, is in the hands of Africans."

Moreover, it will be shown later that while peasant-cultivation by the Uganda Africans is virtually the only form of production in Uganda even today, such is the development of the productive forces in the Protectorate that the only item considered by the Government Agricultural Farm in working out the cost of cultivation is human labour. And while Uganda thus remains purely an agricultural country and her productive forces are so ill-developed, only 4 per cent of the budgeted expenditure for 1951 was ear-marked for "agriculture"[259].

Today, even after more than half a century of British rule, Uganda remains practically bereft of any education. According to official literature, 6 out of every 14 African children attend school at some time or another; but of them less than 1 in 12 complete a four-year course. While such is the officially declared scope of education for the Uganda Africans, it can be worked out from the data supplied by authoritative sources that in Uganda three in a thousand of the population aged between 6 and 15 attend any educational institution[260]. The above figures however include also the European and Asian children who, while giving first priority to the European (mainly British) children, usurp the greater part of educational facilities in Uganda. In 1950, out of the total Government Recurrent Expenditure of £ 566,307 on education, 6 per cent went to the European community composing 0.1 per cent of the total population of Uganda (of whom again a substantial number spend their service period in Uganda without families or leave their children in the mother country or in Kenya for "proper" education), 14 per cent went to the Asians composing 0.8 per cent of the total population, and 76 per cent were spent on the Africans composing 99.1 per cent of the total population[261]. The result is that illiteracy is rampant in the Protectorate. Even so, only 5 per cent of the budgeted expenditure for 1951 was ear-marked for "education"[262].

And then, even from such a paltry allotment, proportionately much more was spent on the non-African (and particularly European) children than on the natives of Uganda. As can be worked out from the official figures, out of the total Government Recurrent Expenditure of £ 639,793 in 1951, the expenditure per head of the 343 European children for primary education only was £ 112; the same per head of 8,791 Asian children for both primary and secondary education was £ 10; and for some 414,528 African students* from the primary school to the university standard at Makerere College in Uganda (and including the secondary and technical schools) the per head expenditure was £ 1[263].

To consider next the physical well-being of the peoples of Uganda, the Africans are ridden with horrible diseases even though the British rulers have been there since a long time. Over half the population suffers from tuberculosis, 2 per cent are infected with leprosy, yaws and infestation by worms are widely prevalent, cancer of the liver is a common complaint, Kwashiorkor (malignant malnutrition) is widespread, and the death rate for pneumonia and venereal diseases is very high indeed. The following extracts from a British Colonial Office publication give some idea of the health position of the Uganda Africans today[264]:

> "Malaria is the most prevalent disease, occurring throughout the country. It is endemic among the native population and accounts for a high proportion of all cases of sickness treated in Government hospitals and dispensaries. Its indirect effect upon health and efficiency is generally regarded as considerable, although precise informationis lacking. . . . Venereal disease has continued to spread, . . . Skin diseases, including scabies, fungus infections, ulcers, and abscesses, are prevalent and lead to much disability. Respiratory diseases, including pneumonia, pulmonary tuberculosis, and bronchitis are extremely common. . . . Trypanosomiasis (sleeping sickness) has been held increasingly within check in recent years, though reports of the further spread of the insect vectors emphasise the constant potential danger. . . . Relapsing fever is endemic in Ankole and Masaka Districts. The spread of the tick northwards and eastwards has been aided by the movement of the Banyaruanda whose migration route lies through these districts. . . . Intestinal disease including typhoid and enteric fevers, and dysentry, are endemic throughout the Protectorate. . . . Cerebro-spinal fever was formerly rare, but has increased steadily throughout the Protectorate; incidence is greater in Northern and Eastern Provinces which account for some four-fifths of all cases notified. . . . Malnutrition is widespread even amid abundance of food. Most Africans have an unbalanced diet including too much starchy food (cheapest available — RKM) and too little fat and high-grade protein. Diseases resulting directly from this diet deficiency are not uncommon but more serious is the resulting diminished resistance to infection and proneness to complications."

While such is the position now in once healthy Uganda where the abundance of food in the pre-British period of her history was available to the Africans, today medical facilities in the Protectorate are infinitesimal. In 1949, in the whole Protectorate, there were only 145 Registered medical practitioners and 86 Licensed medical practitioners[265]; and there is no reason to suppose that since then the situation has markedly changed for the better. Moreover, whatever medical facilities there are, are concentrated in the towns where very few Africans live[266]. In the "bush" hardly any medical attention can be received by the Africans, and in the administrative centres in the rural areas treatment can be had only from some scattered dispensaries which are again very few in number. In 1949, there were only 28 Government hospitals with 3,242 beds and 5 Mission hospitals with 777 beds in the whole Protectorate. This means that even if these beds were available only for the Africans

* The total number of African students, as given above, is only approximate because exact figures are not available. However, a difference of a few hundreds or even a thousand will not materially change the above picture of gross discrimination.

(which they are certainly not, for a large number of beds are reserved firstly for the European and then for the Asian patients), there would have been on the average 1,224 of the total African population of Uganda per hospital bed. In 1953, as stated in the Annual Colonial Report, all told there were about 6,200 beds in Uganda in hospitals, dispensaries and maternity centres maintained by Government, the Missions and some employers of labour. Of these again, only 2,600 were in rural dispensaries and maternity centres, which means that only these few beds were available to the general mass of the Uganda Africans[267]. Finally, it is worthy of note that there is roughly only one physician available for 23,000 inhabitants of Uganda[268]. Such is the position in the Protectorate as regards the care for the health of the people. Even so, only 6 per cent of the budgeted expenditure for 1951 was ear-marked for "medical"[269].

To continue with such characteristics of colonial expenditure in the light of the minimum demands of the Uganda Africans to live like human beings, while the peoples of Uganda suffer from gross malnutrition and it is so marked even among the better-paid railway workers that in a recent report to the Colonial Office Dr. C. Northcott made it quite clear that such "malignant malnutrition", due to starvation in childhood, is "probably incurable"[270], and while compared to the British in Uganda, who present death rates lower than those for corresponding age-groups in the United Kingdom[271], death rates among the Uganda Africans are very high and infant mortality is reported to be one of the highest in the world, besides the paltry contribution of 6 per cent of the budgeted expenditure on "medical", 3.7 per cent was ear-marked by the Colonial Government for "Colonial Development and Welfare Schemes" and 0.28 per cent for "Labour"[272]. That the benefit of "colonial development" of Uganda will go essentially to the foreign capitalists will be explained later, and that the benefits of the "welfare schemes" are largely enjoyed by the non-natives living in urban areas has been noted before. Nevertheless, even if the total share of 3.7 per cent was spent on welfare of the Uganda Africans it could not have been anything more than a mere drop in the ocean. Moreover, not merely will only a part of the amount be so spent, but "welfare schemes" for the Africans essentially mean organising social clubs, etc., in order to keep them away from politics. Regarding this policy something has been said at the beginning of this study, and more will be said later. And, as regards the expenditure of 0.28 per cent of the total budget on "labour", it hardly deserves any comment if it is borne in mind that, as revealed from a sample survey of African labourers in Kampala in 1950, they spend 70 per cent of their incomes on food and clothing, and the nourishing articles, like meat, consist of only 17.3 per cent of the total expenditure on food, the rest being essentially starchy substances[273].

Such then are the results of the civilising mission of the British in Uganda where after more than half a century of British rule the people lead a sub-human existence.

And while such is the care devolving upon the African subjects from the British rulers, the bulk of the colonial expenditure is for item-heads like Administration, Police and Prisons, Pensions and Public Debt, and for the so-called "Urban Services and Public Utilities" and "Development of Natural Ressources". The underlying meaning of Public Debt and Development of Natural Resources will be clear when the character of the colonial economy will be discussed on the following pages. Pensions and urban services are obviously enjoyed by the foreigners and their local allies, who are virtually the only people entitled to such claims; the millions of Africans living in the "bush" having no access to such benefits. And as regards Administration, Police and Prisons, one need not elaborate further. Indeed, so important are the institutions of police, prisons and defence to the colonial government that the budgeted expenditure on these three heads in 1951 was more than double of that on agriculture, nearly two-times that on education, more than one and half times that on medical, and nearly 35 times that on "labour"[274].

It is further of importance to note that in the post-war period of 1946—51, while the expenditure on education increased only by 80 per cent, that on defence was more than thrice increased. Similarly, the expenses on "medical" increased by less than two and half times, but those on prisons by more than five times. Likewise, while the expenditures on agriculture and police marked the same rate of increment, by six times, that on police consistently received more attention than on agriculture, even though the latter is virtually the only source of livelihood available to the Africans. In 1946, the budgeted expenditure on agriculture was 78.1 per cent of that on police; the corresponding proportion for 1951 was 82.9 per cent[275].

It should also be borne in mind how little is spent from the treasury of Native Administration Governments for the socio-economic development of the people, and how such budgets are ear-marked mainly for the emoluments of the native bureaucracies, and then to maintain local police and prisons. There the argument is that looking after the social and economic needs of the people is the responsibility of the Protectorate Government. But how admirably the Protectorate Government discharges its duties in this respect has been explained earlier.

Finally, a comparison of the budgeted expenditure of Great Britain with that of Uganda may not be out of place here, for although there can be no doubt that there is a very big difference between the standard of civilisation and social amenities available to the people in the metropolitan country and in their colony, it would be of interest to examine how this gulf is further widened year by year by the attention paid by the metropolitan and colonial governments to the welfare of their respective peoples. According to the budget estimate for the year 1950—51, in Britain £ 1,243 millions were spent on "Education, Housing subsidies, and other Local services; Health and Food subsidies". As compared to that, in Uganda in the year 1951, £ 1,998,694 was spent on "social services", viz. education, medical facilities, labour, public relations and co-operation. Per head of population in the two countries, the figure for Britain was £ 24—11s.; for Uganda, 8s., or less than 2 per cent of the former figure. To look at these two figures from another aspect, even ignoring the fact how little such expenditures on social services really benefit the African peoples of Uganda, the amount accounted for only 15.5 per cent of the total budgeted expenditure, while the corresponding figure for Britain was 31.8 per cent. More revealing is the fact that while Britain, a mighty industrial country, could spend £ 48 millions on agriculture and fisheries for the year 1950—51, Uganda spent only £ 546,923 for that purpose in the year 1951. And this is so in a colony which is of the same size as the metropolitan country, and where moreover while almost all the native population live on agriculture (excluding for the sake of extreme accuracy some 250,000 who at one time in the year work as wage-labourers), in the metropolitan country only about 5 per cent of the total population have the same form of livelihood.

It should further be noted that while 15.5 per cent of the Uganda budget was to be expended on "social services" in 1951, expenditure on "Urban Services and Public Utilities" amounted to 33.9 per cent[276]. The former might to some extent benefit the Uganda Africans, but that the latter was meant primarily for the foreigners in the colony there can hardly be any doubt. In a country where the native population is carrying on a sub-human existence, Public Utilities included item-heads like "civil aviation", which evidently only the foreigners can indulge in.

Lastly, a look into the mechanism of Public Finance of the Protectorate shows that the per capita incidence of "ordinary revenue" on the total population of Uganda, which as explained before is obtained primarily from the African peoples, is always higher than the per capita "public expenditure", of which a little may go back to the people through the schemes of "social services", however poorly developed they are. Thus, in 1925, the per capita incidence of "ordinary revenue" on the total population of the Protectorate was £ 0.415, while the corresponding figure for "public expenditure" was £ 0.413[277]. In 1951, as can be worked out from the available data, the former figure came to £ 2.827, while the latter came, to £ 2.481[278]. It will moreover be noticed from the above that during this period of 1925 to 1951, while the per capita incidence of "ordinary revenue" increased by 681 per cent, the per capita "public expenditure" showed an increment of 601 per cent. This gives some idea of the nature of Public Finance in the Protectorate.

These facts and figures should be read in the light of assertions made by the British Government that they are still remaining in colonies like the Uganda Protectorate only to bear the White Man's burden of Trusteeship for the backward peoples who are "not yet able to stand by themselves under the strenuous conditions of the modern world"[279].

11. Colonial Economy

It may however be argued that in spite of the fact that the Uganda Africans had to bear the burden of colonial rule, the British did one great thing for them, namely to bring them within the orbit of world civilisation. Railways were brought to Uganda from the port of Mombasa, thereby connecting this region of the "Dark Continent" with world trade routes and bringing her closer to other countries of the world. Roads were built in Uganda, so that people could travel from one part to another of hitherto inaccessible regions in this territory. Above all, the peoples of Uganda were brought within the fold of a cash economy, and the introduction of cotton production provided the people with sufficient resources to buy cotton goods and sundries from the retail shops which are to be found in remote parts of the Protectorate today. Hence, instead of praising the British for their great achievements in Uganda, why should one blame their colonial rule ?

It could of course have been said in answer to such questions that, as was shown from the analysis of the Uganda societies in the pre-British period of their history, such progress could have been attained by the Uganda Africans in their normal course of development, and for that they need not have lost their sovereignty and independence. However, besides that point, in the following pages the true character of the colonial economy will be discussed to indicate who actually profit from the introduction and flowering of a "modern" economy in the Protectorate. Later, it will be shown how the colonial "developments" were demanded by this colonial economy and why without such "developments" the British capitalists could not have made their millions of profits in Uganda.

There is the special reason for discussing the character of Uganda's economy at some length, for the economic enslavement of colonial and dependent countries is the main content of colonial exploitation. Indeed, it was for this purpose that the colonial administration was set up, the local bureaucracies were created, and the peoples of Uganda were thenceforth made to pay for running the colonial system. It was for this reason that while the stage of development in Uganda at the time when the British arrived called for the eradication of the oppressive feudal despots and a democratic organisation of all societies, the British not only further stabilised the interests of these native rulers but imposed similar elements on top of other societies which had not yet come under such control; and over all these societies they imposed their own rule for which they were not only not answerable to the peoples of Uganda but also not to their local rulers. It was because colonial exploitation was the central aim of the British that their appearance did not lead to any progressive development of the Uganda societies in their political sphere; on the contrary, autocracy passed from the weaker rulers of the scattered societies to the centralised authority of the supreme one, and to the previous burden of the peoples was added the even more oppressive burden of an efficient colonial rule.

Therefore, in a study of this kind, attention should primarily be directed to the economic development of the territory, for the colonial policy of the metropolitan country, whether the territory under consideration is described as a Protectorate or a Colony, is operated in the direction of arresting the development of productive forces in such areas. This leads to several points of advantage to the occupying power. First, it makes it possible for the bourgeoisie of the metropolitan country to utilise the cheap, or virtually free, labour-power of the colonial people; second, the colony in question can then be kept as a dumping ground for the merchandise of the metropolitan country; and third, the very backwardness of the colony is favourable to the foreign rulers because it facilitates the possibility of exploitation by retarding the development of an anti-colonial national liberation movement and prevents the creation of the pre-requisites of economic independence of the colony by depriving it of the possibility of developing industry which would then be capable of bringing goods on the market competing with the goods produced by the metropolitan country.

In order to expose fully the character of the colonial economy, the following discussion will be undertaken under several sub-headings: — (a) *Economic Subjugation* of the Uganda societies, a pre-requisite for the introduction of a colonial economy; (b) "*King Cotton*", that is, the emergence of cotton production as the basis of the colonial economy after some initial experiments on colonial exploitation; (c) *Policy of Commodity Production* of crops (mainly cotton) by the Uganda Africans, which transformed the character of the economy of the pre-British Uganda societies from subsistence to production for the external market; (d) *Britain's Control over the Economy,* as is evident from the commercialisation of crops and the one-sided development of the colonial economy depending overwhelmingly on cotton export; (e) *Fallacy of Britain's Economic Aid,* indicating how the import of machines and their accessories into Uganda in recent times is meant primarily for a more determined exploitation of Uganda's resources and labour in order to realise the *maximum profit* for the British bourgeoisie; (f) *Dynamics of the Colonial Economy,* revealing how all-embracing is the control of the British capitalists over the economic life of Uganda; (g) *Economic Drain on Uganda,* resulting from an adverse balance of trade, — a typical feature of colonial economy; and (h) *Internal Economy of the Protectorate,* an analysis of which reveals how the Uganda Africans must work in order to fill the coffers of foreign monopolists and their local agents and how even the small sum which is formally given to them as primary producers of the raw materials needed by the foreign capitalists is taken back, almost entirely, through the administrative and commercial channels controlled by the foreigners.

11(a). Economic Subjugation

With the occupation of Uganda, the foreign rulers set about the task of subjugating the whole area politically and economically. The former aspect of the policy has already been dealt with. To consider the second, at the beginning the essential job of the colonial administation was to prepare Uganda for exploitation by the British capitalists. Along with Uganda they had also occupied Kenya. While Lugard had advocated Uganda also as a favourable spot for White Settlement, this idea was discarded by his compatriots. At that time the intention was to prepare a Kenya suitable for White Settlement with an African labour force at hand to fulfil the demands of the British settlers, while Uganda would be "developed" to supply raw materials from her fertile lands as demanded by the bourgeoisie of Britain. Therefore, although Uganda had been declared a Protectorate, without any consultation with the local people or their native rulers, portions of her territory were hacked off and joined to the Kenya Colony in order to suit the needs of the White Masters. In 1902, during the administration of Sir Charles Eliot as the Commissioner of the East Africa Protectorate, the Eastern Province of Uganda was ceded to the new province of Kenya, "thus bringing all the area suitable for White Settlement under one administration, and in addition introducing a small non-colonisable district containing some million Africans who would be ready to hand as a labour force"[280]. And, in 1905, "the administration of British East Africa and Uganda was transferred from the Foreign Office to the Colonial Office, and in 1906 the two Commisioners got the rank of Governor"[281]. There remained no doubt that Uganda had been turned into a colony, although it was called a Protectorate. Henceforth it was her role to supply the foreign rulers with precious gifts from her rich land.

That this was the key-note of British aims in Uganda, in contradistinction to their avowed aims of bringing peace to the country (by abolishing slavery and stopping civil wars) and civilising the "black savages", has been discussed before. Indeed, this was particularly evident from Stanley's encouragement and Lugard's advice to the British ruling class to annex Uganda, and the very explicit report of the London Chamber of Commerce in 1893 on the prospects in Uganda. Some of them went even further. They advised the Uganda Africans to orient their economy in order to produce in accordance with the needs of the British bourgeoisie. In this respect Stanley's letter stands out as a memorable document. Along with Apolo Kagwa, Zakaria had also "profited largely" from the land distribution under the 1900 Agreement, and due to the kind favour of the British rose to great power as a Regent of the State. Now in this letter, dated 25th June 1900 (that is, written three months after the Agreement had been signed), Stanley directed him how to set about preparing the country for the benefit of the British bourgeoisie. To quote from the letter[282]:

> "We have just voted a great sum of money to finish the railway to the lake, and to put more steamers on the Nyanza, that the communications between you and us should be rapid and certain. There are many people in this country who still doubt whether the Waganda [previous spelling of Baganda] are as willing to work as other people, and grudge giving more money until they are sure that there is any trade to be got with Uganda. They say, if Uganda produces nothing that the British want, is it worth spending money upon it. Now I want you, and the other Regents, to do all you can to induce the people to work, and show that we travellers have spoken the truth about Uganda. The English want coffee, skins, ivory, rubber, gums, and such things. Those who live with you should never be able to say that food is scarce, therefore there should be plenty of bananas, corn, potatoes, vegetables, fish and meat. In fact all kinds of eatables, for it is better to have too much than too little. Everything you can produce. . . . You have good soil, you have plenty of rain to make things grow. . . . It will not be as in the old days when the white people did not know there was such a country as Uganda, or were surrounded by ignorant black people; but you will have a railway, and white people will be coming and going continually to do all manner of business with the Waganda. What have you got to trade with them? That is the question the Regents should ask of one another, and if Uganda has nothing to sell to the white men, then we travellers who have done so much for Uganda will be utterly ashamed."

In response to such precise orders from Stanley, who had already done "so much for Uganda", Zakaria and other newly-created henchmen of the British began "to induce the people to work" — in order to provide whatever the British needed, ranging from the production of plenty of food for the British masters in the colony to the extraction of precious raw materials for the metropolitan country, from looking after the comforts of the foreigners and paving the way for the flowering of the colonial economy by building roads, trade-centres, etc., to working for the transportation of the resources of the country out of Uganda, In short, the peoples of Uganda had to undertake all labouring jobs connected with colonial life.

In this way, the foundation of the colonial economy was laid, first of all in Buganda where British rule was then most consolidated and later over the whole of the Protectorate. Simultaneously, as noted before, a money economy was introduced and work and production in all forms were henceforth geared to the needs of the British monopolists. The upshot was that the subsistence and independent character of Buganda's economy was gone forever. Soon the other provinces of the Protectorate followed suit.

11(b). "King Cotton"

Concurrently with thus demanding the labour of the Uganda Africans in order to build up the colonial economy, experiments were conducted by the British bourgeoisie in Uganda in order to find out what would be the best means of profiting from the rich lands. Referring to this period of experimentation, Thomas and Spencer have noted[283]:

> "The years between 1905 and 1909 were a time of bustling activity. . . . dozens of applications to tap for wild rubber, over sections of various size of these (Uganda) forests, were received from European syndicates or individuals between 1905 and 1907. Most of these applicants were ready only to wring what they could out of the forests without care for their preservation. In the face of the Government's insistence of specific working conditions, and hampered by the spread of sleeping sickness in the lake-shore forests, the local rubber boom had faded out by 1909. Only one undertaking established itself and remains in existence today to evidence this early enthusiasm. *Funtumia elastica* had been discovered in the Mabira Forest in 1904, and this led to the formation by the Uganda and East Africa Exploration Syndicate, Ltd. . . . of the Mabira Forest (Uganda) Rubber Co. Ltd.; and to the latter company the Mabira Forest was leased from 1st June, 1906. This, the first public flotation in England of an Uganda undertaking, was to have a chequered career. It exists today primarily as a coffee-plantation company, its forest and wild rubber interests having been surrendered in 1924. . . .
>
> Among other schemes, mostly on a grandiose scale, which were afoot at this period were the extraction of fibre from papyrus to be cut in the swamps around Lake Victoria; the collection of sansevieria; of raphia fibre; and the harnessing of the Ripon Falls and the Sezibwa Falls. . . . (A proposal) was concluded by the freehold grant to the Uganda Company Ltd., of about 5,000 acres, originally intended for cotton growing, in Bulemezi County, Buganda, which land is still held by the Company. Of an early urge to acquire land in Busoga, the freehold Lumbuye Hill Estate granted on 4th August, 1905, is now the only existence."

It will be noticed that the central aim of these schemes was not to improve the state of productive forces in Uganda, which was the foremost demand of the hour but to make money quickly by tapping the natural resources which were still in their virgin state in the colony. It will also be noticed how the British bourgeoisie by "public floatation" or otherwise were spreading their tentacles in the Protectorate during these early years of British rule in Uganda.

Soon, however, the "other schemes" faded out, and those interested in the production of rubber shifted their attention to German East Africa. Brode noted[284]:

> "The sudden rise of the rubber prices in 1910 produced a sort of fever among English businessmen; the high dividends which were paid to rubber shareholders in Ceylon and in

the Straits Settlements caused the foundation of numerous new companies in London, which tried to acquire plantations wherever they could. So many of the German East African plantations came into English hands, and those who had started them realized high prices, which enabled them at once to establish new plantations. The fear has been expressed in some German quarters that the rubber of the colony would become an object of English speculation, and that an English ring might be formed to dictate the prices of the article."

Without going into the question of German monopolists in this situation, and concentrating upon Uganda one finds that British interests in the Protectorate were directed more and more to the production of cotton, and to a lesser extent of coffee, tobacco and tea.

This concern to grow cotton in the newly-established colony was dictated by the international situation. As Brode noted[285]:

"The American War of Secession in the sixties of the past century caused a great decrease in the production of cotton, and latterly the production of this article has not kept pace with the needs of the world market. Since 1889 the consumption of America has risen by 60 per cent, whilst production has sunk by about 12 per cent compared with the average of the last ten years, a fact which led to wild speculations, with the result that the prices were raised by more than 50 per cent. There was every reason to fear that the whole European textile industry would fall under the tyranny of an American trust, and therefore the mill-owners of the various European nations started to take joint steps to open new fields for the cultivation of this important product. In German colonies the Kolonialwirtschaftliches Komitee, in British territories the British Cotton Growing Association, have made successfull efforts in this direction. In May, 1904, an international congress of cotton manufacturers accepted the necessity of a joint struggle for the cultivation of cotton."

Even earlier than 1904, Stanley had informed the British capitalists of the possibilities of the production of cotton in Uganda, of which something has been said before. Lugard also gave the same advice[286]. But at the beginning the British capitalists were more interested in schemes which could be easily introduced in the colony and were spectacular in character. They did not immediately want to take the trouble of initiating a Protectorate-wide cotton economy. Probably they were also not aware that this could be done fairly easily. However, the demands of imperialist economy soon decided their colonial policy. Brode noted[287]:

"The experiments in German East Africa were followed with great concern in the interested British quarters. In December, 1902, the Chamber of Commerce at Manchester declared a sample grown in that colony to be equal to the best Egyptian produce, and in the next year, in a report about the general market of the cotton industry, the Vice-President of the B. C. G. A. (British Cotton Growing Association) referred to the German success, in order to encourage the production of the article in British East Africa and Uganda."

The experimental phase in British Uganda was thus coming to a decisive end. First rubber, and then cotton, together with some other crops like coffee, tobacco and tea, were considered suitable products to be obtained from this colony. Indeed, production of cotton soon outstripped all other commercial crops, and assumed the pivotal position in Uganda's economy. As can be worked out from official figures, out of the total acreage under cash crops in Uganda in 1951, cotton (which is entirely African-grown) accounted for 84 per cent, coffee and tobacco (very largely African-grown) for 13 per cent, and other cash crops, such as rubber, tea, sugar, and sisal (grown mainly by non-natives) for 3 per cent[288].

This point is well worth remembering, for it is often declared by British spokesmen today that one of their great achievements in the colonies is the introduction of cotton production in the Uganda Protectorate. That it was dictated entirely by the imperialist demand of the day, like almost all such so-called progressive measures, should not be lost sight of.

How important became this production of cotton in Uganda would be evident from the following brief account of its career in the Protectorate. It will be remembered that cotton seeds were introduced in the Kingdom of Buganda during Mutesa's reign, but before its production could be seriously taken up by the people, the country was overrun by the British. After that, the commercial cotton plant was first introduced in 1903[289]. In 1904, for the first time, the Agricultural Department "imported a large quantity of seeds for distribution throughout the country"[290]. Great progress in cotton growing was reported from all parts of the country during the next few years[291]. From a modest beginning of 54 bales in 1904—05, the production of cotton rose to 480,000 bales in 1938; from the average yearly figure of 59,000 metric tons of ginned cotton during 1934—38, it rose to 66,000 metric tons during 1948—51[292].

Cotton began to dominate the economy of the Protectorate. While since 1938 six to seven million acres of Uganda's land have been under crops, cotton accounts for one and a half million acres[293]. In 1946 the Agricultural Department of the Protectorate Government estimated that while the total crop production of the Africans by means of their primitive peasant-cultivation, and of the European and Asian planters by improved methods of farming, accounted for about £ 7,895,020, African-grown cotton and cotton seeds alone accounted for £ 4,236,422, or 53.7 per cent of the total crop value[294]. The 1951 Economic Survey of Uganda, published by the British Colonial Office, stated[295]:

> "The majority of the population work on the land as peasant cultivators. Both food and cash crops are grown. Sale of cotton is the main source of money income though recently other crops such as coffee have been increasing in importance. Only small surpluses of food crops are marketed."

And, like all other commercial crops grown in the Protectorate, virtually the total production of cotton is ear-marked for export. Thus cotton became the main item of domestic export from Uganda. The export value of cotton lint f. o. b. Mombasa, that is, excluding the export of cotton to other parts of British East Africa (Kenya and Tanganyika), rose from the small figure of £ 326 in 1904—05 to £ 354,146 in 1914—15; then, with a few slight drops in 1915—16, 1921, 1922, and 1926, it rose to £ 4,268,283 in 1927[296]. In the post-war years before 1950, it was of the order of 5 to 7 million pounds sterling[297]. In 1951, it reached the figure of £ 28,741,604[298]. The percentage value of the exported lint cotton related to the total domestic exports during 1915—16 to 1947 gives an average value of 73 % $\pm$2 % (with a coefficient of variation of 14), excluding the years 1924, 1939, and 1940, for which no figures were available when making the above calculations[299]. In 1951, the corresponding proportion still remained as high as 61 per cent, even though the export of coffee and other crops, as well as mineral resources of the country, was largely increased in the post-war years[300]. Furthermore, it is worthy of note that data on the export value of cotton seeds (including derivatives, in oil and cake) were not available when making the above calculations; otherwise the importance of cotton in Uganda's economy would have been further emphasised. Thus, for instance, the percentage value of exported cotton lint and cotton seed related to the total domestic exports was 65 per cent in 1913—14; it rose to 94 per cent in 1925; in 1947, in spite of the increase in the production of other exportable goods and a smaller production of cotton in that year, the proportion remained at 63 per cent; and in 1951, the proportion was 64 per cent[301]. Dalal wrote in 1939[302]:

"The significance of the cotton industry to Uganda cannot be exaggarated; in fact, cotton is the life-blood of this Protectorate. To an average resident in Uganda, or for the matter of that, in East Africa, the importance of cotton production is generally well-known, but only few realise how much Uganda is dependent on its cotton crop."

"King Cotton" thus became the central figure in Uganda's economy. In Uganda it is also poetically described as "white gold".

11(c). Policy of Commodity Production

From the beginning, production of this "white gold" was left to the Africans to be done by means of peasant cultivation. They were indirectly forced to grow commercial crops because of the introduction of a money economy and because of money-taxes and other burdens imposed on them. Moreover, in order to facilitate extension of the cultivation of commercial crops to those areas where the people were not well versed in the techniques of agricultural production, the Baganda "agents" who went to many of these "simpler" societies to establish British rule therein were also entrusted with the task of teaching the people how to grow cotton and rubber.

This schooling was also needed in the kingdoms of Bunyoro-Kitara, Ankole, and Toro, for there the pastoral overlords were either averse to agricultural production or were not yet well-versed in its technique. Therefore, the Baganda who were exported also to these areas as "chiefs" had this mission entrusted to them.

No doubt, these faithful Baganda did not fail their masters. Thus, for instance, it has been noted earlier that the Banyankole were brought into submission with the help of Baganda "chiefs". But this was not the only role these Baganda played in their society. The Banyankole finally accepted their fate under British administration, but, as noted earlier, pastoral traits were very strong in their society. Therefore, the "innovation that chiefs should have an interest in agricultural pursuits militates against all the sacred rites of the pastoral people. Cultivation under the pastoral custom could not be undertaken without prejudice and injury to milk institutions; the latter forbade any man keeping cows to dig or to grow vegetable food, because it was considered detrimental to the health of the herds"[303]. So Roscoe noted[304]:

"The introduction of Baganda for this purpose may possibly be the method which the people have adopted to overcome what appeared to be an insurmountable difficulty."

Whether this policy was ordained by the "people" or by the interests of the foreign rulers it is no longer necessary to discuss. But the upshot was that "there are a number of chiefs in charge of the agricultural districts, seeking to develop the land, but these men are Baganda who have been brought into the country to superintend the cultivation of cotton and rubber"[305]. Nothing could be more explicit; the interest lay chiefly in cotton and rubber cultivation in which the British monopolists were most interested at that time. Furthermore, Roscoe noted[306]:

"Chiefs belonging to pastoral clans were unable to supervise agricultural pursuits, and so capable Baganda have been placed in charge of certain districts to supervise the labours of the agricultural clans. Some fifteen years ago the king of Ankole would not have tolerated any Muganda, as a chief, in his kingdom, but would have been afraid of the influence over his people; now, however, he welcomes them, to instruct his peasants in the art of cotton and rubber growing, in order to satisfy British officers, who, alas! know little of the pastoral prejudices and superstitions."

Evidently, within a short time the British had consolidated their position in the kingdom of Ankole with the help of the Baganda Agents, and thanks to their assitance this territory was also brought within the orbit of colonial economy.

As in Ankole, the Baganda sent as "chiefs" to the Teso country were also entrusted with the same mission. Roscoe noted[307]:

> "The British Government placed Baganda chiefs over these Bateso (Iteso) some years ago when the country was occupied by the English, and a more systematic form of government was established by these chiefs. It was the task of the principal Muganda chiefs to bring the country into peace and harmony with other parts of the Protectorate, and they have done their work well. They have gained the confidence of the people and made friends with them, while they have taught them to cultivate cotton and rubber, to make roads, and to do other work for the cultivation of the district, thus uplifting a tribe and redeeming a country, in addition to bringing about a sound system of government, and making it safe to pass through a people hitherto wild."

Here one finds the reason why "peace" was brought into these societies, the "tribes" were "uplifted", and the countries were "redeemed". Everywhere the Baganda Agents went, they taught the people to cultivate cotton and other commercial crops, for which "peace" was required in the country and a "sound system of government" as a part and parcel of the colonial administration. How these Baganda Agents fulfilled this task in the Kigezi country and other societies like that of the Bagishu and Bakedi has also been mentioned earlier.

Soon the economy of cotton production engulfed the whole of the Protectorate, and thenceforth the peoples of Uganda began to live by means of commodity production of crops, — mainly cotton, and, in addition, coffee, sugar-cane, tea, tobacco, etc. According to the estimates of the Agricultural Department of the Protectorate Government (which has been referred to earlier), in 1946 the commercial crops, like cotton, coffee, sugar-cane and tea, accounted for more than 88 per cent of the total value of crop production in the Protectorate. Not only as before the advent of the British the African peoples went on living on plantains, maize and some vegetables, but with the growing demand of Uganda's agricultural produce for foreign market the subsistance sector of her economy shrank appreciably. In 1950, while out of the total value of "net Geographical Products" in Uganda agriculture accounted for 54.4 per cent, the latter proportion was composed as follows: — "African production for sale" — 26.7 per cent, "Non-African" production — 2.6 per cent, and "African subsistence" production — 25.1 per cent[308]. The economy of Uganda societies was thus transformed from subsistence production for themselves to commodity production for the external market.

11(d). Britain's Control over the Economy

The commercialisation of crops in this way was to have a far-reaching effect on the future of the Uganda Africans. To note the three main features of the colonial economy arising out of this situation:

Firstly, Uganda's agricultural production henceforth depended overwhelmingly on the external trade which has been entirely in the hands of foreign interests. Simultaneously with political freedom, the Uganda societies thus lost their economic independence.

Secondly, by basing Uganda's economy mainly on the production of cotton for external trade, she was made to depend decisively on the vagaries of the world capitalist market, and thus tied to the apron-string of British imperialist economy. Her one-sided economy, dictated primarily by the inner-imperialist rivalry between the British and American monopolists over the demand and supply of cotton, decided her unstable position and her chronic dependence on the metropolitan country. Her other forms of production, though of much less importance than cotton, were also dictated by the interests of the British imperialist economy. This further strengthened the position of Britain in the colony and in the world capitalist market, and correspondingly whatever economic power the Uganda societies might possess it was now reduced to practically nothing.

Thirdly, the economy of the Protectorate, thus based on the production and export of cotton (and other less important commercial crops as well as some minerals, etc.) satisfied the conditions of colonial trade. This is undoubtedly a fundamental objective in possessing a colony as it enhances in all respects the position of the dominating foreign bourgeoisie in the world capitalist society and reduces the dominated people to abject submission.

Of the above three consequences, the first two are still in full play, for the Uganda Africans have virtually no control over the Protectorate's external trade, and cotton (as noted before) dominates the colonial economy. As regards the third, the conditions of colonial trade in Uganda were fully satisfied in the earlier years of British rule, but in recent times the emergence of rival capitalist interests from other countries and the growing weakness of Britain in the capitalist world have tended to undermine them. Even so, the system has not lost its fundamental importance, and, furthermore, since the last war Britain is again asserting herself in Uganda in this respect. Because the system of colonial trade plays a very important rôle in the process of colonial exploitation, for Uganda it has been discussed below in some details.

One of the two main conditions of colonial trade, viz. the guarantee of inelastic foreign demand (that is, preference for Home goods in the Colonies), remained assured by the simple fact that the industrialisation of Uganda is virtually nil, and agricultural production is "the predominant asset of the country", as Hailey has recently noted. Out of the total value of "net Geographical Product" in 1950, "Manufacturing industry" accounted for only 3.7 per cent[309]. Hence, almost everything, from candles, matches, hardware, etc., even to textile goods, has to be imported into the Protectorate for consumption by the peoples of Uganda. Although cotton production has had such tremendous growth in the Protectorate since 1904, not a single weaving or textile factory has yet been opened in Uganda. Therefore, for all manufactured products Uganda must depend on imports. Needless to say, all these goods come almost entirely from Britain, with the exception that in the last few decades the Indian capitalists have to a certain extent penetrated into the textile market, and Japan became a serious rival in the thirties. Even so, British interests have not seriously suffered, mainly because even this unavoidable concession to non-British capitalists could be utilised in favour of Britain's imperialist economy as a whole. For instance, the market for British goods in Uganda was first disturbed by Japan entering the field during the 'twenties of this century. In 1929, 39 per cent of cotton piece-goods (which comprise the most important item of import in Uganda)[310] came from Japan; in 1935 the figure rose to 88 per cent, with the United Kingdom (and British India) as the second and practically the only other exporting country in the list[311]. But, as Hailey noted[312]:

> "Japan has entered the world's markets not only as a seller but also as a buyer. From Great Britain and the British Empire she imported in 1936, according to the Japanese import statistics, about £ 51,000,000 of goods, and exported to Great Britain and the British Empire about £ 41,000,000. Thus the British countries as a whole have a favourable balance of about £ 10,000,000."

Thus, although aparently it seemed that the mother country had lost its "preferential" marked in Uganda, she really gained in an over-all balance.

Moreover, since the last war Britain has fully re-asserted her position in the Uganda market. During those post-war years of 1944—47, when India was still under the complete control of Britain, the trade reports indicated that the United Kingdom and India were the "principal sources of supply" for all imports including "cotton piece-goods, cotton blankets and cotton thread" to the Uganda Protectorate[313]. According to the latest available figures, in 1951, 46 per cent of Uganda's net imports came from the United Kingdom, and "some two-thirds of total imports were from Commonwealth countries"; the share of other principal countries from which goods were imported was 10 per cent from Japan, 6 per cent from Italy, 4 per cent each from Western Germany and the USA[314]. In 1953, more than half of Uganda's imports came from the United Kingdom[315]. Britain's national export to Kenya-Uganda in 1938 was 3.03 million pounds sterling; in 1951 it rose to 32.34 million pounds[316]. Thus, an increase of nearly eleven times has been effected since the last year before the Second World War. Britain has, evidently, fully re-established the *preference* for Home goods in the "colony".

As regards the second condition of colonial trade, viz. the export of raw materials from the colony to the mother country, in the early years of British rule in Uganda this was the law without exception. Production of sugar-cane was not then in full swing, but the other commercial crops of secondary interest, such as coffee, went (and still goes in bulk) to the United Kingdom either to be consumed there or for re-export to those countries from which the British monopolists could obtain the desirable profits. And other precious resources of Uganda (such as gold, tin, and wolfram) were extracted from her mines and went straight to Britain. As regards cotton, Carter noted[317]:

> "Up till 1917, the principal, and almost the only, market for Uganda cotton had been that of Liverpool; but in that year, owing to the war and the difficulty of maintaining shipments to England, Bombay entered the market as a serious competitor, and has ever since, with Japan which entered the market later, taken a large proportion of the crop."

It is, however, worthy of note that Bombay functioned essentially as a centre for re-export of cotton, mainly to Liverpool and Japan, as Contractor noted in 1926 and Brendel in 1934[318]:

> "Bombay is the biggest market in the world as far as *Cotton Arrivals* are concerned. Bombay has been the market for import and export, attracting all the varieties of Indian Cotton, also East African, some American, Egyptian, Persian, and Afganistan, which are all received in Bombay, from where distribution takes place to all other ports afterwards".
>
> "Interessant ist es, daß von Bombay aus, dessen Bedeutung als Baumwollmarkt nicht unterschätzt werden darf, ein Teil der Ugandabaumwolle wieder nach Liverpool und Japan ausgeführt wird."

Thus the emergence of Bombay as a "serious competitor" in the world cotton market did not materially affect the conditions of British colonial trade.

With Japan, however, the situation was different. Hailey noted regarding the intervening years between the two world wars[319]:

> "In 1936 Japanese imports from Kenya and Uganda, mainly cotton from Uganda, were almost equal to exports to those territories, according to value at the Japanese ports."

Even so, because of her large empire the situation was not too bad for the British, as has been explained before. Besides that, to these non-British capitalists Britain could dictate the price, after having secured the needs of the Home monopolists. For instance, during 1924/5 to 1936/7, the price of East African cotton in the Liverpool market was consistently higher than the corresponding price of the American Middling; the average price of Uganda cotton being 15 % $\pm$ 2 % higher than the American cotton[320].

Moreover, since the last war, when the British Power has become seriously weakened in other parts of the world, she has re-asserted her position in Uganda. Britain's "General Imports" from Kenya-Uganda in 1938 were 2.55 million pounds sterling, in 1951 the figure came to 23.55 millions; that is, an increase of nearly ten times[321]. According to the latest available information, in 1951, 30 per cent of Uganda's domestic exports went to the United Kingdom, 26 per cent to India, 11 per cent to Western Germany, 6 per cent to France, and 5 per cent to the USA; 65 per cent of the total domestic exports were to the Commonwealth Countries[322].

In this way the conditions of colonial trade have been properly maintained in Uganda, and thereby the well-being of the African peoples have been left at the mercy of the British capitalists who hold all the resources of the country under the octopus-like grip of the colonial economy.

11(e). Fallacy of Britain's Economic Aid

As opposed to the above conclusion, claims have been made in British Government quarters that Britain had, even before the last war, exported more than she imported from her East African Protectorates, that in recent times this process has become quite obvious, and that this shows how consistently Britain endeavours to develop the economy of her colonies. In favour of such an argument it can be said that in 1938, for instance, the excess of Britain's "national exports" to "general imports" from Kenya-Uganda was £ 480,000; in 1951, the figure came to £ 8,790,000.

However, while examining such an argument it should first be noted that the inclusion of Kenya in the above data confuses the situation, for the White settlers in this colony depend to a large extent on imports from their mother country with respect to both their production (machines, etc.) and consumption (special food, clothes, and such). Therefore, even in pre-war years, the excess of Britain's exports to imports in regard to Kenya-Uganda does not necessarily suggest a policy to better the life of the Africans.

Moreover, for Uganda alone, an examination of the types of articles imported gives an insight to the nature of "help" coming from the metropolitan country to the colony. Thus, an analysis of the net imports to the Uganda Protectorate in 1951 by the nature of articles imported shows that more than a quarter of her total import-value was under the heads of "machinery, apparatus and appliances", "road motor vehicles and tractors" (including spares and parts), "tyres and tubes, pneumatic", "petrol, gas oil, fuel oil, and lubricating oils". In all, they accounted for £ 6,134,052 out of the total import value of £ 22,435,674[323]. This feature of Uganda's external trade, assuming of course that most of the goods enumerated above were from the metropolitan country (which is no doubt a reasonable assumption), is said to indicate that Britain has seriously taken up the task of progressively developing her colonies under the "Colonial Development and Welfare Schemes". But is this contention justified?

In the next chapter it will be discussed for whose benefit such schemes have been launched, how they are paid for, and who own and utilise these imported articles. But here it may be worth stating that these machineries and their accessories are meant primarily for a more determined exploitation of Uganda's resources and labour as called for by the ever-aggravating economic crisis of Britain whereby her bourgeoisie are pushing for the *maximum profit* wherever they still have a stronghold. As it will be evident from a later discussion, most of these machines sent to Uganda in recent times are for the extraction of her mineral wealth (which had not been taken up seriously in the pre-war years) or for the primary processing of her agricultural products; and it will also be noticed that the other articles mentioned above are in one way or another associated with building and running such extractive industries or to transport the raw materials (or their primary products) from the production-centres to the ports for export. The export of these machines, etc. thus does not show any significant change in the character of colonial trade, but to a certain extent it indicates the critical phase of Britain's parasitic economy today, about which more will be said later.

Indeed, this feature of Britain's economy accounted to a large extent for her slight excess of exports to Uganda even in the pre-war years when in Uganda's external trade she was forced to give way, to an extent at least, to non-British monopolists. This feature has been further intensified at the present time when, in order to earn dollars for the metropolitan country, the ingenious scheme of "triangular trade" has been launched whereby a substantial part of exports from British colonies is now ear-marked for the United States of America.

These two factors, viz. the present policy of Britain to draw the *maximum profit* from her colonies and her scheme of the "triangular trade", account for the greater divergence in Britain's export and import to Kenya-Uganda and stress that instead of a change of policy towards her colonies Britain is bent upon living as before (in fact, more ruthessly) on the resources and labour of these territories and utilising them to meet her "dollar shortage". Therefore, like other colonies, Uganda lives on an adverse balance of trade — with a much higher value of exports than imports. This, a typical feature of the colonial economy, will later be discussed in detail.

11(f). Dynamics of the Colonial Economy

Indeed, how all-embracing this control of the British capitalists over the economic life of Uganda remains must be evident from the following account of how the peoples of Uganda fare under the colonial economy in regard to production of economic goods. The economic relations in the Protectorate can be classified under three categories according to the different stages in the operation of the colonial system of production. They are as follows.

The primary stage of production naturally rests on the Africans who have undergone very little development under colonial conditions, and therefore have to live on cultivation alone. The system of production is peasant-farming, and the instrument of production in most parts of Uganda is the hoe. According to the estimate made by the Agricultural Department of the Protectorate Government in 1946, which has been mentioned before, out of the total crop production valued at £ 7,895,020, production from the African peasant-farming accounted for 78 per cent, and within the African production again cotton and coffee accounted for 87 per cent of the total value of crop production. According to the study of "net Geographical Products" of Uganda in 1950, the non-African sector of the agrarian economy accounted for less than 5 per cent of the total valuation of agricultural products[324]. Further to note, the entire production of cotton is from such primitive farming, and of coffee also only about 17 per cent of its total produce-value is from the European farms equipped with better means of production. Thus the position of the Uganda Africans as producers of raw materials for the foreign monopolists, and that also by their own capital, is the essential characteristic of the society. Thereby they represent the lowest rung of the economic structure of Uganda society today.

The second stage in production, viz. processing the commercial crops for export, rests mainly on the Indians. Coffee, tea and tobacco are processed mainly by British-owned farms, but although of some importance they do not account for the bulk of Uganda's export; altogether (including the production of cigarettes) they accounted for 29 per cent of the total value of domestic exports in 1951[325]. Sugar is produced and processed almost entirely in an Indian-owned farm, and ginning and baling of cotton also rests today almost entirely on the Indians; cotton export alone accounting for 61 per cent of the total value of export in 1951[326]. Before the First World War, in the heyday of Britain as a world power, ginneries in the Protectorate were owned by three British concerns[327]. But the proportion of ginneries owned by the Indians gradually increased since then; it was 77 per cent in 1930, and after the last war the figure rose to 90 per cent[328]. There are thus very few British ginners in the Protectorate today, and only one African ginning factory has been started in recent years in Buganda after a good deal of agitation and struggle. Thus, in relation to the most important production in the Protectorate, the Indians are placed in the middle rung of the economic structure of Uganda society, whereby they buy raw cotton from the African growers, gin and bale the cotton lint, and hand it over with cotton seeds to the British interests for export.

The last stage in the operation of the economy, viz. selling the crops in the export markets, is virtually entirely in the hands of the British. Coffee, tea and tobacco have always been in their hands. As regards the main crop, cotton, at the early stage of cotton production in the Protectorate, the crop was traded in a free export market, viz. the Liverpool Cotton Exchange, which has been mentioned before. Then followed an intervening period of free export conditions in which both British and Indian interests prevailed, but the former interest was decisive, for the banking institutions and export concerns in the Protectorate were and are controlled almost entirely by the British. Finally, since 1942, the Uganda Government with the help of British banking institutions and export concerns, and in collaboration with the United Kingdom Government, has taken over the entire control of cotton-sale. (And the same is true for the sale of African-grown coffee.) Thus the top stratum of the economic structure of Uganda society today is represented by the British interests.

How domineering is this top stratum and how it controls the entire cotton economy primarily for its own benefit would be evident from the following. To take the relation between the British and the Africans first, it can be shown that greater and greater official control over the cotton economy has meant a continually worse position for the Africans and mounting profits for the ruling interest. Thus, when the growers benefited by free competition between the ginners, they used to receive the highest share of the export value of the crop that they have ever-realised. Then a Statutory Pool was established by the Uganda Government in 1934 which controlled the price to be paid to the growers but ultimately the price was determined by the world market price[329]. The result of the beginning of government interference was that the growers' share of the export value of cotton was significantly reduced. Next the Cotton Exporters' Group was set-up by the Government in 1942 and the price paid to the growers was fixed by this organisation; in consequence the African growers faced a still worse situation. Finally, the establishment of the Uganda Lint Marketing Board in 1949, in order to "control the entire marketing of the crop from producer to foreign buyer", has not changed the situation in favour of the Africans; on the contrary, it has become still more critical. These succeeding changes can be seen from Table IV.6 in the next page. This table has been prepared from the available data, in which the phases of 1942—48 and 1949 onwards have not been shown separately, for in both phases the type of control on pricing the crops remained much the same[330].

The Table VI.6 shows that while the African growers' share of the total value of cotton has declined in successive periods of more and more government intervention, the export value itself has not only not fallen in the first two periods, but has trebled during the entire period.

Thus, while the African growers are being given a still lower *price for their labour*, the foreign interests are accumulating fortunes. This gives an idea of who really prospers from the cotton economy of Uganda, and how the British Colonial Government looks after the interests of its African subjects.

Table IV.6

Period	Characteristics of the period	Percentage of export value of cotton as received by African growers	Export value of cotton (in £) per c.w.t.
(1)	(2)	(3)	(4)
1924—33	Free competition between ginners to export	76 ± 3	3.86 ± 0.52
1934—41	Price paid to growers controlled by Statutory Pool	60 ± 3	3.01 ± 0.15
1942—51	Government control of cotton-sale and fixation of price paid to growers by the Cotton Exporters' Group (1942 to 1948) and by the Lint Marketing Board (1949—51)	38 ± 2	10.09 ± 1.61

Notes: (1) The figures for 1939 and 1940 were not available for the table.
(2) The calculations in col. (3) do not take account of the value of exported cotton seed, and of the value of cotton sold to other countries in Africa, such as Kenya and Tanganyika, however small these quantities may be. Therefore, the calculated share of the growers' of the marketable value of the produce is somewhat higher than the actual figure, and is certainly the maximum.
(3) There are significant differences between the three figures in col. (3), as *t* test shows by comparing them in pairs.
(4) In col. (4) there is no significant difference between the values for 1924—33 and 1934—41, but the value for 1942—51 is significantly different from the other two values, as is evident from *t* test.

In dealing with coffee also the situation is not different. Since 1942 the Protectorate Government has assumed control of the African-grown coffee, while that grown by Europeans (British) falls outside the scope. Every year the price for high-quality Bugishu coffee is fixed by a Statutory Board of Management and for other classes of African-grown coffee by a Control Committee. The result has been that while the export value of coffee has soared very high since the last war, the price paid to the African growers lags remarkably behind. This can be seen from the following table which gives relevant figures for the years 1947—1951[331]. Unfortunately, data for earlier years is not available.

Table IV.7

Year	Return to African growers per ton of coffee crop		Export value of hulled coffee per ton	
	(in £)	Index	(in £)	Index
(1)	(2)	(3)	(4)	(5)
1947	21.08	100	73.77	100
1948	21.75	103	86.07	117
1949	28.37	135	121.07	164
1950	36.33	172	261.48	354
1951	48.81	232	312.99	424

It is, however, argued that from the profits accruing to the Uganda Government from the sale of cotton and coffee a reserve fund has been established in order to help the Uganda Africans. It has been said that part of the reserve fund would be used to pay the African growers a fair price in the case of a sharp drop in the world price of cotton and coffee. But available reports do not indicate that such assistance has yet been given for the production of coffee, while in mid-1953 the Coffee Fund stood at £ 14 million, and for cotton only £ 200,000 was so used in the first instance in 1953 while on the 31st July, 1953, the Cotton Fund ammounted to £ 24.8 million[332]. This argument therefore does not hold water when it is seen that while the African growers are deprived of their legitimate gain in the post-war world in which the prices of cotton and coffee are so high, the colonial government is holding on to a colossal sum obtained by depriving the poverty-stricken Africans.

The other plea for holding on to the accumulating profit is that the bulk of it would be used for the "Development of Uganda"[333], for the purpose of which about £ 30 million have already been expended from the Cotton and Coffee Funds. But, firstly, such "development" of Uganda would be of course along the lines the foreign rulers would like, and, secondly, as will be shown in the next chapter, the "Development Plan for Uganda" is meant primarily for the prolongation of British interest in Uganda and more intensive exploitation of her wealth and labour by the British and allied monopoly concerns. This plan, therefore, does not envisage any serious change in the character of Uganda's colonial economy.

Moreover, as regards other benefits to be received by the Africans, it is sufficient to state here that according to the principal schemes launched by the Development Plan under different heads of expenditure, during 1947—56, (a) out of £ 705,900 to be spent on Education, while £ 357,000 was ear-marked for expenditure on aided schools (which would benefit some of the 5.3 million Africans and about 50,000 Asians, according to the estimates of population in 1953), £ 200,000 was kept for European education (benefitting the thinnest stratum of only 7,000 souls in all in the present-day Uganda society) and for "provision for technical and commercial education"; and (b) out of £ 4,225,300 to be spent as Public Works Extraordinary, while £ 900,000 was ear-marked for African housing, £ 966,000 was kept for European housing, £ 387,000 for Asian housing, £ 636,000 for town-water supplies (to be enjoyed virtually by the non-natives only), and £ 819,000 for "Communication"[334]. Thus, per head of total population in the three communities in 1953, the expenditure on housing is £ 0.1 for the Africans, £ 7.7 for the Asians, and £ 138 for the Europeans (British). Only these two instances of expenditure on education and housing, therefore, make it obvious that no declaration that the Uganda Africans are given a very low price for their crops for their own eventual benefit can hide the true role of the colonial government as serving foreign interests at the expense of the well-being of the native population.

In this connection, the history of the Cotton and Coffee Funds would be of some interest. The Cotton Profits and Hard Coffee Control Funds, set up in 1942, amounted to £ 10,550,000 on the 30th of September 1948. The bulk of the amount was then allotted to African Local Governments and Development and Welfare Funds, and the remaining belance of £ 3,925,000 was paid to new funds called the Cotton and Coffee Price Assistance Funds. Since then surpluses from each year were paid to this fund, which at the end of 1952 stood at more than £ 22 million and £ 13 million for cotton and coffee, respectively. On the 31st of July, 1953, the Cotton Fund stood at £ 24.8 million (including £ 5 million capital of the Lint Marketing Board, but excluding £ 5 million already transferred to the "Development Fund"), and a further £ 5 million was due from 1951/52 receipts. Subsequently, £ 9½ million was transferred to the "Development Fund".

The Coffee Price Assistance Fund, established in 1948, amounted to approximately £ 11 million at the end of 1951/52 marketing season, and in mid 1953 to £ 14 million. Of this, £ 5 million was declared to be retained for price stabilisation and most of the balance to be transferred to the "Development Fund"[335].

Thus it appears that the main purpose in establishing these reserve funds was to take away a large share of the legitimate gain of the African growers for the so-called "Development of Uganda". This conclusion is further supported by the fact that as regards the European (British) coffee planters, the Non-native Coffee Marketing Board does not accumulate trading surpluses and there is no so-called "price stabilisation fund".

Considering now the relation between the British and the Indians in the working of the colonial economy, it is seen that while in coffee the monopoly interest of the British is fully assured, the same interest is given support by the government control of the cotton economy at the expense of the Indian interest which acts as intermediary between the British and the Africans. Surveying the situation it can be shown how in course of increasing government-control of the cotton economy the Indian interest has been undermined, so that today mainly the British interest prevails.

Thus, in the first phase of the cotton economy in Uganda, when the ginners directly profited by the sale of cotton in the export market, there were many British ginners; in fact, practically all the ginners were then British. And there were of course the British export concerns and banking institutions. Notable among them were the Barclays Bank — Colonial and Overseas Branch, the National Bank of India (which should not be mistaken for representing any Indian interest), and the Standard Bank of South Africa. All these concerns are controlled by British capital (with a mixture of South African capital), and have always controlled practically the entire banking business in the Protectorate. In this phase of cotton economy, therefore, besides the lowly position of the African growers, virtually the only interest in the economy was that of the British.

In the second phase, when, with the establishment of the Statutory Pool, large profits were indirectly assured to the ginners, and of course also to the exporters[336], there were still 23 per cent non-Indian (or British) ginners, and as before there were the British exporters and banking interests.

Finally, during and after the last war, while ginning of Uganda cotton virtually became the monopoly of the Indians, the Uganda Government set the up Cotton Exporters' Group and the Lint Marketing Board, in which British interest prevailed. The result was obvious. Even as early as in 1949, while the world prince of cotton lint per pound was 2.65 shillings, the government-fixed price received by the ginners was 1.15 shillings. There remained no doubt whose interest the Colonial Government intends to serve.

The dynamics of the colonial economy thus shows that by the monopolisation of the external trade of Uganda the British Government fulfils another basic demand in the mutual relations between the metropolitan country and her colonies, viz. to subordinate to its control and regulation all the channels which connect the colonial economy with the world market. Thereby, while the national markets of the Africans are hardly developed at all, Britain's monopoly interests thrive on the adaptation of the scattered internal colonial trade to the needs of export, and on the "bleeding" of the national wealth of Uganda.

11(g). Economic Drain on Uganda

Indeed, how severe this plunder of Uganda is would be evident from an examination of the main articles of Uganda's export and import and of her balance of trade. Even as late as in 1951, that is, after the British had "developed" Uganda for more than half a century, the principal articles of domestic export from the colony was "raw cotton", which accounted for 61 per cent of the total export value in that year. Next in importance was coffee, accounting for 28 per cent of the total export value. Then came "hides and skins (undressed)", accounting for 2.5 per cent of the export value; and the remaining articles in the order of decreasing importance consisted mainly of tea, salted fish, sisal fibres, chillies, precious ores and minerals (like tin, copper, gold, tungsten, wolfram, etc.), cigarettes, and sugar. And while Uganda thus parted with her raw materials, she received as import mainly textiles, which in 1951 (even though machines had begun to be imported into the Protectorate in the post-war periods for a more efficient extraction of her mineral wealth) accounted for 21 per cent of Uganda's net import value. In that year, household utensils and such metal manufactures accounted for 14 per cent of import value; and then came the machines and their accessories as described above[337]. It is obvious that none of these imported goods help Uganda to develop her economy to the point from which she, with her own raw materials, could manufacture those goods which she is now forced to buy from the foreign monopolists.

Evidently, as always under colonial rule. Uganda remains a producer of precious raw materials handed over to foreign monopolists, and a consumer of manufactured goods received from the same class of monopolists. And thus living under a foreign rule her external trade consistently registers an adverse balance, she being forced to part with a much greater value of her wealth than she receives by import. In 1926, the Protectorate exported goods worth £ 5,097,215 and imported goods worth £ 2,788,365[338]. There was thus an export-surplus of 182 per cent. In 1951, domestic exports amounted to goods worth £ 47,197,150, and net imports were £ 22,435,674[339]. The export surplus had thus increased to 210 per cent. Of equal importance to note is how this form of external trade affects the individuals in the Protectorate, of whom the overwhelming majority are Africans. In 1925 Uganda's import and export values per head of her total population were £ 0.888 and £ 1.620[340], respectively. In 1951, the corresponding figures were £ 4.308 and £ 9.077[341]. Thus, while the value of imports has increased by less than four times from 1925 to 1951, the value of exports has gone up by nearly six times during the period. This is how the colonial trade is balanced in the Protectorate.*

* That the situation is no better for other colonial countries and that the economic relations between Britain and her colonies primarily serve the needs of Britain's parasitic economy will be evident from the following analysis of her external trade figures for 1950 and 1951. Firstly, it can be seen from the table below that while Britain had an import surplus both from the "Commonwealth countries and Irish Republic" and the foreign countries, as compared with 1950, in 1951 this went up by ten times in the case of the former countries (from £ 26 million £ 261 million), but for the rest of the world the increase was about two and half times (from £ 365 million to £ 966 million).

Table 1

Source and Destination	External trade during 11 months ended November (all figures in £ million)					
	Import		Export		Import surplus	
	1950	1951	1950	1951	1950	1951
Commonwealth countries and Irish Republic. . .	1068	1566	1042	1305	26	261
Foreign countries	1305	2036	940	1070	365	966

The above table thus tends to indicate that Britain's parasitic economy had become more pronounced and that Britain lived through 1951 by a greater exploitation of her empire than in 1950.

This point is further proved by the following table which gives the types of goods involved in Britain's external trade with the empire countries. The table shows that, except for the export of manufactured textile and clothing and other manufactured goods for human consumption or for the extraction of colonies' raw materials (as mentioned before), the import figures are higher than the export figures, and that between 1950 and 1951 the most marked change took place in the import of raw materials which had nearly doubled in value since 1950.

Furthermore, it will be seen from Table 3, which gives the data on Britain's trade relations in 1951 with the colonies and other countries in the Commonwealth (such as Canada, Australia, New Zealand, Union of South Africa, Irish Republic, India, Pakistan, Ceylon, Burma), that the colonies mainly bear the burden of maintaining Britain's parasitic economy and that, quite contrary to the pious claims made about the equal share of the burden and prosperity enjoyed by the members of the Commonwealth, there is a significant dif-

But even such chronic impoverishment of a country like Uganda by the Foreign interests is sometimes regarded as the signal of a healthy economy of the Protectorate. Thus Hailey noted[342]:

> "The investment of capital in the development of African territory contemplates, as it has been seen, the subsequent creation of an export surplus from which to provide for interest payments. It has been one of the difficulties of colonial territories to make the transition from the stage of investment, when imports tend to exceed exports, to the stage in which a sufficient exports surplus exists to meet these charges."

In Uganda higher exports than imports certainly provides for such "interest payments", but does this operation free the peoples of Uganda from constantly handing over their resources and labour for the benefit of foreign monopolists and consuming the manufactured goods sent by these monopolists to the colonies? In 1936 ores and minerals accounted for

---

ference in Britain's trade relations with the colonies and those countries in the Commonwealth with which Britain has to deal more and more in terms of parity.

Table 2

Types of goods involved in Britain's external trade with the Commonwealth countries and the Irish Republic	Monthly average in £ million				Percentage to 1950 of 1951	
	1951		1950			
	Import	Export	Import	Export	Import	Export
Food, drink, tobacco	108.2	13.4	85.8	11.2	126	111
Raw materials	142.9	7.9	83.1	8.8	172	90
Manufacture: Metal	—	100.1	—	89.4	—	112
Textile & clothing	—	44.7	—	34.6	—	129
Others	—	44.6	—	32.9	—	136

Table 3

Source & destination of Britain's external trade within the Commonwealth	Monthly average (in £ million)		Import surplus (in £ million)
	Import	Export	
Colonies	45.9	28.3	17.6
Others	96.6	90.4	6.2

Finally, Table 4 shows that the exploitation of the colonies in supplying (a) food, drink and tobacco, (b) raw materials, and (c) partly processed goods, which came under the heading of "manufactured goods" (such as metals after primary extraction from the ores, wood and timber made into furniture, textile yarns, etc.), was further intensified in 1951 as compared with the situation in 1950. On the other hand, the change in regard to other countries (that is, the remaining Commonwealth countries, the Irish Republic, and the foreign countries) was much less, as will be evident from the figures for "all countries" including the colonies.

Table 4

Imports to Britain from	Percentage of the value in 1950 to that in 1951		
	Food, drink, & tobacco	Raw materials	"Manufactured goods"
Colonies	123	179	176
All countries	123	171	150

Thus it is evident that, except for some machineries and accessories sent for a more skillful extraction of their wealth, to Britain the colonies have remained, as before, the supplier of raw materials and the dumping ground for her merchandise.

only 3 per cent of the total export-value of the Protectorate[343]. Now, therefore, some machineries and accessories are imported for a more efficient exploitation of the mineral and other so-far untapped resources of the land, such as iron, cobalt, copper, timber, etc. And for this purpose the colony has to provide for what is called "interest payments", in order that it can be further impoverished. This is how the "bleeding" of the national wealth of the colony goes on, while even the tools for such operations are supplied from the resources and labour of the same colony.

11(h). Internal Economy of the Protectorate

Furthermore, this is not all that happens within the colonial economy. It is necessary to mention that besides aligning the whole economy to the needs of imperialist Britain and guaranteeing monopoly profits to the British banks and export business organisations in the Protectorate by the previously described method of dealing with the commercial crops of Uganda (viz. cotton and coffee), the colonial economy being based primarily on cotton production directly serves British interest in the internal economy of the Protectorate from another aspect as well. It is interesting to observe that according to the 1931 census of population 68 per cent of the European population in Uganda (of which 82 per cent were British) were gainfully employed, and 76 per cent of the European community consisted either of officials or missionaries or their families[344]. This suggests that not many of them settle down in the Protectorate. They come at a working age, with or without family, and leave the country after a prosperous career. Since the 1948 census of population there has been a high rate of European immigration, and by 1951 the European population had reached the figure of 6,624, or nearly double the figure for 1948. But such increments are partly due to the fact that some Poles refused to be repatriated to Poland after the last war and preferred to remain in Uganda, and partly it is because of the recent import of large numbers of British specialists and assistants (with or without families) in order to (i) advise the Government on Uganda's "development", (ii) run these "development" schemes, (iii) organise a more efficient extraction of Uganda's natural resources, and (iv) run the administration more efficiently in the face of the post-war upsurge of the African peoples for freedom and democracy. These individuals do not form a lively component of the Uganda society today, for most of them are birds of passage who have come to make money in the colony. Also those who have settled down do not consider themselves as belonging to Uganda, but as mere overlords of the benighted Africans. And thus living in Uganda, these "Europeans" are in control of the government and large business organisations and spiritual welfare of the people. In the 1931 census 88 per cent of the "occupied" Europeans were returned as "workers", of whom half were in governmental occupations and the rest presumably in professional and non-governmental jobs. Many of these non-governmental highly-paid officers can be maintained in Uganda because of her "prosperous" cotton economy, and the same is also true for the government services. Dalal, who has painted a very rosy picture of Uganda on account of her cotton economy, noted[345]:

> "The almost spectacular prosperity enjoyed both by the people and government of Uganda is largely due to 'King Cotton'. . . . Cotton is responsible not only for the huge surplus balance which the Protectorate Government has been able to build up, but has also enabled the government to establish and maintain public services involving large amounts of expenditure."

To what extent the benefits from these "public services" go to the people has been discussed earlier. Since, however, the salary of the "European" officers forms a large share of this expenditure, and in 1948 such officers employed in the government departments received an average yearly salary which was 23 times that of the African employees and 3 times that of the Asians (as has been shown in Chapter II), government control of Uganda's cotton economy thus directly serves the interest of the ruling class in Uganda. And it should also be borne in mind that in the commercial and banking institutions of Uganda the salaries of the European officers are still higher than in the government departments, and that the cotton economy has a significant bearing upon their prosperity.

No doubt, within the total structure of the colonial economy, this is one way "to provide for our ever-growing population", as Lugard had noted in 1893[346], and thus to buy up a section of the British middle class.

Moreover, it should be borne in mind that Britain's full share in Uganda's colonial economy is not realised only by means of the enormous profits from the sale of cotton and coffee, maintenance of the system of colonial trade, and providing a comfortable living for a section of the British middle class. Direct and indirect taxation of the Africans by means of imposing additional taxes and raising the existing ones because of "increasing prosperity" of the native population through the cultivation of cotton and coffee, differential freightage-rates of the state-owned railways, and differential tariffs of the administration, all these manoeuvres are closely linked up with this colonial economy. All these points have been touched upon in the foregoing pages. To take now a comprehensive view, firstly, it is appropriate to quote what Dalal wrote while elaborating on the *Importance to the Protectorate of Its Cotton Industry*[347]:

> "Customs and Excise and Licences and Internal Revenue ... form over 75 per cent of the total revenue of the Protectorate and it is clear that both these have a direct or indirect bearing on cotton."

Secondly, it will be remembered that as the Government Treasurer of the Protectorate wrote in his report: "In Uganda the bulk of the taxation is paid by large numbers in rural areas and the amount paid by each individual represents a very large proportion of his money income"; and that this money income comes to the Uganda Africans from the sale of cotton and other cash crops of lesser importance, such as coffee, sesame, etc. Thus in order to survive in a cash economy, the peoples of Uganda must sell their crops and spend the money obtained thereby to buy the most essential goods. But this means adding to the government's share in the colonial economy in its consumption aspect, which, as has been explained in the discussion on colonial budgets of expenditure, does not eventually flow back to the people but is employed for furthering the interests of the foreign rulers. To quote Dalal again in this connection[348]:

> "The export and sale of cotton gives us our spending power, which, in turn, stimulates imports. Thus, the customs duty realised on imports has a close connection with cotton production. The purchasing power of the African population mainly depends upon this and the revenue from this source would at once suffer if we were to experience any large set-back to our cotton crop. ... The importance of the cotton industry of Uganda is not confined to Uganda alone. It brings large benefits, direct and indirect, to Kenya and Uganda Railways and also to the harbours. Piecegoods merchants at Mombasa have to take special account of the figures of the year's crop of cotton before they decide to place their indents in foreign countries."

Thirdly, it will be borne in mind how the differential freightage and tariff rates affect the Africans adversely and the British interests favourably. All these in total, therefore, explain how foreign interests profit also by importing manufactured goods for consumption by the African peoples of the Protectorate.

And now if one recalls the words of the Government Treasurer that in Uganda "taxation is the principal incentive to labour and production", the conclusion becomes irresistable that, thanks to the policy of the colonial government, all Africans must work to fill the coffers of foreign monopolists and their local agents and that even the small sum which is formally given to them as cotton and coffee growers is taken back, almost entirely, through administrative and commercial channels.

This in essence is the benefit the Uganda Africans derive from the colonial economy introduced in their country by the British rulers with the ostensible purpose of uplifting the "primitive" economic life of the "tribal" peoples.

12. Colonial "Developments"

Still it may be thought that even though the British received the lion's share of the benefits from the colonial economy, the Uganda Africans were indirectly benefited by the "developments" which were undertaken along with the introduction and flowering of the colonial economy. Such developments, which the British undertook from the beginning of their rule in Uganda, were connected with linking Uganda with the outer world by building railways and a net-work of roads penetrating into the remote corners of the Protectorate; with building towns and trade-centres within the Protectorate; and with providing electricity and other amenities of urban life (which, however, as noted before, are enjoyed almost entirely by the foreigners living in the towns). In the following pages, therefore, an attempt will be made to examine how this form of colonial development has benefited the African people.

With reference to the character of the colonial regime of Uganda, as discussed before, there cannot be any doubt that in order to drain away the wealth of the country and to dump British merchandise in the colony, both for the benefit of the British bourgeoisie, an effecient transport and communication system was the *sine qua non* of the colonial economy. A good road system was necessary to bring the fruits of African labour from the fields and homes to the export-centres, and then railways were indispensable for a quick and efficient transport of these goods to their ultimate destinations. And the same channels from the other end were needed to bring foreign merchandise to the Africans. Therefore, businessmen, administrators, explorers, and politicians, all with one voice raised the demand for building roads and railways in Uganda. Of Stanley's part in this, a good deal has been noted above. It has also been noted how the Directors of the Imperial British East Africa Company worried about building the railway in order to connect Uganda with the coast in Kenya, and that the administrator Sir Gerald Portal also strongly stressed the immediate building of railways. The politician Mr. Winston Churchill (now Sir Winston) was also not far behind. In 1907, when he described the newly-founded Uganda Protectorate as "the jewel of the East African Empire and confidently anticipated its becoming one of the most valuable of our tropical possession"[349], he further noted that "the Railway was vital to its development, and to its extension urgent"[350]. Thus, roads and railways were then the first demand of the British bourgeoisie in Uganda, and therefore this was major a form of "development" introduced in the Protectorate.

But, typical of the epoch of imperialism, private capital was shy of pioneering but eager for safe investment. In their private capacity the British monopolists were not willing to invest money in order to pave the way for colonial exploitation. It has been noted earlier that the Imperial British East Africa Company could not raise sufficient finance to build the railways, and this was one of the reasons why the Company became "anxious that Government should relieve them of their burden". Brode noted in this connection[351]:

"The Company could not afford to make the railway themselves, because, running the first part of the way through a sterile country, and having few lateral feeders, it could not be expected to be remunerative for some time. Finally, in June, 1895, after Lord Salisbury's return to office as Premier, it was resolved that the line should be constructed by the Government itself. At the end of the year the first rails were laid, and the first train reached the lake [the Lake Victoria adjoining the Eastern and the Buganda Provinces of the Uganda Protectorate] in December, 1901."

So, from the beginning, money came from the Home Government in order to "develop" Uganda for the benefit of the British monopolists.

The returns were of course very pleasing to the British bourgeoisie. "In 1889 the value of goods transported from Uganda to Mombasa amounted already to £ 30,272"; in 1902—03, the export from Uganda was of £ 32,179, and in 1903—04 of £ 67,375[352]. The upshot was that "the tendency of the railway to develop the country soon made itself felt"[353]. Brode noted[354]:

"Thus Mombasa soon began to be served by direct steamers from Europe. After 1899 the intermediate steamers of the Deutsche Ostafrika Linie touched at the port; in July, 1900, the Austrian Lloyd from Trieste commenced running: first bi-monthly, later on every month. In July, 1901, chartered steamers of the British India Steam Navigation Company began calling each month, coming from Europe, and proceeding from Mombasa by Zanzibar and Delagoa Bay to India. In addition, the regular ships of this line used to call at Aden on their way to Bombay."

Further activities in railway building were in progress to connect the Uganda Protectorate with the civilised world. An expert arrived in 1907 to prepare a technical report on railways and steamer lines[355]. Stores and materials brought from Europe for the building of the Kenya-Uganda Railway rose to the figure of 305,000 tons at a cost of £ 2,306,300; from British India 41,000 tons valued at about £ 150,000[356]. Trade began to flourish for the benefit of the foreign monopolists. In 1925 the value of the total exports from Uganda was £ 5.1 million, and the commercial imports £ 2.7 million[357]. Evidently the British bourgeosie were having their way, while the Uganda Africans "benefited" from the railway by means of differential freightage-rates of exports and imports as noted before.

Besides its major economic motive, there was another reason for building railways in the colony. This was political. Brode noted[358]:

"The great political events which then took place — the European crusade against China in 1899 and the rebellion in German South-West Africa in 1904 — enlarged the ideas of the home people, awakened interest in matters abroad, and taught them the great importance of railways in the colonies."

This importance of building the Kenya-Uganda Railways was quickly realised, as "for the suppression of the German East African rebellion in 1905, the existence of the Uganda Railway proved also of great assistance. When riots broke out near Bukoba and Muansa, a naval detachment used this quick conveyance, and so, in a very short time, were enabled to restore order"[359].

Such was another benefit enjoyed by the Africans from the "development" of their land by the introduction of railways.

In addition to railways, building of roads and trade-centres were also of prime importance to the colonisers for a successful emergence of a colonial economy. So money was also spent to build in Uganda roads, towns, etc. Referring to such activities of the colonial government in the early years of British rule in Uganda, Thomas and Spencer noted[360]:

"During these years also, the establishment of stable civil administration and a quiet expansion of trade were leading the Government to investigate developmental projects of general utility in particular the improvement of roads, and to under-take the planning of the various districts stations with a provision in each for a trading centre."

"The increase in the prosperity of the Eastern Province owing to cotton led to a rapid extension of both trade and Government activities and, between 1912 and 1916, H. B. Thomas planned and marked out a number of new townships in this province, Bugondo, Soroti, Njanji, Lira, and Kumi being dealt with."

But the repercussion of such "developments" on the peoples of Uganda was that in regard to the maintenance of motorable roads for the transport of cotton it has been officially commented:

"A disadvantage of this method (motor transport) is its expense and the consequent reduced price to the grower"[361].

Uganda was thus "developed" by the Home Government in order to pave the way for the British monopolist interest in the Protectorate. Up to 1936, £ 46 millions were invested in Uganda, of which the proportion of "public-listed capital" to the total was 68.36 per cent[362]. Of this roughly £ 31. million of "public-listed capital", until 1934, £ 23 million were spent as capital expenditure on the railways[363]. In addition, between 1900 and 1926 the Uganda Government received "assistance" of £ 2.2 million to build roads and export-centres, etc.[364]. Further capital flowed since then, and is still flowing in the post-war years in order to draw away more efficiently the precious raw materials of Uganda while keeping her as "primarily a country of peasant producers"[365]. Of this new scheme more will be said later, but here it should be noted what a terrible economic drain it is on Uganda, for which she has been and is still being "developed". As mentioned earlier, in 1925 the total value of exports of the Protectorate was 1.5 times that of imports; in 1951, the value of exports was 2.1 times the value of imports.

It is of equal importance to note that the capital which thus flowed (and is still flowing) to further prepare Uganda for exploitation by the British imperialist interests and to safe-guard their possessions in this "jewel of the East African Empire" is being raised from the Uganda Africans by means of the system of Public Debt, ostensibly contracted to "develop" Uganda.

How great a burden this public debt has been on the chronically improverished peoples of Uganda is evident from the following. By 1926, after having shared the "assistance" given to the Kenya-Uganda Railway between the Kenya and Uganda Governments, the total public debt for Uganda, "including extinguished debt", came to the colossal figure of £ 6,233,970[366]. Per head of Uganda's population at that time, it worked out at 2 pounds sterling, or, as it will be seen in the following, according to the British Government rate of exchange for African labour, at about 4 to 10 months' slave-labour from each Uganda African. The standing burden of Uganda's indebtedness in that year was £ 1,174,000 (with interest of £ 58,864, or 5 per cent) in the form of loans, and £ 2,548,512 of "grants-in-aid". The burden of total Public Debt in that year was estimated by Buell at 24 shillings per head of the population in the Protectorate[367]. The ratio of interest to revenue as estimated by him was 4.5[368]. These figures give an idea of the load the peoples of Uganda had to bear from the beginning of British rule for the "development" of their country.

In 1936, Public Debt amounted to £ 2,235,600, which Hailey, an expert on Africa, considered as not very big. The Uganda Government expended £ 137,598, or 6 per cent of the amount, as debt charges for the year[369]. What an economic drain it was from the country is evident from the fact that only the "debt charges other than railways" accounted for 8.4 per cent of the budgeted expenditure of 1936—37 of the Protectorate[370]. The people went on paying. The incidence of public debt per head of Uganda's population at that time was 12s. 7d.[371], evidently a lower figure than a decade earlier.

But in the next two decades which witnessed the shrinkage of British Power in the world, the situation was different. By 1951 the Public Debt reached the figure of £ 5,950,000, of which 2 millions were at 5 per cent and the remainder at $3^1/_2$ per cent Inscribed Stock. The per capita burden of public debt, in spite of an one-third increase in population, now shot up to 23 shillings[372]. The people had to pay more. In 1946, out of the total budgeted expenditure of the Protectorate, 5 per cent was ear-marked for public debt; in 1951 the proportion of debt charges to the total budgeted expenditure rose to 13 per cent, an increment by 8 points within a period of five post-war years[373].

It is of further interest to note that the amount expended on debt charges in 1951 was £ 1,731,666, which was only a little lower than the total amount of 2 million pounds expended under the heads of "agriculture", "education" and "medical"[374]. And this was so in a country which still depends and will so depend in future, as declared by the present rulers (to be quoted later), entirely on an agrarian economy, where illiteracy is rampant, and where terrible diseases and high death rates are ravaging the African people. And this is so in a country where the huge drain of her resources by export without compensating import has been hailed by the Africa-expert Hailey as a sign of her healthy economy[375].

It is equally worthy of note thatwhile in such a poor colony 13.4 per cent of the budgeted expenditure for 1951 was spent on Public Debt, in the mother country of Great Britain the proportion of expenditure on National Debt to the total budgeted expenditure for 1950—51 was 12.6 per cent.

Thus, while the British monopolists in their private capacity avoided the burden of preparing Uganda for colonial exploitation, although it was for their own benefit, their government financed the projects and recovered the exported amounts (evidently more than what they lent out, for to this was added the interest charges, etc.) from the peoples of Uganda by the mechanism of Public Debt. The process is still going on, and it goes under the spectacular banner of "colonial developments".

Moreover, the "development" of Uganda has been effected not only by exporting some capital, as noted above, but in various forms the labour of the Uganda Africans has been directly exploited from the beginning with the same purpose in view. Notable among them is the introduction of forced labour from the onset of colonial administration in Uganda.

As mentioned earlier, previous to British rule, in the semi-feudal kingdoms of Uganda the people had to give compulsory free labour for making roads and royal buildings. This custom was most developed in Buganda where the British first entrenched themselves. Now under the rule of the defenders of "western civilisation", this custom was pursued with renewed vigour to further the cause of the British bourgeoisie. It has been noted how important was the building of good roads and trade-centres for the efficient functioning of the colonial economy. Hailey noted[376]:

> "Uganda is little served by railways, and the growth of cotton cultivation has given a special importance to the existence of a good road system."

Also, although the railways had not penetrated very far into the interior, there was the need for workmen to maintain the lines, to build stations and other establishments, etc. These were not overlooked by the British rulers, and they were searching for the cheapest way to solve the issue. For instance, Stanley wrote in his letter to Zakaria which has been mentioned earlier:

> "The Waganda are very clever, I know personally, but they can be cleverer, and when the railway reaches the lake there will be a greater demand for all kinds of workmen to build houses, stores, shops, churches, schools, etc. .... I should like to hear that your people are preparing themselves for the great change that is coming."

Hence, to see to it definitely that the "people prepare themselves for the great change that is coming", while direct slavery was abolished after the British flag flew over Uganda, the system of forced labour was kept in force. Not only that. Previously this was restricted only to some parts of the present boundary of the Protectorate, but, under the administration of a more efficient ruler than of Buganda or Kitara, it was extended to cover the whole of the Uganda Protectorate. Henceforth, every able-bodied male African was required to furnish thirty days' free labour in a year, "or six days more than in Kenya"[377]. Railways and new towns were built thereby as centres of trading business, and, in the words of Hailey, Uganda could boast of "a system of roads that is generally regarded as being among the best in Africa"[378].

In this way, from the beginning the Uganda Africans were forced to contribute to the import of "western civilisation" into their country which finally bloomed through the cotton economy, the pivot of colonial exploitation under British rule in Uganda.

This form of slavery for "communal purposes" was not, however, retained only for a short while during which the foundation was laid for the colonial administration and the colonial economy. Even when the transport system had been well-organised and a number of towns and trading centres built, the Africans were not allowed to get rid of this form of unpaid forced labour. On the contrary, it was perpetuated in a cash economy, by allowing them to commute this communal or *luwalo* labour into a regular tax. Hailey noted[379]:

> "The proceeds of the commutation of luwalo are paid to the native administrations; about one-third of the adult males avail themselves of commutation in Buganda, and large numbers in the other provinces. Persons who, having opted to commute, fail to meet their liabilities by the due date, may be required to pay a fine of 10s. and to work for a period up to two months, failure to work being punishable with imprisonment; neglect or refusal to discharge the penalty within two months of the due date is punishable by a fine of 20s., and the defaulter remains liable to work for two months."

Thus, under a *civilised* administration, slavery took a new form.

In later years the *luwalo* tax was further increased. In Buganda it was raised from the former rate of 10 shillings to 14 shillings in 1947[380].

Moreover, the native bureaucracy laid additional burdens on the people in lieu of their labour "for communal purposes". The Native Administration of Bunyoro and Toro imposed a shilling education rate [381], although how extensive and deep educational facilities are has already been noted. Needless to say, most of such ill-gotten money was spent for the welfare of these bureaucrats. Much has already been said on this point. Even so it may be of interest to note here that from the commutation of *luwalo* the native government of Buganda received in 1925 a sum of about £ 29,000, of which £ 17,400 were spent at headquarters and the remainder in the districts. Of the £ 17,400 retained at the headquarters, about £ 5,000 were expended on "personal emoluments" and about £ 7,000 as the out-of-pocket expenses of the native bureaucracy in order to maintain motor cars, for entertainment, and as "alimentary allowance for certain of the Kabaka's relatives"[382]. It is also worthy of note that from this collection every year about £ 1,500 used to be spent "in maintaining the Lukiko police and the Lukiko prison"[383]. In 1935, the proceeds of the commutation of *luwalo* in this province were £ 75,853, out of the total estimated income of the native government of £ 128,638[384]. Comments are unnecessary. There is no wonder that the newly-created native bureaucracy remained docile to the British regime.

In addition to this forced unpaid labour, the Africans were also obliged to undertake "compulsory paid labour for certain public purposes". These "public purposes" centred round the colonial administration and activities ancillary to the introduction of a colonial economy. Buell noted[385]:

"In addition to obliging the natives to perform a month's unpaid labor for communal purposes, the Uganda Government also resorts to compulsory paid labor for certain public purposes. Befofe 1921, practically all the labor required for government services was secured by a form of compulsion, under a system of *kasanvu* labor. No serious effort to obtain voluntary labor was made; men were called out under the Native Authority Ordinance under which they could be required to do work 'of a public nature' for not more than sixty days in any year, unless employed for a period of three months in any other occupation. It is estimated that twenty thousand men were annually obliged to submit to this obligation. The situation was described by a government circular as follows: 'At present, but a small proportion of the unskilled labour employed by Governmnet departments is voluntary. There are many reasons for this, not the least of which is that officers, being assured of a continous supply of compulsory labour, have not found time to give that "porsonal atteniont to the requirements of their labour and to causes for discontent amoug them which is essential if a supply of volunteer labour is to be attained. Head-men also have not encouraged voluntary labourers as they must be handled with tact and patience — qualities which need not be exercised in dealing with Kasanyu labourers, who are not free agents'."

Thus, even though some remuneration was paid, this form of forced labour was no better than slavery, and perhaps even worse than the domestic slavery prevalent in some of these areas before; for, in that case, at least a bond developed between the master and the domestic slaves as has been so often reported by Roscoe and others, whereas for the Kasanvu labourers no "tact and patience" were necessary for they were not "free agents".

And how big was the remuneration paid to these working-slaves? Many of them were employed on the railways as porters, etc. The importance of railways in developing the colonial economy has already been noted. The Africans, being thus employed for "public purposes", received wages of Rs. 2 to Rs. 3 per month, "and this made it possible, in spite of the cost of transport, for ground-nuts to be cultivated on the Lake at a paying rate, which was impossible on the coast on account of the high price of labour"[386]. Was this form of forced labour anything better than slavery?

However, this "compulsion worked in a vicious circle". Buell noted[387]:

"It also resulted in all the dullards being swept into government employ, while the intelligent natives, to escape this requisition, sought more popular employment with European coffee planters. Such a form of "compulsion, though nominally for public purposes, indirectly benefited private employers."

Thus, even though the government labour force was not up to the mark, evidently because the people resented this form of veiled slavery, it helped the British in another way, by indirectly selecting efficient labour force for the planters.

But it was at the same time leading the country to desolation, and the African economy to rack and ruin. Even though the colonial administration might not have been seriously concerned with the devastation of the peoples' life and economy caused by this pernicious system, they had to make some ammendments to it when it led to a "severe labor shortage". Buell noted[388]:

"Only steady men, having a stake in the land, men with families cultivating shambas (the native word for farm), could be caught, as the unmarried men would scoot off at the rounding-up time. These young men were reluctant to marry and settle down upon a shamba, and became wanderers, escaping all obligations, including those imposed by the tribal authority. The institution was thus destroying native society. The system itself was breaking down because of the difficulty of catching the men. Between 1917 and 1920, the government experienced a severe labor shortage. These considerations led to the abolition of this system in 1923."

This measure for Uganda's "development" thus helped in the disintegration of the societies and in turning the settled populations into vagrants — wandering about to escape the agents of slave-drivers.

Moreover, it should be noted that although Kasanvu labour was partially abolished because it could no longer serve the interests of the colonial regime, the institution of indirect slavery of the Africans remained in force. Only it now took a slightly different form. An amendment was made to the Native Authority Ordinance, according to which this Kasanvu labour "can not be employed, except for government porterage, with the prior consent of the Secretary of State"[389]. The consent was of course readily given "for railway or road construction", the two prime concerns in building the colonial economy.

By this amendment, "the entire male population within a radius of five miles on each side of the road" were called out in installments to contribute their share to "communal labor", the new sign-board for Kasanvu labour. "During 1923, 14,097 men were called out under the communal system. In 1924, the number was 19,900, each working an average of 23.4 days"[390]. Buell noted that these Africans were paid the "market wage"; but he also noted[391]:

> "In 1924, nearly seventy-five miles were built at a cost of about one hundred and seventy-five pounds a mile, compared with a cost of six hundred pounds a mile under a system of regular voluntary labor, where men work under departmental officers for a long period rather than under their chiefs and political officers."

Slave-labour thus still remained the key to Uganda's "development".

This, however, could not go on for long. Unpaid or "paid", the people evaded this form of slavery as far as they could, and if finally roped in by the local agents of the colonial administration they did not put their heart into the work. Consequently, in spite of the use of "communal labor", the government departments began to face a serious labour shortage, "particularly in regard to transport and building". The government then required about 1,500 men a month in the Public Works Department and 5,500 on railway construction[392]. But direct slavery had become outmoded; how then to find this labour force to pave the way for greater and greater colonial exploitation?

Finally, the administration decided to recruit only, or mainly, "voluntary" labour. But "voluntary "labour could no longer be found on a large scale from Buganda and the Eastern Province where the foreign rulers had the most intimate contact in the beginning and therefore the Africans there had already been made interested in cotton production. Hence, the African labour reserves which were not yet employed to grow cotton, such as in the Northern Province and elsewhere (from the outlying district of West Nile at the Belgian Congo border, for instance) were now tapped to "develop" Uganda. Buell noted[393]:

> "Unable to obtain labor by compulsion and to find volunteers in the cotton growing areas of Buganda and the Eastern Province, the government in 1923 established a Labour Bureau for the purpose of recruiting government labor in the non-producing provinces, primarily from the Ruanda district in the south and from the West Nile in the north — areas where native production has not yet been developed."

This Labour Bureau, headed by a Labour Commissioner (of course British), employed native recruiters to decoy Africans from the "simpler societies". These recruiters received six pence for every recruit. There were also two British "conductors" to supervise the operation.

How "voluntary" this labour-force was cannot be discussed for want of relevant information. But it may be noted that in a similar way labour was recruited in India for the British-owned tea plantations in Assam and Bengal. With high hopes and false promises, given by the recruiting agents of the planters, the impoverished peasantry from the "civilised" villages as well as the people from "tribal areas" were decoyed to the tea-gardens where, once they entered the estates, they were not allowed even to go beyond the garden areas lest they should escape from the inhuman treatment meted out to them. When it was urged that the "slave law" for tea-gardens should be abolished, the British India Government declined to do this, under direct pressure from the British planters[394]. And thus remaining under the clutches of slave-drivers, most of these "voluntary" labourers could never get out of their bondage. They had to borrow from the estates, for the wages they received could not keep body and soul together; and thereby they entered into the scheme of "life imprisonment" in the estates[395].

In Uganda also the Africans thus removed from their homes were to sign four-months contracts at twelve shillings a month. If they stayed for another two months, their wages were increased by two shillings[396]. In this way, in 1925, "the Bureau recruited about twenty thousand men, including twelve thousand who came from Ruanda and six thousand from the West Nile"[397].

And how were these "primitive" Africans treated by the members of a superior civilisation? Buell noted[398]:

> "Natives were obliged to walk most of the way without adequate provision being made for their wants. Sanitary and cooking arrangements in rest camps were particularly bad. Apparently, the medical examination was inadequate, since many natives carried tick or spirilium fever from one end of the country to the other. According to the medical report on this 'imported' labor, the annual death rate was one hundred and eighty per thousand, of which fifty deaths per thousand were on account of dysentry. The treatment of labourers returning home after the completion of their contracts was particularly bad. They were simply given rations at the beginning and then left to shift for themselves, with the result that many died."

Such were the aftermaths of the much-vaunted abolition of slavery and bringing "civilisation" to Uganda by the British.

Without a penny or for a mere pittance, the peoples of Uganda were thus forced to work in order to "develop" Uganda as required by the British monopolists; and after this slavery had ravaged the country for decades, peoples from less "developed" regions were lured into the death-camps by offering them a comparatively higher wage-rate. British Uganda was thus built on the life-blood of the "primitive" Africans; and this is the development that the experts and propagandists now boast of before the world public.

13. "Prosperity" of the People

Yet it may be argued that, in whichever way it took place and for whatever ultimate purpose, by providing Uganda with modern towns and an efficient transport system Britain has benefited the Africans. But whether the Africans can avail themselves of these amenities of civilisation obviously depends primarily on their prosperity. Therefore, of still greater importance than the above account of the plundering of Uganda's wealth and labour behind the facade of civilising the Uganda Africans and "developing" their socio-economic life is the question whether the British improved the state of productive forces in the Protectorate. This being fundamental to a better future of the Uganda Africans since the time the British appeared on the scene, all forms of British oppression could have been at least partly forgiven if they had initiated progress in the economic life of the Africans. But did the foreign rulers do that at least during their rule of more than half a century? The answer to the question is in the negative.

From the very onset of their rule in Uganda, the consistent policy of the British monopolists has been not to improve the state of productive forces in the African societies, although as stated at the beginning this was virtually the only way to uplift these societies and put the people on the way to prosperity and progress.

It has been noted before that during the first experimental stage the central aim of the British capitalists in Uganda was to make money quickly without large investments of capital. Later also, even when in the face of inter-capitalist rivalry production of cotton became their supreme policy in Uganda, they did not undertake the task of improving the agrarian economy. A few farms with somewhat better means of production were established for growing coffee, tobacco and tea, but in the main the aim of the British was to persuade the Africans to produce commercial crops for the benfit of foreign interests by means of their traditionally primitive form of peasant cultivation. Thereby the British monopolists could avoid large investments of capital in primary production, ultimately destined for their own good, and also could keep the colony as a producer of raw materials by means of the capital and labour of the subjugated peoples.

Indeed, this remained such a consistent policy with the British that as late as in 1946 the Agricultural Department of the Uganda Protectorate Government estimated that while the crops produced by the Africans by means of peasant cultivation were valued at £ 6,152,281, those produced in the better-equipped farms of non-Africans were valued at only £ 1,742,739. What is of greater importance to note is that cotton was produced entirely by African peasant-cultivation. Evidently, production of raw materials from the colony by native capital and labour has remained the dominant note of colonial policy in Uganda. In 1951, while some 71,000 acres were under cultivation by non-Africans, the total area under peasant cultivation by the Africans amounted to 6.9 million acres[399].

And while the foreign monopolists thus insist on making the Africans produce the crops without themselves taking any interest in production, virtually no assistance is forthcoming from the government for the better equipment of productive forces of the African peasants. It has been noted before how little is allotted to the account of "agriculture" in the Protectorate budget. Furthermore, even in the loudly advertised "Development Plan for Uganda" (of which more will be said later) only 5.3 per cent of the total expenditure during 1947—56 has been ear-marked for "agriculture"[400]. And this would be used mainly for "experiments" which means mainly on the high salaries of British officers conducting such "experiments".

The Africans also by themselves cannot change the state of affairs so long as they remain victims to colonial and feudal exploitation. Because of the very low incomes which they draw from peasant-cultivation, as will be seen later, they cannot afford to invest money in capital equipment to mechanise or even partially improve the state of productive forces in agriculture. The upshot is that the state of productive forces in Uganda still remains almost as primitive as it was when the British first arrived. Very little attention has been paid during these many years to improving the mode of production. The hoe is still the principal tool used to prepare the field, and human energy the only motive power available for production. Because of the "ever-advancing menace of the tsetse fly which has already swallowed up more than one-third of the land area of Uganda"[401], even though Uganda has been under British rule for nearly six decades, cattle cannot be used as motive power to a large extent. Therefore, even plough cultivation, which would have been a little advance in the agrarian economy, is practically non-existent if the whole of the Protectorate is taken into consideration. And, consequently, even as regards the main economic activity in the Protectorate, the 1951 *Economic Survey of Uganda* reported[402]:

> "Seed cotton is produced in Uganda entirely by African peasant farmers, usually on small plots of 1—3 acres. . . . The preparation of the land for the crop is done largely by hand, as is the cultivation and picking."

The result is obvious. Even though the few British and Indian owned farms in Uganda do not always employ the most modern methods of production, their rate of productivity is many times higher than that of the African peasants, as can be worked out (very roughly of course) from the available data. Quoting the figures supplied by the Agricultural Department of the Uganda Government (which probably referred to 1946), Hailey noted that non-native cultivation in Uganda was restricted to 48,662 acres, and that in 1946 the British and Indian farms produced crops of the total value of £ 1,742,739[403]. On the other hand, in the same year the African peasants in Uganda produced crops of the total value of £ 6,152,281, and the total area under their cultivation has been estimated at about 5 million acres at that time[404]. Therefore, even by such a rough calculation as dividing the total crop value of the non-natives by 48,662 acres and that of the Africans by five millions, one finds that the rate of productivity per acre of land is £ 35.6 for the non-natives and £ 1.2 for the Africans. The difference in the two rates of productivity is no doubt very revealing, and gives a fitting answer to the loudly proclaimed "prosperity" of the Uganda Africans because of the colonial economy of cotton production. Indeed, as it can be calculated from the figures supplied by the Uganda Government, while the rate of productivity of British-grown tea in 1946 was £ 88.1 per acre, that of African-grown cotton was only £ 1.9 in 1947 and £ 6.2 in 1953*.

The situation as a whole can now be formulated as follows. The "prosperity" of Uganda on account of the cotton economy is not due to an increase in the *constant capital*** employed for the production of the crop by investing money in improving the state of productive forces. The increase in production is almost entirely due to an extensive use of the *variable capital***, viz. the cheap labour-power of the native population. It may be of interest to note that to work out thecost of production of cotton per acre of land, the only item of importance which the Government Agricultural Farm at Ngetta considered worthy of note is the man-power[405]. It is also worthy of note that Carter reported quite candidly that the spectacular growth of cotton production in Uganda, centering on which the entire colonial economy was built up, was effected not by raising the productivity of land but only by bringing in more and more land under cultivation[406]. In other words, the foreign interests thrived mainly by plundering the natural resources and labour of Uganda without putting in anything substantial in return.

This weak *organic composition of capital*[407], that is, a high share of the *variable* to the *constant* capital, is no doubt a typical feature in colonial countries[408]. It keeps Uganda, like

* For tea, Hailey has noted that the acreage was of 4,335, and the value of production, £ 381,932 (cf. *loc. cit.* 134, p. 3). The value of the yield of cotton per acre in 1947 is given in Table IV.8 in page 196, and that for 1953 is calculated from the figures supplied by the Uganda Government, viz. the cotton-area in 1952—53 was 1,611,000 acres and "the return to growers was estimated at a little over £ 10 million" (cf. *loc cit.* 266, p. 43).

** Karl Marx, who was the first to use the concept of constant and variable capital, explained it as follows (cf. *Capital: A Critical Analysis of Capitalist Production,* Vol. I, George Allen & Unwin Ltd., London, 1949, pp. 191—192):

"The means of production on the one hand, labour-power on the other, are merely the different modes of existence which the value of the original capital assumed when from being money it was transformed into the various factors of the labour-process. That part of capital then, which is represented by the means of production, by the raw material, auxiliary material and the instruments of labour does not, in the process of production, undergo any quantitative alteration of value. I therefore call it the constant part of capital, or, more shortly, *constant capital.*

On the other hand, that part of capital, represented by labour-power, does, in the process of production, undergo an alteration of value. It both reproduces the equivalent of its own value, and also produces an excess, a surplus-value, which may itself vary, may be more or less according to circumstances. This part of capital is continually being transformed from a constant into a variable magnitude. I therefore call it the variable part of capital, or, shortly, *variable capital.*"

other colonies, a backward territory, and thus makes a decisive mockery of the rulers' claim of introducing progressive technique in the colony's economy. And simultaneously it underlines the true meaning of the so-called "prosperity" of the Uganda Africans, for it does not allow the productivity of labour to develop. Indeed, such is the situation in this colony that, in order to examine the comparative position of the cotton growers vis-a-vis other types of working Africans in Uganda, it has been calculated that in 1947 the income of a cotton grower was twenty or twenty-one shillings per month, which according to the wage-rates prevalent in Uganda in that year was less than the wages received by any grade of semi-skilled labourer, and was even less than the wages of an unskilled labourer in the Sugar Estates or the building industry or the Public Works Department of the Protectorate Government[409]. This can be seen from the table below prepared from the official data[410].

Table IV.8

1.	1947 yield of seed cotton per acre of land	204 lbs.
2.	1947 value of seed cotton per acre yield at 18.45 shillings per 100 lbs.	37.64 shillings
3.	Equivalent daily wage-rate of cotton grower at 45 man-days work per acre of land	84 cents
4.	Equivalent monthly wages of cotton grower at 26 working days a month	21 shillings

Note: — In the above calculation the only item of cost of production is regarded as "human labour". I a hoe is to be regarded as an extra expenditure, at the average price of 3.23 shillings (cf. *Uganda Blue Book 1945*, Entebbe, Uganda Protectorate), then the monthly wage of a cotton grower would be 20 shillings.

Since it has already been noted how miserable is the existence of even the better-paid railway workers in Uganda who also suffer from malignant malnutrition, no further comment on the vast mass of African peoples who eke out a living as "traditional" peasants is necessary. It should, however, be borne in mind that even the life of a wage-labourer is denied to the bulk of the people because of the lack of industrialisation of the country.

Evidently, the "economic development" of the Uganda Protectorate by the British did not really lead to the much-advertised prosperity of the African people. It was a shrewd method of exploiting the resources and labour of the country. The essential connection between colonial trade and imperialist exploitation has been fully maintained, and the cool calculations of the British monopolists have been fulfilled. According to their "uniform experience ... (Uganda did) amply repay the first expenditure ... both by extension of trade and shipping and in the growth of national power and status". And, in consequence, as mentioned earlier, half a century after British rule the Baganda patriots declared that "Great Britain is still fond of slaves".

Furthermore, the author of this publication, *Buganda Nyafe*, stated that "the modern slave trade includes children and women and English laws put yokes round our necks even though we think we are in our native home". The truth of this statement is also inescapable to any one who is seriously interested in undertaking a scientific examination of the *problem of Uganda*. For, although the process of exploitation is so cleverly executed that it leaves a mirage of prosperity and progress in the country, it cannot hide the real poverty of the African people.

Having been brought within the fold of a money economy, it has been imperative for the Uganda Africans to become sellers and buyers. Hence, after paying all taxes and tributes from the cash they receive in the cotton season by selling their produce, the rest of their money income they spend in the *dukhas* (small retail shops run mostly by Indian petty traders) to buy clothes (mostly cheap calico prints) and such essential articles. This is their "prosperity" which has been repeatedly stressed by the Government spokesmen and even by several scientists. For instance, a bicycle is an article coveted by them, because in a country where the transport system hardly at all caters for the Africans living in the villages, this becomes a necessity. But their purchase of bicycles is regarded by the experts as a mark of great prosperity. Hailey noted in 1950 while explaining the reason for heavy taxation of the Buganda Africans[411]:

> "It will be seen that the total taxation paid by individual native taxpayers in Buganda is noticeably higher than that in neighbouring territories, though this is held to be justified by the relatively high production of marketable crops, and the general standard of living in Buganda. It was stated a few years ago that no less than 60 per cent of the taxpayers owned bicycles."

No doubt, therefore, there cannot be any argument against such heavy taxation as is imposed on the Baganda! But such is their prosperity that even after six decades of British rule they remain tied to the stagnating colonial economy as "sons of the soil" without any possibility of developing the productive forces, for they have no resources and no opportunities to break away from it. Even the demobilised soldiers of the Second World War came back to put on this yoke of slavery. As the Colonial Annual Report of the British Government on Uganda admitted in 1947[412]:

> "The majority of ex-soldiers returned to the agricultural pursuits in which they were engaged before joining the forces."

The trades that they had learnt in wartime service were of no use to their motherland. The Governor of Uganda declared in 1946 that "Uganda is and must remain for the present at all events primarily a country of peasant agriculture"[413]. And the British expert on Africa, Hailey, noted in 1950[414]:

> "Agricultural production is therefore the predominant asset of the country and this, with the exception of the sugar and tea estates and a number of coffee estates, is in the hand of Africans."

Thus, even though the Uganda Africans have been in contact with one of the foremost protagonists of "western civilisation" for more than half a century, they still live in the "bush" as the immemorial tillers of the soil.

And how does their standard of living compare with that of the foreigners who live in the towns and enjoy the benefits of Uganda's "developments"? In the villages and in the outskirts of towns the Africans live in "traditional" mud-huts; while in the towns the Europeans have modern houses and bungalows with extensive gardens, and the Asians also have brick-built houses or apartments in a large building, such as in Kampala. In the towns the foreigners enjoy electric lighting or in remote places kerosene lamps; while, as noted by Leys, "hardly any East African can afford artificial light". The foreigners in urban areas have a regular water supply from central reservoirs or from tube-wells; while the Africans in the villages do not always have earthen wells for their supply of drinking water, and, if they have any, they are not always kept in working order, as the author himself found in many places during his sojourn in Uganda. And this is so, even though medical authorities have often asserted that the Africans contract several serious diseases from impure water. Like the supply of pure water, there are no organised sanitary arrangements in the villages; and whatever medical and educational facilities are available in the Protectorate they are mainly centred round the urban areas, whereby they can best be enjoyed by the foreigners. Moreover, as noted before, the proportionate expenditure on these accounts is higher for the foreigners than for the Africans.

In this way much could be written to show that in Uganda there are two sharply differentiated standards of living. One, which applies particularly to the Europeans (British) and to some extent also to the Asians, is modern and comfortable, — to the point of being luxurious for the ruling class. The other, which applies virtually to the Africans alone, is shabby, miserable, and is taken to be "traditional". And this so-called "traditional" way of living of the Africans has been made good use of by the Government spokesmen. As recently as in 1954, a Government report on the question of Social Security of the Africans stated[415]:

> "It is policy whenever practicable to leave the care of the destitute and disabled in the hands of the tribal clan and family organizations which have traditionally accepted this responsibility."

Thus, on the one hand, the foreigners live well, ride in the upper classes of the railway trains, and make use of the excellent road conditions with their cars and trucks, and, on the other, the Africans, to whom this country belongs, lead a sub-human existence, can hardly aspire to ride in railway trains (and then also not higher than in the third class), and likewise only a small proportion of them (when they are so wealthy as to possess bi-cycles) can make some use of the good road system which they have themselves built by means of slave-labour. In short, while some amenities of modern civilisation have been strewn over Uganda and the spokesmen of foreign rulers make a great song about it without, of course, ever mentioning what it meant in terms of African life and labour, these are almost entirely usurped by the foreigners, and among them again mainly by the Europeans (or British).

Such are the true meanings of "developments" of Uganda and of "prosperity" of the Africans under British rule. There the Africans have been brought within the pale of "civilisation", but they remain in the dark while holding the lamp for the "white" standard-bearers. Their lot is suffering and misery, such as they had never experienced before; while the light of "western civilisation" shines over Uganda, and all amenities of the modern world are enjoyed by the foreign rulers who, whether living in the mother country and clipping coupons or staying in the colony for a while to make a fine career, fatten on their toils and the riches of their land.

14. Retrogression of Uganda

Along with the foreign rulers, the foreign or local middlemen thrive in colonial Uganda. The role of the Indians as middlemen between the British and the Africans has been mentioned before. Besides them, there are the African middlemen — the *traditional* "kings" and the *traditional* or *British-made* "chiefs", who enjoy their privileged and exalted position by fulfilling the demands of the colonial regime and live on the labour of the common people. For the people in the "developed societies" this burden of native interest has been even more oppressive, because of the introduction of landlordism. This course of retrogression, based on imposed feudal relations within the colonial regime, is not only having a far-reaching effect in the "developed" societies, but is also spreading into the "simpler societies". To explain the full implication of British rule in Uganda, it is therefore necessary to examine this colonial-feudal characteristics of Uganda today.

The foundation to this course of retrogression was laid in Buganda where the native rulers were made into a landed aristocracy by the 1900 Agreement. This was the kingdom in the pre-British period where agriculture had flourished most in this part of the "Dark Continent". Now, in order to consolidate British interest in this most-developed region of the Protectorate, besides collection of taxes and tributes, further incentive was given to the native rulers for profiting from land. It has been explained before how the landed aristocracy developed in Buganda. Here it is worthy of note that in the subsistence economy of pre-British Uganda landownership by itself could not have seriously affected the life of the people. But the simultaneous appearance of the foreign monopolists and their agents in the kingdom had a profound effect on the society. The landowners began to understand the value of their possession. Mair noted about this time[416]:

> "The concurrent [with the 1900 Agreement] adoption of a money economy brought with it the idea of land as a source of profit and their heirs in perpetuity resulted in the almost complete dissociation of rights over land from governmental functions."

Profiting from possession of land became an important source of income to the native rulers. At the beginning, however, they did not realise the full worth of their possession. Therefore when they were offered what they considered as high prices they began to sell their holdings to the British speculators who had come either individually with a view to settle down as planters or as representatives of monopoly concerns to prospect for fortunes in the Protectorate. The scramble for land began. In this scramble while the British reaped the first benefit, the Indians who had first come as indentured labourers or as petty traders also followed suit. Thomas and Spencer noted about the situation[417]:

> "It was not until the advent of European settlers began to give value to plantation land that there was any realisation that a vital and permanent alteration in the old land-holding customs had been effected. ... Alienations by natives to non-natives soon began to take place, first in freehold, but after 1916 in leasehold only, and despite the investigation given by the British Government to each case, the cupidity or supineness of native landowners did on occasions lead to clan lands and their customary tenants passing under the control of European planters. ... Native landowners in favoured situations began to derive large incomes from rental of plantation land, and later from cotton ginnery sites or from tenancies by petty Indian traders who tended to congregate in considerable settlements on native land."

This, however, happened at the very initial stage of British rule in Uganda. Soon the policy of the foreign rulers was crystalising into making the Africans employ their own capital and work for the benefit of the foreign capitalists. Cotton had then been introduced to become eventualy the pivot of the colonial economy. The declaration of the Secretary of State in 1915, that "no more land — whether Crown or Buganda mailo [allocations by the Uganda Agreement of 1900] — should be sold in freehold to non-natives"[418], may not therefore be quite fortuitous.

But that declaration did not do away with the incentive given to the native rulers as landowners; on the contrary, it bore fruit in producing for the colonial market. Occasionally they still alienated their lands to non-natives in lease-hold, as noted in the above extract, but rapidly they learnt that their profit lay in growing cotton themselves, or more in persuading other Baganda and immigrants to Buganda (such as from Ruanda-Urundi) to grow the crop as their tenants. The first was easy for they had plenty of land, and the second was also not a difficult task. Traditionally the Baganda are an agricultural people. So with the introduction of a cash economy it was no problem to make them interested in the cotton economy in order to pay their taxes and from the remainder, if there remained any, to buy "interesting things", like printed calico and metal trinkets, of which Stanley had told the British bourgeoisie some time before and which they had now brought to Uganda in great quantities. The Baganda thus took to the cultivation of cotton at the cost of their own industries, and in this economy the possession of land, the *primary means of production,* became the governing objective of the upper stratum of the society. Consequently, the declaration of the Secretary of State in 1915 did not put a stop to alienation of land. It began to work *within* the Buganda society.

Those Baganda who had made money by serving the British in exploiting their own country and those who were made into landlords by the 1900 Agreement (mostly the same persons) now began to buy more land. Mair noted[419]:

> "Almost every native who has been in European employment for any length of time has managed to acquire a small piece of land, while in more prosperous time people bought land with the proceeds of economic crops."

The Uganda Agreement initially created about 3,700 original allottees, although according to the clause laid down there should have been only 1,000 landowners[420]. In about 1921 the number, through succession and purchase, rose to 10,000[421]. In 1931, the number of landowners in Buganda was estimated at 16,000[422]. In 1950, Hailey noted that there were 25,510 registered titles of privately-owned plots and that "there are now two to three times that number still unregistered"[423].

Such *differentiation* of the people according to property-inequality in land opened up a vista of prosperity for the landlords. They began to earn a good living by their parasitic existence on the labour of the "peasants". Rents were imposed on the tenants which, after 1918, were "generally fixed at 10 shillings a year for each peasant's garden, occupying usually one or two acres, but signs are not wanting in succeeding years of a desire on the part of certain landowners to increase this figure", as Thomas and Spencer remarked in 1938[424]. A rough calculation, based on the available figures, shows that during 1942—47, in spite of the relatively higher prices of crops in this period, which should benefit the grower if the land rent is fixed, 16 per cent of the peasant's gross income went to his landowner. This can be seen from the following table prepared from official data[425].

Table IV.9

1.	Average yield of seed cotton per acre	330 lbs.
2.	Average price of 100 lbs. of seed cotton	13.85 shillings
3.	Gross income of a peasant from an acre	45.71 shillings
4.	Average land rent for an acre at 10 shs. for a plot of one to two acres	7.50 shillings
5.	Percentage of land rent to peasant's gross income	16 per cent

The above calculation of the landlord's profit at 16 per cent of a peasant's gross income is, however, likely to be an under-estimate and is certainly the minimum, for, as remarked by Thomas and Spencer, "signs are not wanting in succeeding years of a desire on the part of certain landowners to increase this figure" of land rent from 10 shillings for a peasant's plot of usually one or two acres.

Moreover, the actual extent of the landlord exploitation is not realised only from the proportion of agricultural income which accrues to the landowner from the individual peasants. It also depends upon the volume of exploitation, that is, on the number of tenants under a landowner. Thomas and Spencer noted[426]:

> "A few thousand fortunate landowners were entitled to the rents of several hundred thousand tenants.... [In about 1921] 10,000 persons out of a total population in Buganda of about three-quarters of a million were concerned as individual owners of interests in land."

Evidently, the landowners began to enjoy a great deal of "prosperity" under the protective wings of the colonial system. Mair noted[427]:

> "The ownership of land now makes the essential difference between the rich and the poor. ... As happens inevitably with the introduction of a money economy and the immensely increased possibilities of material acquisition which it brings, the difference in

material conditions between rich and poor in Buganda has enormously increased in recent years, and the establishment of freehold tenure has been the most important single cause contributing to this result."

Thus the creation of landlordism in Buganda, besides stabilising the interest of the native rulers in the colonial regime, provided a great incentive to the growth of commodity production in the society, resulting in the sharp difference in "material conditions between rich and poor". Indeed, this institution of landlordism was found so useful that, as noted before, when the cotton economy was rapidly spreading beyond the borders of Buganda, it was extended to other kingdoms and to Busoga which was also known as a flourishing agricultural country next to Buganda and where also a class society had definitely emerged in the pre-British days, so that there also it was necessary to appease the upper stratum of the society.

But what happens to that part of the exploited labour of the common Africans which remains in the country as land rent? Is it employed to improve the productive forces in agriculture? The answer is in the negative; firstly, because its scope within the colonial economy is small, and, secondly, as allies of the foreign rulers, the landlords prefer to thrive on rents without taking any interest in the actual process of production. Thus referring to Buganda, where landlordism had its classical development in the Protectorate, Mair noted[428]:

"The trouble is that the modern landowners are acquiring tastes beyond what the offerings of his faithful followers can satisfy. He has learnt to appreciate life in the capital and to think the country dull; he generally manages to find some work under the native government, thanks to the education which his father's rent have made possible; he has been taught to be businesslike and applies his knowledge to the exaction of his dues, merely visiting his land from time to time to see that the shillings are coming satisfactorily."

No doubt, this state of affairs is not limited only to Buganda. It has spread all over the Protectorate where landlordism was introduced. In all such places, unter colonial conditions which do not give scope to invest money for greater production, the accumulated wealth of the landowners is not used for that purpose. Even after six decades of rule, the Colonial Government has not been able to cope with the Tsetse fly, so that even a large-scale introduction of plough cultivation is out of the question in the Protectorate. Moreover, apart from agriculture, investment in industry or trade is also strictly limited. Even now the Africans are not allowed to gin their own cotton. This is done almost entirely by Indian interests, except one ginnery leased recently to the Uganda Growers Co-operative which is a Government-controlled African Co-operative Society. And the same applies to coffee-hulling, where the processing industry is in the hand of the British. Investment of African capital in industries, even of the primary processing variety, is thus very limited, although assurances have been given recently that a few more government-controlled African co-operative ginneries will be opened in the future. Retail trade, in the form of peddling and keeping small shops, is virtually the only avenue open to them to invest their money. But here again, firstly, it has long ago been usurped by the Indians settled in Uganda, and, secondly, its scope is not growing further because of the growing distress of the mass of the African people who can be their only customers. In short, within the limits of the present-day colonial life, whatever opportunities for industry and trade there are in the Protectorate are already utilised by foreigners. The upshot is that while living on a substantial profit from land rent, very few of the landlords can or will invest money for large-scale or mechanised cultivation and for business activities. Their parasitic existence in society is thus conditioned by colonial conditions.

Simultaneously, because the common people have to bear the double burden of colonial and feudal interests and there are no possibilities to develop their productive forces, they become more and more impoverished, and in increasing numbers they cannot even maintain the mere reproductory rate of the economy. This has been noted also by authorities like Whitely and Mair from 1930 onwards[429]. Consequently, when their holding is not enough to provide for them, they become lease-holders of the landlords by paying a bigger land rent, or, if available, take to labouring jobs. Those who still have some land prefer to remain in the village, but most of them can do so only as lease-holders of the landlords, that is, by paying more land rent. Some can become hired labourers in the farms of better-off peasants, but, as will be explained in the next chapter, in the present state of agricultural production in the colony such a possibility is very limited. Others might prefer to go to neighbouring towns or to mining and such industrial concerns to work as wage-labourers, but because of the lack of industrialisation of the Protectorate not all of them can take to even such a course of livelihood. Many of them, of course, work seasonally in the cotton ginneries which early in the year employ a labour force of about 21,000, but in general their peasant-existence within the sphere of the primitive peasant-economy is conditioned by the colonial life. Therefore, even though "unpeasantised", whether they like or not they remain victims to landlord oppression and colonial exploitation.

While mere pauperisation thus remains the prospect for the mass of the people under colonial rule and imposed feudal land relation, for the upper stratum of the African societies landlordism remains the only scope for making a fortune. Therefore, because their profit depends primarily on the extraction of more and more land rent, they prefer to settle their lands on temporary lease-holds of one or a few years and live on the prospect of successively increasing the land rent. And, thanks to the colonial rule, the demand for leasing their land is always there. Indeed, because the colonial system ravages not only Uganda but also her adjacent territories, besides from the local impoverished peasantry, the supply of such temporary lease-holders is very big. A steady stream of poor people arrive every year in Uganda from the neighbouring countries to work on the lands of the landlords. The 1947 Annual Colonial Report of the British Government on Uganda estimated the number of such immigrants from Ruanda-Urundi alone at about a hundred thousand every year. The 1953 Annual Report recorded a smaller figure for all immigrants, but even so the figure is not an insignificant one. To quote from the report[430]:

> "Migrant labourers come mainly from two directions: the north-west, which includes West Nile, the Belgian Congo and the Sudan, and the south-west which includes the Kigezi and Ankole districts and Ruanda-Urundi. From the south-west also come a number of immigrants from Tanganyika and there is a smaller number from the Nyanza Province of Kenya who enter across the eastern border. During 1953 a total of 64,070 immigrants entered Uganda by the south-west route, compared with 68,952 in 1952. The number coming to central Uganda by the north-west route was 27,118 compared with 27,797 in 1952. Only a relatively small proportion of these immigrants enter the employment of non-Africans: the majority either grow cash crops on a leased plot or work for African farmers. Most of them stay for periods ranging from six months to one or two years and then return to their homes."

Just for their survival these immigrants thus lead a homeless life, and the root cause of their degradation — the colonial rule — gives the landlords the very best opportunity to exploit them. But this callous utilisation of human labour, ultimately for the benefit of foreign monopoly interests, is paraded by the admirers of colonial rule as an indication of the exploiting instinct of the Africans who are therefore declared to be no better than the Europeans! To quote Huxley[431]:

> "As we drove north we overtook party after party of them [Banyaruanda] streaming up the road. ... Some were on their way to the big Indian-owned sugar plantations near Jinja, but most work for Baganda landowners and farmers. This great movement of population is based on the employment of African by Africans, not on the intrusion of foreign capital. ... Exploitation is not a European invention.
>
> However rich a man in Buganda is — and some landowners are very rich, with fine houses, motor cars, servants and incomes running into four figures — he pays no more in direct taxation than the poorest labourer. Uganda has an income tax, but only Europeans and Indians pay it; Africans are specifically exempted. This is racial discrimination with a vengeance — in favour of Africans."

Such apologies for "European" exploitation and for favourable discrimination to the local lackeys of the British by exempting them from paying income tax cannot conceal the obvious fact that colonial and feudal interests (all for the final benefit of the British Power) devitalise the "developed societies" of Uganda. And it should be borne in mind that these "developed societies" include the whole of the Buganda Province, the districts of Bunyoro, Ankole and Toro in the Western Province, and the district of Busoga in the Eastern Province. According to the 1948 census, in all this region accounted for 46 per cent of the total land-area of the Protectorate and for 53 per cent of the total African population.

Furthermore, landlordism today is not restricted only to those societies where it was established by the *agreements* signed between the native rulers and the colonial government. Although, for reasons mentioned earlier, it was not directly introduced in other societies in Uganda, nevertheless in many of them, and especially in those which were adjacent to those where landlordism was planted, the institution blossomed forth with the flowering of the colonial economy. Production of cotton having been made the backbone of the economy all over Uganda, wherever there was any possibility those who amassed some wealth invested it in land and aspired to become a landlord. Thus, referring to the Sebei people (Basabei of Roscoe), Hailey noted in 1950[432]:

> "They, like the Bagishu, are becoming land conscious, though their area is at present more sparsely populated, and the individualisation of holdings has not proceeded to the same extent as in Bugishu."

It will be remembered that when the British came to Uganda, both these people were probably still in their tribal stage; but now here too land has become a possession of great value.

In this way, wherever private property in land and the right to alienation of land have been established, landlordism is developing, just as it developed in India even in those areas where landlordism was not directly planted by the British[433].

Colonial and feudal have thus become the essential characteristics of Uganda, especially in the more developed areas. And it means that the British have not only arrested Uganda's normal course of development, but have also effected retrogression in her societies. Instead of leading the peoples of Uganda to progress and prosperity, the British colonial rule has endeavoured to turn back the wheel of history and impose more organised feudal exploitation of the peoples in their economic life than they had ever experienced before.

15. A Blind Alley

The consequence of such colonial and feudal characteristics in Uganda is that they are constantly devitalising the people without giving an opening to break away from the vicious circle. In this sense the situation in Uganda under British rule is similar to what Marx observed about India after she became a British colony. Marx wrote[434]:

> "All the civil wars, invasions, revolutions, conquests, famines, strangely complex rapid and destructive as their successive action in Hindostan may appear, did not go deeper than its surface. England has broken down the entire framework of Indian society, without any symptoms of reconstruction yet appearing. This loss of his old world, with no gain of a new one, imparts a particular kind of melancholy to the present misery of the Hindoo, and separates Hindostan, ruled by Britain, from all its ancient traditions and from the whole of its past history."

The same has taken place in Uganda, for both were under the same system of colonial domination. There were, of course, differences in the stages of development in India and Uganda in the pre-British periods of their history, and they have left marks on their present and future course of change. But in both colonies the old basis of the societies disintegrated under British impact, and newer and newer systems of oppression were imposed one above another, with the result that while the societies exhibit a glitter of "modernisation" and "prosperity", there was no progressive course of development[435]. Therein lies the secret of the poverty of the colonial countries.

In Uganda, the disintegration of societies is spreading over all facets of life. To take the Acholi, for example, who were one of the last to come under colonial bondage, gone are their old sense of values along with their freedom and normal course of development; but new and progressive values have hardly emerged among them. Thus, in the old days the young people were not allowed to drink except on certain ceremonial occasions, such as dance festivals or funerals. It will be remembered that in the stage of *military democracy* the Acholi had always to be prepared for a "war", so that they could not allow their young people to get drunk in case of a sudden attack. In the "peaceful" life of present times there is, of course, no need for such a taboo; but since their life today is one of dejection and misery, with little possibility to develop a new conception of living, the most common form of recreation is for them to return from work and take to drinking. If a father does not approve of his son drinking, he cannot do anything. The traditional sanctions of the society are gone, while new ones based on a progressive socio-economic life have not emerged. The upshot is that if the father beats his son drinking, according to the "civilised" law he would be fined; but the "civilised" administration does not look after the socio-economic needs of the son. Therefore, at his present cultural level, and without any future scope for development, the son also sees no reason to foreswear the only source of enjoyment left to him, and so he gets drunk to escape reality. Not only the young men, the women also have taken to excessive drinking, which was never prevalent before. Frustration is the order of the day. Even Huxley, the great admirer of British Colonial Rule, noted about the Acholi[436]:

> "Community hunts have vanished with the march of progress; so have tribal wars, cattle-raids, witch-hunts and even dances. An unforeseen result of civilised rule has been that the life of the ordinary "bush" African has grown duller and duller."

Huxley could not diagnose the disease, although she noted the symptoms. It is not "progress" and "civilisation" which have produced this dull life, and how could the results have been "unforeseen"? They are the logical consequence of a colonial life in which the people are "preserved" as "peasants" or are transformed into this category to produce for the foreign interests, while their traditional life loses its significance. A new conception of life to replace the old one is not produced by the interaction of the old "tribal" life and the present "civilised" mode of living, because the old form of living in the "bush" is carefully retained. As a result, lacking cultural development and opportunities to improve their life in the future, the primary instincts of food and sex play the most dominant role in Acholi life today.

To add to this, the "tribal" sanctions on sexual behaviour have been loosened in a "civilised" administration, but the people are not given opportunities to concentrate their energy in other productive pursuits. Educational facilities are extremely limited, cultural activities are non-existent, even the old forms of entertainment like dances only take place occasionally, while the adulterer, instead of being punished by death as it was before, can now get away by paying a small fine. The result is, as the District Commissioner of the Acholi District reported, adultery is one of the serious social problems among the people.

So the old form of life is disintegrating very rapidly, but a new form to replace it at a higher level is not forthcoming.

Needless to say, this feature is not limited only to Acholi. All over Uganda it is the same sad picture. In some places it has gone to such a length that Huxley reported, as she was informed by the authorities, sometimes husbands encourage their wives to commit adultery, so that they can make an income out of the share of the fine which is charged on the culprits!

In the "developed societies" also the situation is not different. In the Buganda society, for instance, it has been noted how British occupation was engineered by splitting the society into factions and even by breaking the bonds of clan and family ties. There also social and moral laxity have resulted in a terrible spread of venereal diseases which now rage all over Uganda. Even Sir Apolo Kagwa, who, as noted before, always remained extremely loyal to the colonial regime, commented upon the difference between the rule of the African despots in a free Buganda and that of the British in the colony. Referring to the slaughter of the common people by the king, about which Speke and others painted very lurid pictures, Sir Apolo commented[437]:

> "This wholesale slaughter did not, however, decrease the population alarmingly, as did several religious wars and newly introduced diseases."

The author also found during his stay in Uganda that although the towns like Kampala do not have any African habitation within the town-area, African prostitutes strolling in the evening in front of the big hotels are an everyday occurence. This has been one of the many contributions made by the colonial regime to the Uganda societies.

Likewise, theft and robbery have grown, while the sympathetic European travellers of the nineteenth century, like Rowley, had testified to the honesty of the Africans and the safety of living among them. Today burglary and violence in the towns is not an uncommon occurrence in Uganda. The 1953 Colonial Annual Report on Uganda stated[438]:

> "Offences against property in general increased by 9.5 per cent over the previous year, thefts increased by 9.7 per cent and breakings by 5.7 per cent. An increase of nearly 10 per cent in offences against property following an increase of 8.6 per cent in 1952 is a substantial one and indicates a lowering of the general standard of honesty and the need for increased police protection for the public. Juvenile crime, though not yet of serious proportions, more than doubled in comparison with 1952 and indicates the problems attending growing urbanisation."

Many such instances could be given to indicate how the moral fibre of the Uganda Africans has degenerated under colonial rule. But this degeneration of the Uganda societies is paraded by the spokesmen of the foreign rulers as evidence of why they must remain in the colony in order to "civilise" the Africans. The missionaries speak of low instincts of the Africans; the sympathetic foreigners speak of them as "children who have not yet developed proper sense of values"; and the propagandists dwell elaborately on the depravity of African character. None of them, of course, admits that this shocking state of affairs has been brought about by the putrid stagnation of the societies.

Furthermore, the retrogression of Uganda is seen not only in the disintegration of the life of the mass of the people in all its aspects and in the perpetuation of feudal exploitation under colonial rule. It is also seen in the obstruction put forward by the colonial regime to the emergence of new social forces among the Uganda Africans. Based on the rudimentary or developed class structure in the Uganda societies in the pre-British period, a class of *nascent* bourgeoisie, as "entrepreneurs" and "intelligentsia", has emerged in most of them from their upper stratum. This class, by initiating progress in the country's economy and by introducing social reforms and at least some sort of democratic political organisation, could have played a progressive role in Uganda today. But, as explained earlier, their course of development is blocked in the colonial society. This leads to frustration of many of those belonging to this class, and helps in stabilising the retrogressive features of Uganda societies under British rule.

To take, again, the Acholi society as an illustration, for this was one of the less-developed societies before the introduction of British rule in Uganda, even among them these tendencies can be observed. Thus, in 1949, the author found that the leading figure in the Acholi Bus Company, a rival concern to the Indian-owned Bus Company, was the Head Master of the Church Missionary Society's School at Gulu, the main town in the Acholi District. An ex-Rwot (County Chief) is a part-owner of the company; he also employs his capital to lend at a profit to poor Indian shopkeepers. One of his sons is a Sub-Inspector of Police, another the Head Master of a Secondary School, and another is studying at Makarere College; the daughter is a teacher in a Secondary School, and the son-in-law is the first qualified doctor among the Acholi, who is now serving as an Assistant Medical Officer in the Protectorate. The Acholi doctor is again a near relation to the Secretary of the Acholi Bus Company.

To continue with this growth of the class of *nascent* bourgeoisie among the Acholi, a young man, who once served the Government as a clerk and was also in the Armed Forces during the last war, was found in 1950 to be an important trader at Pajuli — a place in the interior of the Acholi District. He told the author of his desire to oust the local Indian trader within a few years' time. Another ex-Rwot was found to have a son who had previously held a high position in the Native Administration, and was then a Rwot himself. An ex-Jago's (Sub-County Chief's) son was contemplating to go to the Union of South Africa or to the United Kingdom for higher studies; in 1950 he was a government employee. An ex-priest of the Church Missionary Society was seen to be a shop-keeper, who was also engaged in the cotton industry as a buyer of raw cotton from the African growers and supplier to the Indian-owned cotton ginneries.

Several such instances can be cited to show that the growing bourgeoisie in the Acholi society have their roots in the upper class of the old society It is also interesting to observe that they are now marrying among themselves, obviously, as the author was told by one of them, to maintain their social status. Thus, in course of time, they will certainly grow as a class both in the social and the economic sense.

But the colonial character of this *nascent* bourgeoisie is revealed by the fact that, because of the lack of opportunity and the necessary capital, not only it is quite out of the question to become industrialists, but also not many of them can thrive in trading activities. An Acholi Traders' Co-operative helps the small shopkeepers to run their business, but because of the lack of necessary funds it cannot expand properly. As a result, retail trade remains in the hands of the Indians, and the growing Acholi bourgeoisie is forced to turn to other means of improving their economic and social position. And this, in the present context, means identifying themselves with the colonial bureaucracy by taking up government jobs. This is therefore seen to be the ambition of many of the young Acholi belonging to this class.

Of course, all these young men cannot be absorbed in the administration, and moreover as a class they are sure to fight on their own to have a voice in the economic and political life of their country This is only too apparent in their somewhat critical attitude to the colonial administration (especially in relation to running the Native Administration by appointed chiefs) and to the usurpation of economic opportunities by British and Indian interests. Nevertheless, it remains a fact that under colonial conditions their development is fore-shortened.

This feature is naturally very well marked in the "developed societies", and particularly in Buganda. Here, on the basis of a share of colonial tributes from the people as well as from the land rents, has developed a class of *nascent* bourgeoisie which is constantly complaining of lack of opportunity for their development. Although, as noted by Mair, some of those belonging to this class have in frustration cast their fate in with the government by identifying themselves with the colonial bureaucracy, yet signs are not wanting that as a class they are far from satisfied with the present state of affairs. Some of their complaints have been noted in Chapter II. In recent years, through great agitation and struggle, in which they had active support of the masses, they have secured an African cotton ginnery. But this government-controlled ginnery is to small to accommodate all of them. Besides they want to control the economic life of their country as a whole, and also demand a democratic political organisation. But in every aspect they come across hidden or open opposition from the foreign rulers and the foreign intermediaries. As noted earlier, they find it very difficult to develop as merchants and industrialists. In the Government and commercial organisation also they are discriminated against. Higher posts are still reserved for Europeans and then for Asians, and even if they get a reasonably high post their salaries according to their qualifications are found to be much less than those of Europeans with the same or even lower qualifications. The following extract from Mr. Kartun's recent booklet *Africa*, in which he refers to a specific case mentioned by Mr. E. M. Mulira (an Assistant Lecturer at the School of Oriental and African Studies at the University of London) in his booklet *Troubled Uganda*, is very revealing[439]:

> "Mr. Mulira goes on to present the case of an African from Uganda who studied education and took an Honours degree at Cambridge at the same time as the son of a white planter in Uganda was obtaining a diploma in education. The two students returned to Africa at the same time. The white student was made an education officer at once. The African was told that he would be watched for three years. The white man started at a salary of £ 550 yearly, rising to £ 1,346. The African started at £ 330, rising to a maximum of £ 456. Later, says Mr. Mulira, the African graduate became headmaster of a secondary school under the supervision of the far less highly qualified Englishman."

Mr. Mulira further noted in his booklet referred above[440]:

> "In East Africa they tell us we have not the necessary qualifications (to fill reasonable posts). When we demand the opportunities of securing the necessary qualifications they accuse us of being too fond of letters after our names and then add that degrees are not necessary. They offer us scholarships to the United Kingdom. They send us to do short courses in most cases, and in insufficient numbers, ... When some of us obtain qualifications they tell us we lack character and experience and so on."

Several such cases could be cited to show that the colonial rule puts up severe obstacles to the development of the African bourgeoisie and the African intelligentias. It appears thus that all over the Protectorate the future of the African bourgeoisie is doomed under colonial rule. So the societies stagnate in all their aspects, while the reactionary elements enjoy *their* prosperity.

Such then is the upshot of British rule in Uganda. It has neither initiated progress in the political structure of the Uganda societies nor has it improved their productive forces. All that the Uganda Africans have faced under colonial rule is retrogression, behind glittering "prosperity", from their previous normally-developing socio-economic life. However slow that development might have been, it could ultimately lead them to peace, prosperity and progress. Instead, under British rule, Uganda has entered into a blind alley of disintegration and decay without any scope of future progress so long as the colonial-feudal domination persists. Under the British auspices has emerged the machinery of imperialist exploitation, with the thriving foreign monopoly interests at the apex of the unholy pyramid of the colonial society — and the vast mass of down-trodden people at the bottom, suffocating in abject misery, backwardness and degeneration. In between these two extremes are the host of intermediary parasites, foreign or local, representing either the old decomposing forces or new unscrupulous ones, but all of them subscribing themselves to the mechanism of exploitation for a few crumbs. This is what the British have brought about in Uganda by their "silent" and "bloodless revolution", as Elspeth Huxley acclaimed in 1948.

REFERENCES

1 Speke, J. H. — "Journal of the Discovery of the Sources of the Nile", Everyman's Library, J. M. Dent & Sons Ltd., London, 1922, p. 12

2 Baker, S. W. — "Ismailia", Macmillan & Co., London, 1874

3 (a) *loc. cit.* 1;
(b) Roscoe J. — "The Baganda: An Account of Their Native Customs and Beliefs", Macmillan and Co. Ltd. London, 1911;
(c) Roscoe, J. — "The Northern Bantu" Cambridge University Press, 1915;
(d) Roscoe, J. — "Twenty-Five Years in East Africa", Cambridge at the University Press, 1921;
(e) Roscoe, J. — "The Soul of Central Africa", Cassel and Company Ltd., London, 1922;
(f) Roscoe, J. — "The Bakitara or Banyoro", Cambridge University Press, 1923;
(g) Roscoe, J. — "The Banyankole", Cambridge University Press, 1923;
(h) Roscoe, J. — "The Bagesu and Other Tribes" Cambridge University Press, 1924;
(i) Johnston, H. H. — "The Uganda Protectorate", Hutchinson, London, 1902, Vol. I, pp. 235—592, 716, 834, etc.

4 Buell, R. L. — "The Native Problem in Africa", The Macmillan Company, New York, 1928, Vol. I, p. 259

5 *ibid.*, p. 259

6 Annonymous — "Handbook for East Africa, Uganda and Zanzibar", Stationer's Hall, Mombasa, 1905, p. 2

7 *ibid.*, p. 2

8 Coupland, R. — "East Africa and Its Invaders", Oxford at the Clarendon Press, 1938, p. 52

9 Dobb, Maurice — "Studies in the Development of Capitalism", George Routledge & Sons, London, 1946, Chapter 3ff.

10 *loc. cit.* 8, p. 52

11 *ibid.*, p. 52

12 *ibid.*, pp. 52—53

13 *ibid.*, p. 53

14 *ibid.*, p. 53

15 *ibid.*, p. 55

16 *ibid.*, p. 56

17 *ibid.*, p. 56

18 *ibid.*, p. 56

19 See, Mukherjee, Ramkrishna — "The Rise and Fall of the East India Company", Verlag der Wissenschaften, Berlin, 1955, Chapter 3 ff.

20 Brode, H. — "British and German East Africa: Their Economic and Commercial Relations", Edward Arnold, London, 1911, p. 2

21 *ibid.*, pp. 2—3

22 Palme Dutt, R. — "The Crisis of Britain and the British Empire", Lawrence & Wishart, London, 1953, pp. 72—74

23 quoted in *loc. cit.* 22, p. 74

24 Morton, A. L. — "A People's History of England", Lawrence & Wishart, London, 1951, p. 477

25 *loc. cit.* 19, Chapters 3 and 4

26 *loc. cit.* 6, p. 11

27 *loc. cit.* 20, p. 3

28 Aaronovitch, S. and K. — "Crisis in Kenya", Lawrence & Wishart, London, 1947, p. 21

29 *loc. cit.* 22, p. 75
30 *ibid.*, p. 78
31 *ibid.*, p. 78
32 Lenin, V. I. — "Imperialism, the Highest Stage of Capitalism. A Popular Outline", Selected Works, Foreign Languages Publishing House, Moscow, 1952, vol. i, part 2, p. 445
33 *ibid.*, pp. 453—454
34 *loc. cit.* 28, p. 19
35 *loc. cit.* 32, pp. 459—460
36 *ibid.*, pp. 464—465
37 *ibid.*, p. 477
38 *loc. cit.* 9, p. 300
39 *ibid.*, p. 309
40 *ibid.*, pp. 309—310
41 *ibid.*, 311
42 *loc. cit.* 28, pp. 19—20
43 *loc. cit.* 9, pp. 311—312
44 *loc. cit.* 32, p. 511
45 *ibid.*, p. 519
46 *ibid.*, p. 496
47 *ibid.*, p. 512
48 *ibid.*, p. 510
49 Harris, J. H. — "Africa: Slave or Free ?", Student Christian Movement, London, 1919, pp. 19—20
50 *ibid.*, p. 20
51 Eliot, Sir Charles — "The East Africa Protectorate", Edward Arnold, London. 1905, p. 24
52 *loc. cit.* 6, p. 11
53 *loc. cit.* 51, pp. 24—25
54 Lugard, F. D. — "The Rise of Our East African Empire", William Blackwood & Sons, Edinburgh & London, 1893, Vol. II, p. 591
55 *loc. cit.* 51, p. 29
56 Anonymous — "Uganda 1947", Colonial Annual Report, His Majesty's Stationery Office, London, 1949
57 *loc. cit.* 28, p. 21
58 *loc. cit.* 54
59 *loc. cit.* 3 (i), Vol. I, pp. 230—231, footnote
60 *loc. cit.* 1, p. 379
61 Lucas, Sir Charles — "The Partition and Colonization of Africa", Oxford at the Clarendon Press, 1922, p. 72
62 Jackson, Sir Frederick — "Early Days in East Africa", Edward Arnold & Co., London, 1930, p. 143
63 *loc. cit.* 51, p. 28
64 Quoted by Leo Huberman in *Man's Wordly Goods*, Peoples' Publishing House, Bombay, 1953, p. 208
65 *loc. cit.* 28, p. 22
66 *loc. cit.* 54, Vol. II, pp. 66—67, 119, 452, etc.,
67 *loc. cit.* 51, p. 28
68 Gordon, General, C. G. — "Gordon in Africa", 1893, p. 183
69 *loc. cit.* 28, p. 22
70 *loc. cit.* 61, p. 76
71 *loc. cit.* 54, Vol. I., pp. 381—382
72 *ibid.*
73 Reported in the *Times* of London, October 4, 1892
74 (a) Portal, Sir G. — "The British Mission to Uganda in 1893", Edward Arnold, London, 1894;
(b) *loc. cit.* 3(i)
75 *loc. cit.* 54, Vol. I, pp. 379—380
76 *ibid.*
77 *loc. cit.* 51, pp. 28—29
78 *loc. cit.* 54, Vol. II, Chapter XL
79 *loc. cit.* 51, p. 25
80 *loc. cit.* 20, p. 4
81 *loc. cit.* 61, pp. 91—92
82 *loc. cit.* 51, p. 28
83 *loc. cit.* 20, p. 5
84 Thomas, H. B. and Spencer, A. B. — "A History of Uganda Land and Surveys and of the Uganda Land Survey Department", Entebbe, Uganda, 1938, p. 12
85 *ibid.*, p. 34
86 *loc. cit.* 61, p. 92
87 *loc. cit.* 62, p. 260
88 *ibid.*, p. 251
89 *ibid.*, p. 260
90 *ibid.*, p. 263
91 *loc. cit.* 61, pp. 78—79
92 *ibid.*, pp. 79—80
93 *ibid.*, p. 79
94 *ibid.*, p. 80
95 *ibid.*, p. 83
96 *ibid.*, p. 83
97 *ibid.*, p. 84
98 *loc. cit.* 84, p. 29
99 *ibid.*, p. 34
100 *loc. cit.* 3(i), Vol. I, p. 216;
loc. cit. 54, Chapter VII;
loc. cit. 3(b)—(h).
101 *loc. cit.* 1;
loc. cit. 3(b)—(h);
loc. cit. 3(i), Vol. I, pp. 235—592, 716, 884, etc.
102 *loc. cit.* 62, p. 4
103 Thomas, H. B. — "More Early Treaties in Uganda, 1891—96", *The Uganda Journal*, Kampala, Uganda, Vol. 13, No. 2, September 1948, pp. 173—174
104 *ibid.*, p. 174
105 Thurston, A. B. — "African Incidents", London, 1900
106 *loc. cit.* 84, p. 45
107 *loc. cit.* 62, Chapters XVIII and XIX;
loc. cit. 3(b); *loc. cit.* 3(d); *loc. cit.* 3(f);
loc. cit. 3(i)

108 *loc. cit.* 3(d), p. 103
109 *ibid.*, p. 105
110 *ibid.*, p. 108
111 *ibid.*, p. 108
112 *loc. cit.* 62, p. 251
113 *ibid.*, p. 262
114 *ibid.*, pp. 262—263
115 *ibid.*, p. 263
116 *loc. cit.* 3(d), p. 109
117 *ibid.*, p. 109
118 *ibid.*, p. 109
119 *ibid.*, p. 111
120 *ibid.*, pp. 124—125
121 *ibid.*, p. 111
122 *loc. cit.* 62, pp. 263—264
123 *loc. cit.* 54, Vol. II, p. 17
124 *loc. cit.* 3(d), p. 176
125 *ibid.*, pp. 251—252
126 *loc. cit.* 62, p. 274
127 *loc. cit.* 3(f), p. 2
128 *loc. cit.* 3(d), pp. 176—177
129 see, for instance, *loc. cit.* 19, Chapter 4
130 *ibid.*
131 Muir, Ramsay — "The Making of British India: 1756—1858", Manchester at the University Press, 1917, pp. 339—342
132 Panikkar, K. M. — "Indian States", Oxford University Press, 1944, p. 3
133 Quoted by W. M. Torrens in *Empire in Asia: How we came by it — A Book of Confessions*, Trübner& Co., London, 1872, p. 358
134 Hailey, Lord — "Native Administration in the British African Territories", Part I, *East Africa: Uganda, Kenya, Tanganyika*, His Majesty's Stationery Office, London, 1950, p. 47
135 *ibid.*, p. 47
136 *ibid.*, p. 13
137 *loc. cit.* 3(g), p. 2
138 *ibid.*, p. 2
139 *loc. cit.* 62, pp. 275—277
140 *loc. cit.* 3(d), p. 177
141 *ibid.*, pp. 171—172, 177, etc.
142 Hailey, Lord — "An African Survey", Oxford University Press, 1938, p. 413
143 Mair. L. P. — "Native Policies in Africa", 1936, p. 56
144 *ibid.*, p. 15
145 Huxley, Elspeth — "The Sorcerer's Apprentice: A Journey Through East Africa", Chatto and Windus, London, 1948, p. 287
146 *loc. cit.* 4, Vol. I, p. 560
147 *ibid.*, Vol. I, p. 561
148 *loc. cit.* 142, p. 448
149 *loc. cit.* 1
150 *loc. cit.* 4, Vol. I, p. 576
151 *ibid.*, Vol. I, p. 576
152 *ibid.*, Vol. I, p. 577
153 *loc. cit.* 142, p. 446
154 *loc. cit.* 134, p. 18
155 *ibid.*, p. 17
156 *ibid.*, p. 18
157 *ibid.*, p. 44
158 *ibid.*, p. 46
159 *ibid.*, p. 47
160 *ibid.*, p. 48
161 *ibid.*, p. 49
162 *ibid.*, p. 50
163 *ibid.*, pp. 1—85
164 *ibid.*, p. 55
165 *loc. cit.* 4, Vol. I, p. 575
166 *loc. cit.* 3(b)
167 Mair, L. P. — "An African People in the Twentieth Century", Routledge & Sons, London, 1934, pp. 33—34
168 *ibid.*, pp. 158—159
169 *ibid.*, p. 165
170 *loc. cit.* 84, p. 63
171 *ibid.*, p. 68
172 Reported in the journal *Uganda Notes*, Kampala, Uganda, December 1907, p. 119
173 *loc. cit.* 134, p. 9
174 *ibid.*, p. 18
175 *loc. cit.* 84, p. 48
176 *loc. cit* 54
177 *loc. cit.* 84, p. 55
178 *loc. cit.* 4, Vol. I, p. 565
179 *loc. cit.* 84, p. 56
180 Lord William Bentinck's speech on November 8, 1829, reprinted in A. B. Keith's *Speeches and Documents on Indian Policy: 1750—1921*, London, Vol. I, p. 215
181 Anonymous—"NativeAdministration", Uganda Government Publication, *For Office Use Only*, Entebbe, Uganda Protectorate, 1939, p. 13
182 Gulliver, P. H. — "The Karamajong Cluster", *Africa*, Journal of the International African Institute, London, Vol. XXII, 1952, No. 1, p. 16
183 *loc. cit.* 134. p. 62
184 *ibid.*, p. 62
185 *ibid.*, p. 62
186 *loc. cit.* 182, p. 19
187 *loc. cit.* 134, p. 63
188 *ibid.*, p. 28
189 *ibid.*, p. 31
190 Lawrence, J. C. D. — "The Karamajong Cluster: A Note", *Africa*, Journal of the International African Institute, London, Vol. XXIII, No. 3, 1953, pp. 244—245
191 *loc. cit.* 134, p. 50
192 *ibid.*, p. 63
193 *ibid.*, p. 62

194 *ibid.*, p. 31
195 *ibid.*, p. 31
196 *ibid.*, p. 28
197 *ibid.*, p. 30
198 *ibid.*, p. 61
199 *ibid.*, p. 61
200 *ibid.*, p. 61
201 *ibid.*, p. 61
202 *ibid.*, p. 61
203 *loc. cit.* 181, pp. 9, 11, 13
204 *loc. cit.* 134, p. 63
205 *ibid.*, p. 63
206 *ibid.*, p. 63
207 *ibid.*, p. 30
208 *ibid.*, p. 31
209 *ibid.*, p. 2
210 *loc. cit.* 84, p. 55
211 *loc. cit.* 134, pp. 36, 55, 66—67
212 *ibid.*, p. 26
213 *loc. cit.* 182, p. 1
214 *loc. cit.* 190, p. 244
215 *loc. cit.* 4, Vol. I, p. 564
216 *loc. cit.* 3 (d), p. 234
217 *loc. cit.* 4, Vol. I, p. 561
218 *loc. cit.* 3 (g), p. 78
219 *loc. cit.* 4, Vol. I, p. 561
220 *loc. cit.* 3 (d), p. 209
221 *ibid.*, pp. 234—235
222 *ibid.*, pp. 244—245
223 *loc. cit.* 4, Vol. I, p. 565
224 *loc. cit.* 20, p. 85
225 *ibid.*, pp. 85—86
226 *ibid.*, p. 86
227 *ibid.*, p. 87
228 *loc. cit.* 142, p. 575
229 *loc. cit.* 4, Vol. I, 565
230 *ibid.*, p. 566 footnote
231 *loc. cit.* 142, p. 1442
232 Anonymous — "Uganda 1951", Colonial Annual Report, Her Majesty's Stationery Office, London, 1953, p. 23
233 *loc. cit.* 142, p. 547
234 (a) Brendel, Horst — "Die Kolonisation Ugandas", Buchdruckerei Georg Weigel, Großenhain, Germany, 1934, pp. 76, 82;
(b) *loc. cit.* 142, p. 1442.
235 *loc. cit.* 232, p. 23
236 Anonymous — "An Economic Survey of the Colonial Territories, 1951", Vol. II, *The East African Territories*, Her Majesty's Stationery Office, London, 1954
237 *loc. cit.* 142, pp. 1583—1584
238 *ibid.*, p. 1335
239 Leys, N. — "The Colour Bar in East Africa", 1941, p. 85; quoted in *loc. cit.* 28, pp. 164—165
240 "Report on the Revenue and Taxation in Uganda", Uganda Government Publication, Uganda Protectorate, 1936, p. 14
241 Calculated from the *Trade Reports* of the Uganda Government and *Colonial Annual Reports* on Uganda of the British Government
242 *loc. cit.* 142, p. 1442;
loc. cit. 232, p. 23
243 *loc. cit.* 142, p. 1442;
loc. cit. 232, p. 23
244 *loc. cit.* 142, p. 575
245 *ibid.*, p. 446
246 *loc. cit.* 134, p. 17
247 *loc. cit.* 142, p. 575
248 *loc. cit.* 134, p. 36
249 *ibid.*, p. 55
250 *ibid.*, p. 66
251 *loc. cit.* 134;
loc. cit. 232
252 *loc. cit.* 232, p. 23
253 *loc. cit.* 240, p. 4
254 *loc. cit.* 4, Vol. I, p. 533
255 *ibid.*, Vol. II, p. 980
256 *loc. cit.* 142, p. 1442
257 *ibid.*, p. 1433
258 *loc. cit.* 134, p. 3
259 *loc. cit.* 232, p. 23
260 Calculated from the *United Nations Statistical Year Book 1951* and *Uganda 1951* (*loc. cit.* 232)
261 *loc. cit.* 236, p. 75
262 *loc. cit.* 232, p. 23
263 *loc. cit.* 236, pp. 75—76
264 *ibid.*, p. 76
265 *ibid.*, p. 94
266 Anonymous — "Uganda 1953", Colonial Annual Report, Her Majesty's Stationery Office, London, 1954, pp. 65—68
267 *ibid.*, pp. 67—68
268 Recorded in the *United Nations Statistical Year Book 1951*
269 *loc. cit.* 232, p. 23
270 Mentioned in *loc. cit.* 22, p. 222
271 *loc. cit.* 266, p. 65
272 *loc. cit.* 232, p. 23
273 *loc. cit.* 236, p. 102
274 *loc. cit.* 232, p. 23
275 *ibid.*, p. 23
276 *ibid.*, p. 24
277 *loc. cit.* 4, Vol. II, p. 977
278 *loc. cit.* 232, pp. 17, 22
279 *Report of the Commission on Closer Union of the Dependencies in Eastern and Central Africa*, Cmd. 3234, British Government Publication, His Majesty's Stationery Office, London, 1929
280 *loc. cit.* 28, p. 24

281 *loc. cit.* 20, p. 22
282 Published in the *Uganda Journal*, Kampala, Uganda, 1948
283 *loc. cit.* 84, pp. 20—21
284 *loc. cit.* 20, p. 102
285 *ibid.*, p. 107
286 *loc. cit.* 54
287 *loc. cit.* 20, p. 109
288 *loc. cit.* 236, p. 85
289 Carter, W. M. (Chairman), — "Report of the Commission of Inquiry into the Cotton Industry of Uganda, 1929", Uganda Government Publication, Entebbe, Uganda, 1929, p. 2
290 Dawe, M. T. — "Lecture to Native Chiefs", *Uganda Notes*, Journal published from Kampala, Uganda, June 1907
291 Reports published in *Uganda Notes*, December 1907, April 1909, etc.
292 (a) *loc. cit.* 289;
(b) *United Nations Year Book of International Trade Statistics 1951*, United Nations Publication, New York, 1952
293 *loc. cit.* 236, pp. 83, 85
294 *loc. cit.* 134, p. 2
295 *loc. cit.* 236, p. 82
296 (a) *loc. cit.* 289, Appendix 2;
(b) Dalal, C. P. — "Importance to the Protectorate of Its Cotton Industry", Uganda Teachers' Journal, Published by the Director of Education, Uganda Protectorate, Kampala, Uganda, Vol. I, No.3, July 1939
297 Annual Reports of the Agricultural Department of the Uganda Protectorate Government, Uganda
298 *loc. cit.* 232, p. 41
299 *Trade and Revenue Reports of the Uganda Protectorate*, Uganda Government Publications, Entebbe, Uganda
300 *loc. cit.* 232, p. 41
301 (a) *loc. cit.*. 296(b);
(c) *loc. cit.* 236, pp. 86, 103
302 *loc. cit.* 296(b), p. 195
303 *loc. cit.* 3(d), p. 210
304 *ibid.*, p. 210
305 *ibid.*, pp. 209—210
306 *ibid.*, p. 210
307 *ibid.*, p. 231
308 *loc. cit.* 236, p. 101
309 *ibid.*, p. 101
310 *loc. cit.* 142, p. 1335
311 *ibid.*, p. 1336
312 *ibid.*, p. 1337
313 (a) "Uganda Blue Book 1945", Uganda Government Publication, Entebbe, Uganda;
(b) "Uganda 1946", Colonial Annual Report, His Majesty's Stationery Office, London, 1948;
(c) *loc. cit.* 56
314 *loc. cit.* 236, p. 103
315 *loc. cit.* 266, p. 38
316 *loc. cit.* 292(b)
317 *loc. cit.* 289, p. 3
318 (a) Contractor (Mobedjina), D. B. — "A Hand-Book of Indian Cotton", Examiner Press, Bombay, 1926, p. 16,
(b) *loc. cit.* 234 (a), pp. 81—82
319 *loc. cit.* 142, p. 1337
320 Calculated from *loc. cit.* 289
321 *loc. cit.* 292(b)
322 *loc. cit.* 236, p. 103
323 *loc. cit.* 232, p. 39
324 *loc. cit.* 236, p. 101
325 *loc. cit.* 232, p. 41
326 *ibid.*, p. 41
327 *loc. cit.* 289
328 (a) Annual Reports of the Agricultural Department of the Uganda Protectorate Government, Entebbe, Uganda;
(b) "Government Circular on the Ginnery Sites Approved up to 31st March 1949", Uganda Government Publication, Entebbe, 1949
329 Whitley, N. H. P. (Chairman) — "Report of the Uganda Cotton Industry Commission, 1948", Uganda Government Publication, Entebbe, 1948
330 (a) *loc. cit.* 289;
(b) *loc. cit.* 328(a);
(c) *loc. cit.* 236, p. 86
331 *loc. cit.* 236, pp. 86—87
332 *ibid.*, p. 86 footnote
333 Worthington, E. B. — "A Development Plan for Uganda", Uganda Government Publication, Entebbe, Uganda, 1949, p. 112
334 *loc. cit.* 236, pp. 103—104
335 *ibid.*, pp. 84—87
336 (a) *loc. cit.* 289;
(b) *loc. cit.* 329
337 *loc. cit.* 232, p. 39
338 *loc. cit.* 4, Vol. II, p. 976
339 *loc. cit.* 232, pp. 39—41
340 *loc. cit.* 4, Vol. II, p. 977
341 *loc. cit.* 232, pp. 39, 41
342 *loc. cit.* 142, p. 1348
343 *ibid.*, p. 1392
344 *ibid.*, p. 1392
345 *loc. cit.* 296(b), p. 195
346 *loc. cit.* 54, Vol. I, p. 382
347 *loc. cit.* 296(b), p. 195
348 *ibid.*, pp. 196—197
349 Reported in *Uganda Notes*, Kampala, Uganda, April 1908
350 *ibid.*

351 *loc. cit.* 20, p. 19
352 (a) *loc. cit.* 20, p. 48;
(b) *loc. cit.* 234 (a), p. 60
353 *loc. cit.* 20, p. 20
354 *ibid.*, pp. 20—21
355 *loc. cit.* 84, p. 23
356 *loc. cit.* 20, p. 20
357 *loc. cit.* 4, Vol. I, p. 535
358 *loc. cit.* 20, p. 10
359 *ibid.*, p. 84
For details regarding why "rosts" broke out in German East Africa, see, for instance, Alexander Bursian's *Die Häuser- und Hüttensteuer in Deutsch-Ostafrika*, Verlag von Gustav Fischer, Jena, 1910, Chapter 4
360 *loc. cit.* 84, pp. 22, 43
361 cf. *The 1931 Annual Report of the Department of Agriculture*, Uganda Protectorate Government Publication, I: 8, also quoted in *loc. cit.* 234(a), p. 77
362 *loc. cit.* 142, pp. 1918—1919
363 *ibid.*, p. 1317
364 *loc. cit.* 4, Vol. II, p. 976 footnote
365 *loc. cit.* 232, p. 19;
loc. cit. 56, 266, etc.
366 *loc. cit.* 4, Vol. II, p. 974
367 *ibid.*, Vol. I, p. 534
368 *ibid.*, Vol. I, p. 534
369 *loc. cit.* 142, p. 1441
370 *ibid.*, p. 1433
371 *ibid.*, p. 1441
372 *loc. cit.* 232, p. 25
373 *ibid.*, p. 23
374 *ibid.*, p. 23
375 *loc. cit.* 142, p. 1350
376 *ibid.*, p. 1557
377 *loc. cit.* 4, Vol. I, p. 567
378 *loc. cit.* 142, p. 1557
379 *ibid.*, p. 576
380 *loc. cit.* 134, p. 17
381 *loc. cit.* 142, p. 576
382 *loc. cit.* 4, Vol. I, pp. 586—587
383 *ibid.*, Vol. I, p. 586
384 *loc. cit.* 142, p. 446
385 *loc. cit.* 4, Vol. I, p. 567
386 *loc. cit.* 20, p. 50
387 *loc. cit.* 4, Vol. I, pp. 567—568
388 *ibid.*, Vol. I, p. 568
389 *ibid.*, Vol. I, p. 568
390 *ibid.*, Vol. I, p. 569
391 *ibid.*, Vol. I, p. 569
392 *ibid.*, Vol. I, p. 569
393 *ibid.*, Vol. I, p. 569
394 Dutt, R. C. — "The Economic History of India under Early British Rule", Routledge & Kegan Paul Ltd., London, 1950 p. 268
395 *ibid.*, pp. xiv-xv
396 *loc. cit.* 4, Vol. I, p. 569
397 *ibid.*, Vol. I, p. 569
398 *ibid.*, Vol. I, p. 570
399 *loc. cit.* 236, p. 83
400 *loc. cit.* 333, p. 11
401 Hall, Sir J. (Governor of Uganda) — "Some Notes on the Economic Development of Uganda", Foreword to E. B. Worthington's "A Development Plan for Uganda" — *loc. cit.* 333, p. v
402 *loc. cit.* 236, p. 85
403 *loc. cit.* 134, p. 3
404 Matheson, J. K. and Bovill, E. W. — "East African Agriculture", Oxford University Press, England, 1950
405 Annual Report of the Agricultural Department of the Uganda Protectorate, Uganda Government Publication, Entebbe, Uganda, 1929
406 *loc. cit.* 289
407 As is well-known, this concept of *organic composition of capital* was developed by Karl Marx (cf. *Capital: A Critical Analysis of Capitalist Production*, Vol. I, George Allen & Unwin Ltd., London, 1949; p. 625ff.
408 For British India, for instance, which was evidently better "developed" than probably all other colonial countries, see: — P. A. Wadia and K. T. Merchant's Our Economic Problems, New Book Company, Bombay, 1948; D. R. Gadgil's *The Industrial Evolution of India*, Geoffrey Cumberlege, Oxford University Press, 1950; Daniel H. Buchanan's *The Development of Capitalistic Enterprise in India*, Macmillan & Co. Ltd., New York, 1934; etc.
409 *loc. cit.* 56, p. 17
410 Annual Report of the Agricultural Department of the Uganda Protectorate, Uganda Government Publication, Entebbe, Uganda, 1949
411 *loc. cit.* 134, p. 17
412 *loc. cit.* 56, p. 83
413 *loc. cit.* 401, p. v
414 *loc. cit.* 134, p. 3
415 *loc. cit.* 236, p. 77
416 *loc. cit.* 167, p. 166
417 *loc. cit.* 84, pp. 68—69
418 Notice dated 11 December 1915 in *Uganda Official Gazette* of 15 December 1915
419 *loc. cit.* 167, p. 170
420 *loc. cit.* 84, p. 66
421 *ibid.*, p. 71
422 *loc. cit.* 167, p. 170
423 *loc. cit.* 134, p. 76

424 *loc. cit.* 84, p. 69
425 Calculated from the Annual Report of the Agricultural Department of the Uganda Protectorate Government for the year 1949, Uganda Government Publication, Entebbe, Uganda, p. 5 and Appendix I(a)
426 *loc. cit.* 84, p. 71
427 *loc. cit.* 167, pp. 147, 172
428 *ibid.*, p. 169
429 (a) Whitley, N. H.P. — "Report of the Commission of Inquiry into the Disturbances which occurred in Uganda during January 1945", Entebbe, 1945;
(b) *loc. cit.* 167
430 *loc. cit.* 266, p. 18
431 *loc. cit.* 145, p. 197
432 *loc. cit.* 134, p. 27
433 See, for instance, the "Report of the Congress Agrarian Reforms Committee", published by the All-India Congress Committee from 7 Jantar Mantar Road, New Delhi, 1949
434 Marx, Karl — "The British Rule in India", New York Daily Tribune, June 25, 1853
435 For a study of India under colonial rule, see, for instance, *India Today* by R. P. Dutt, People's Publishing House, Bombay, 1949
436 *loc. cit.* 145, p. 241
437 Kagwa, Sir Apolo — "The Customs of the Baganda", Translated by Ernest B. Kalibala and Edited by May Mandelbaum Edel, Columbia University Press, New York, 1934, p. 80
438 *loc. cit.* 266, p. 90
439 Kartun, Derek — "Africa, Africa!", Lawrence and Wishart, London, 1954, p. 34
440 *ibid.*, pp. 33—34

CHAPTER V.

UGANDA IN PERSPECTIVE

Come, your drum-beats echo in my soul,
the rhythm of my blood rings, Come Back Africa!
From dust of humiliation have I raised my head,
wiped sorrow-crust off my eyes,
Liberated my arms from pain,
And torn the net of helplessness, Come, Africa!
Each crooked ring of fetters which bound me
is now an armoury in firm grip,
I have broken the halter round my neck and
moulded it into a shield.
My spear-heads like the eyes of wild deers,
Surround him in all our dens,
And the dark of night is red with enemy blood,
The very heart of earth beats with my heart, Africa!
Rivers thrill, and forests tremble,
I am Africa now, I incarnate you, Africa,
Like your lions walk,
Come Africa, Come with the stride of a lion,
Come Back Africa!

[*Come Back, Africa*, written by Faiz Ahmed Faiz, translated from original Urdu by Sajjad Zaheer, and published in *New Age* (weekly), Delhi, October 9, 1955.]

What are the future prospects for Uganda? These the author intends to examine in this chapter. A discussion on *Uganda in perspective* naturally follows from the previous discussions on *Uganda before the British* and *Uganda under British rule*, which have shown that while before the advent of the British in this part of Africa the Uganda societies were undergoing a normal course of changes in which the germs of a progressive development were inherent, the British rule meant arresting this normal course of evolution and channelising the entire activities of the societies to the needs of foreign monopolists. The picture which arose from this contact between the Europeans and Africans is very dismal for the peoples of Uganda. Does it mean that the future of the Uganda Africans is doomed for ever? Or, are there possibilities that the British colonial policy will change its course and will lead the Uganda societies towards progress and prosperity, as often declared by the spokesmen of the British Government since the last war? In this context the perspective that is opening up before the peoples of Uganda under continued colonial rule of Britain should be examined.

On the other hand, one cannot miss the fact that, just as it was noted for British India, "the rising contradictions rooted in the social and economic, no less than in the political conditions" of Uganda today, "again and again defeat the attempts at harmony" made by the colonial rulers. Here, amid the ruins of previous cultures, which instead of developing stagnated under the rule of modern conquerors who did not form a part of the normally evolving life in this part of Africa, "the lowest levels of primitive economy, poverty, and servitude exist alongside the advanced forms of finance-capitalist exploitation". "The two levels, of the most advanced and elaborate finance-capitalist exploitation and domination above, and of the lowest level of social misery and backwardness below, are closely intertwined in a network of cause and effect. In between these two levels —betweenthe two opposing extremes of the imperialist exploiters at the apex of the pyramid and the destitute producing masses at the base —exist a host of transitional forms, intermediary parasitism, subordinate mechanism of exploitation, old decomposing forces and new advancing forces". The upshot is that today practically all stages of civilisation and culture within class society, from the primitive to the advanced, exist in the Protectorate. And, therefore, the wide range of social, economic, political and cultural issues find their sharp expression in the conditions of life of the Uganda Africans. All these issues, reflecting in many respects the backwardness and retarded development of a subject country, and intensified by foreign domination, force themselves to the front today. Inevitably, therefore, the rising political consciousness of the peoples of Uganda and the rising economic demands of the hungry masses lead to an explosive situation, which at the same time show the perspective to the Uganda Africans towards freedom and democracy, progress and prosperity, peace and well-being. Thus, while complicating the conditions of the struggle for liberation by their diversities and complexities, all these social, cultural, economic and political issues cannot but make the peoples of Uganda conscious of the central problem of liberation from imperialist domination. What then are the possibilities of such a perspective opening up in Uganda in the immediate future ? This should also be examined in the context of the present state of affairs in the Protectorate in order to see where lies the solution to the *problem of Uganda*. The present chapter, therefore, will deal with *two* perspective for Uganda, viz. the colonial perspective and the peoples' perspective.

1. Colonial Perspective

One who is conversant with the post-war declarations on the British Empire, emerging from official and semi-official quarters in Britain and her allies, would probably shake his head at the discussion in Chapter IV and think that possibly colonial policy was not so good in the pre-war times, as has been admitted by the Government spokesmen themselves, but the British Government has certainly changed its attitude in the new era which has opened out in the post-war world. He will repeat the phrases one hears so often since the last war from the general run of British politicians, such as "Colonial Development and Welfare", "Help the Backward Peoples", "Develop the Under-developed Territories of the Earth", "A World Plan to Combat Hunger and Poverty", and so on and on[1]. He will remind the un-believers of the British Labour Party Programme, *Labour Believes in Britain*, of 1949, declaring[2]:

> "Great Britain and the colonies have gone into partnership to liquidate ignorance, poverty and disease."

He will echo the announcement in the British Labour Party newspaper, *Daily Herald*[3]:

> "It has been left to a Labour Government to develop the economic and human potentialities of the Empire which successive Tory Governments ignored."

And he may refer to similar declarations made by the Conservative Government of Britain which followed the Labour Government in office. It is therefore necessary to examine the post-war perspective of Uganda under British rule.

What is this perspective ? Obviously, as before, this is to find out how best the colony can be exploited, and to implement such schemes. But this general policy, which basically is in no way different from the previous one, has two special features to be taken into account in the post-war period.

The first feature is specific to Uganda and all such countries bearing the yoke of colonial bondage. Since British colonial policy in Uganda up to the last war meant little more than mere loot and plunder of her rich land and the labour of her peoples, the whole economy is facing a serious crisis. This has therefore to be attended to if the colonial interests intend to make further profits from the colony.

The primitive method of exploiting Uganda by means of the cotton economy is now coming to an end. Owing to the present conditions of production, Uganda's cotton economy is reaching the saturation point. Simple human labour at a low remuneration cannot produce more. This was clearly noticed in the years immediately after the last war. For instance, Mr. Low, a Senior European Agricultural Assistant at the Serere Station in Uganda, remarked to Elspeth Huxley[4]:

> "The ordinary peasant can't cultivate more than about three acres of cotton on a family basis. Now, at present prices, three acres of cotton can't normally bring in more than ten pounds a year ... An extra porter (labourer) to cultivate a bigger acreage .. won't produce enough extra cotton to pay his wages of twelve pounds a year. So .. the peasant comes to a full-stop."

Furthermore, human labour without any reward is no more effective. While the pioneers like Stanley, Lugard and Johnston found the peoples of Uganda as active peoples and were all praise for their cleverness and diligence, Hall (the late Governor of Uganda) noted in 1946[5]:

> "Speaking generally, and judging by European standards, the Africans of Uganda are indolent, ignorant, irresponsible and not infrequently suspicious of foreign intervention."

Although the Governor explained the cause of this "inefficient labour force" without any mention of the absence of *real* prosperity and the lack of proper incentive to the people for work, he was fully aware of the seriousness of the situation that Uganda's economy has entered into a period of stagnation. He, therefore, concluded "Some Notes on the Economic Development of Uganda" with the grave warning[6]:

> "If we fail, then the progress of Uganda will inevitably be retarded and her future, social, political and economic, will indeed be black."

New orientations and revision of the old method of exploitation were therefore called for in British Uganda in the post-war period.

The second feature concerns the general crisis of the capitalist system. Because since the last war the ever-mounting crisis of world capitalism has hit Britain very hard, she must tenaciously hold on to the remaining colonies and find ways and means to intensify the rate and extent of exploitation therein. Britain's rich possessions of India, Burma and Ceylon have passed beyond her control; her hold over central and eastern Europe through the monopoly combines of Unilevers, etc., was lost with the rise of the People's Democracies; her influence in Western Asia is steadily being lost either directly to the local rulers and the people or indirectly to the United States; her domination over China is gone for ever with the establishment of the People's Republic of China; and in the remaining parts of her colonial and semi-colonial empire her position is tottering in the face of the mounting anger of the oppressed peoples, which is particularly noticeable in Malaya, Kenya, Cyprus and Guiana. Britain is thus no longer the supreme colonial power in the world, and whatever power she possesses today it is decreasing steadily.

Moreover, since the last war the single world market has disintegrated, giving rise to the socialist world market which is developing while the capitalist world market is shrinking in consequence.

Release of central and eastern Europe and a large part of Asia from the grip of world imperialism has deepened the crisis of the world capitalist system, while the "Marshall Plan" and such other measures introduced by the United States instead of strangling these areas have given further incentive to the development of the world socialist market, in which, in addition to the socialist states, also other important countries of the world, such as India, Burma, Ceylon and Egypt in the Old Hemisphere and several Latin American countries in the New Hemisphere are increasingly participating. This also has a serious effect on Britain's economy and her position in the world, for Britain is already playing second fiddle to the United States. Her political and economic life are becoming increasingly subservient to the demands of the U.S.A.*; and, consequently, while the colonies are passing out of her hands, in order to meet her "dollar shortage" she must look for greater colonial spoils[7].

* That Britain is giving way to the United States in the world-scale will be obvious from the table below which notes their comparative position from the period just before the First World War to a few years after the Second World War.

Table 1.

Year	Private foreign investments (in billion dollars)	
	United States	Britain
1913	3	17
1930	17	19
1949	19	12

The full implication of the above table will be realised if it is borne in mind that this great expansion of the USA capital has been mainly at the expense of Britain's colonial interests. Thus, firstly, it will be seen from Table 2, which gives Britain's capital investments in some of those countries which were or are under her control or in her direct sphere of influence, that since the last war Britain has become a weaker power in the world-scale.

Table 2.

Country	Nominal capital of British overseas investments (in £ million)	
	1938	1948
Canada	420	162
Australia	520	397
New Zealand	134	66
India & Pakistan	388	69
Burma	20	14
Egypt	10	7
Ceylon	28	26
Sudan	17	11
East Africa	24	15
Central Africa	73	79
West Africa	55	47
Malaya	67	64
West Indies	21	21

(Source: *Commerce Weekly*, 6 March, 1950; *U. N. Directory of International Trade*, January-March, 1950; and *U. N. Economic Survey of Asia and Far East, 1949.*)

Secondly, it should be realised that while the *total* British overseas investments, according to the above sources, went down from £ 3,545,000,000 in 1938 to £ 1,960,000,000 in 1948, the *Times* of London reported on the 24th of December, 1952:

„The Department of Commerce [of the United States] announced yesterday that the first result of a new census of American direct investment abroad established a book valuation of $ 11,800 m. at the end of 1950 which is an increase of $ 4,000 m. over the last census taken in 1943. Since 1950, the Department said in answer to a query, there had been a further rise of at least $ 2,000 m., which would mean a gain of 77 per cent. in American private investment abroad in nine years, bringing the total present figure to at least $ 13,800 m. . . .

Among the countries outside the western hemisphere in which large amounts have been invested are the United Kingdom ($ 840 m.) and the countries of the Persian Gulf area — in which $ 726 m. was invested at the end of 1950. Others included France ($ 285 m.), Australia ($ 198 m.), South Africa ($ 140 m.), Egypt ($ 40 m.),

Therefore, while Britain's economic policy towards the colonies was outlined in the King's Speech of March 1950 as: "My Government is actively promoting the economic and social development of the colonial territories, and the Colonial Development Corporation is proving a useful instrument to this end", Lord Trefgarne, Chairman of the Colonial Development Corporation, made it clear to the public what is the main objective behind such a policy. Adressing a group of Liverpool businessmen, he stated in 1950[8]:

> "The United Kingdom has an annual dollar deficit of £ 500 million — that is the background against which the productivity of colonial territories must be viewed. If the Colonies could raise their overall productivity during the next ten years by £ 200 million a year, that indeed would be a mercy twice blessed.
>
> The reason why we look to the colonies is that their products, food and raw materials, are more acceptable to the United States than manufactured goods. The total value of imports of manufactured goods into the United States in 1947 from all sources amounted to £ 250 million. The total imports of food and raw materials were more than four times as great: thanks to tin, rubber, cocoa, etc., the colonial territories overall are playing a good part in the dollar-sterling balance. Obviously, therefore, it is sound policy to aim at greatly increased dollar exports of colonial products."

As is evident from the above, the colonial crisis and the dollar crisis of Britain have become intertwined, thus expressing the rapidly disintegrating stage of her parasitic economy. The role of the British bourgeoisie is therefore becoming more and more difficult and even desperate to an extent. Hence their endeavour to secure the maximum capitalist profit through (1) "the exploitation, ruin and impoverishment of the majority of the population" of the mother country (resulting in series of strike-waves in Britain in recent years), (2) militarisation of Britain's national economy (in 1952, for instance, war expenditure in Britain was equivalent to 38.5 per cent of the total budgeted expenditure), and (3) continued enslavement of the colonies even by declaring wars against the colonial peoples, as in Malaya and Kenya, in order to intensify their systematic robbery in these territories.[9]

As regards the third point raised above, it may be of interest to note that according to the *Times* of London, dated November 5, 1951, the surplus of the British colonies in dollar

and India ($ 38 m.). . . . Investments in France and England increased in value by over $ 400 m., and the increase would appear larger were it not for the depreciation of the dollar value of assets as a result of currency devaluation. . . .

In less developed countries, the report continues, investments in manufacturing appears to follow the development of more basic resources, after the latter have raised incomes sufficiently to create the necessary demand."

The above facts indicate that in the post-war world the United States not only began to push Britain aside from her „spheres of influence", like the Persian Gulf area, etc., and not only extended its influence by successively increasing private investments also *within* Britain, but it also began to make in-roads into Britain's colonies in order to skim off the cream of colonial exploitation by investing for „the development of more basic resources".

Thus, in Africa, while the colonial manoeuvres are outwardly undertaken by the direct occupying Powers, the United States is everywhere behind the scene, and endeavours to secure the prize of their toils for itself. For instance, the USA directly or through the ECA contributes to „defence preparations" and for strategic minerals, extension of railways, etc. in the colonies; it supplies military equipment to the Union of South Africa; its „observers" attend meetings on „African Defence"; and its private business capital is increasingly flowing into Africa in diverse enterprises and in colonies held by different European Powers, such as in the British colonies of Rhodesias, Nyassaland, Uganda and Nigeria, in the Belgian Congo, in the Union of South Africa, in the Gold Coast, etc. Recent estimates of the USA public and private investments in Africa are of nearly £ 200 million each. There can, therefore, hardly be any doubt that in Africa and elsewhere Britain is already playing second fiddle to the USA in the economic sphere.

As how the United States dominates over Britain in the political sphere is evident from the following rueful comment of the London *Times* in its editorial of the 29th August 1951:

„Over two-thirds of the globe, along the great arc stretching from Europe to Japan, no treaty can be signed, no alliance can be forged, no decision can be made without the support and approval of the United States Government."

trade, which was around \$ 150 million during 1945—49, had risen to \$ 376 million in 1950 and was already \$ 314 million for the first half of 1951; and simultaneously the Sterling Balance held by the colonies had risen from \$ 510 million in 1947 to \$ 908 million in June 1951. Here one sees one of the primary needs of Britain to hold on to her colonies in present times.

Thus, the above two features, viz. (a) the need to attend to the ever-growing economic crisis in the colonies if Britain intends to make further profit from them and (2) the need to "solve" the colonial crisis in order to solve her "dollar problem", which together reflect the crisis of Britain and her empire, dictate the policy of the foreign rulers in Uganda and other colonies in the post-war period.

But such a policy has to be introduced tactfully; otherwise, even in the least disturbed colonies the peoples would rise in unision to drive the colonial power out of their lands. So the plan implementing such a policy must give the impression of a progressive development of the colonies and thus indicate a *change of heart* on the part of the rulers in order to maintain their rule in the colonies in face of mounting opposition of the dominated people; and at the same time the plan must initiate schemes to draw maximum profit from the colonial lands.

This, in essence, is the central idea behind the much-advertised *Colonial Development and Welfare Schemes* which were put into operation in right earnest by the British Labour Government in the post-war period, and which are being diligently carried on by the Conservative Government in recent years, and which the latter can claim as their own, for the sign-board of *Colonial Development and Welfare* was first painted by the Conservative Government in 1940 although the execution of the schemes drawn up under its head were seriously taken up after 1946.[10] Simultaneously there emerged, and are still emerging, post-war declarations from official and semi-official quarters of Britain, notifying the colonial peoples and the world public of the genuine desire of the British Government to lead the backward peoples towards prosperity and progress, of which some mention has been made in Chapters I and II.

In Uganda, in particular, the plans which began to be discussed just after the last war for the "Economic Development of Uganda" were part and parcel of these Colonial Development and Welfare Schemes, and as elsewhere were meant to work out details for the colony in question. Eventually it took shape in the blue print prepared by E. B. Worthington in 1946. This Plan received official blessing, and is now being put into effect in a revised version. "The 1948 Revision of the Plan" by Sir Douglas Harris was concerned mainly with a revision of the cost of its execution, from £ 16 million according to the original scheme to £ 34 million, in order to meet the higher cost of living of the officials and other allied details[11]. Since mainly British officers were to be employed for implementing the plan, the interest of the ruling community within Uganda was further assured by the revision of the plan.

But who would pay for this plan for "development" of Uganda? According to the original scheme, this plan was to be financed by 41 per cent from the Protectorate Revenue, 8 per cent from the Surplus Balances the Government has built up from the sale of cotton and coffee, and 22 per cent from the Cotton and Hard Coffee Control Fund or their profit from 1946—47[12]. In the 1948 revision of the plan also the resources of the Protectorate Government remained the same. In other words, the plan is to be financed mainly from "Local Resources", as it is labelled in recent official literatures, which means that the money extracted from the Uganda Africans is being employed to finance the plan, for as has been discussed earlier all the above sources of government income have a direct or indirect bearing upon the exploitation of the peoples of Uganda.

Even the Colonial Grants received from the metropolitan country to finance a much smaller part of the plan are in fact only a fractional return of the colonial tributes collected by Britain over a number of years, and specially during the last war. This is best realised if the Home Grants to all the British colonies for "Colonial Development and Welfare" are considered together, for Uganda is merely one of the areas where the schemes to usher in a "new era" under Britain's leadership are implemented. Thus, while discussing the figures of "Actual Issues from the Colonial Development and Welfare Vote during the Twelve Years ending March 31, 1952", Palme Dutt noted[13]:

> "Thus over the period of twelve years, from the inception of the scheme to 1952, the actual payments under the Colonial and Development Welfare Acts amounted to £ 66,427,976. This had to be divided between some forty-six colonial territories with a population of 81 millions. A simple sum in arithmetic will show that this is equivalent to a total amount of 16s. 4d. per head for the entire period of twelve years, or *an annual average rate of 1s. 4d. per head*.
>
> Even if the period of five years under Labour Government from 1946 to 1951 is isolated, in order to review only the accelerated expenditure since the 1945 Act, the total of £ 41,588,576 for the five years 1946—51 under the Labour Government, is equivalent to an annual rate of 2s. 0½d. per head of the population of the colonial territories, or *less than ½ d. per week per head*....
>
> On the other hand it is necessary to see certain features in the account which require to be set against the figure of ½d. per week per head before a final balance is struck. The total sterling balances of the colonies, representing goods and services extracted from the colonies during the war and after, payment for which has beeen covered by 'frozen' balances in London (thus equivalent to a type of forced loan) amounted to £ 1,042 million by the end of June, 1952. It will thus be seen that the total amount paid over under the Colonial Development and Wefare Acts during the twelve years of their operation is only equivalent to *less than one-fifteenth part* of the sterling balances owing to the colonies for goods received and not yet paid for. If this one-fifteenth part of the sterling balances had been released, the sum actually spent under the Acts would have accrued to the colonies without any of the humbug and pretence of free gifts and philanthropic grants.
>
> It is further worth noting that these sterling balances have actually increased in the most recent period at the same time as the grants were being paid under the Development Acts. To this point it will be necessary to return, as it is of cardinal importance for the real process of intensified exploitation which has been hidden behind the mantle of "development and welfare". Thus the sterling balance of West Africa alone increased during the year 1948 by no less than £ 20 million or more than three times the total amount paid out to all colonies during the year 1948—9. This is indeed to take out a pound with one hand in order to return a few shillings with the other and call the procedure philanthropy."

This point was also emphatically noted by Professor W. Arthur Lewis, a Member of the Colonial Development Corporation, in an article, entitled "The Colonies and Sterling" which appeared in the London *Financial Times* on the 16th January 1952. In this article he said:

> "Britain talks of colonial development, but on the contrary, it is African and Malayan peasants who are putting capital into Britain."

So much with regard to the financing of the "new vision" of British empire.

The question which is more important is: for whom would this "development" be — for the natives or for the immigrants? The foreword to the Plan, written by the Governor of Uganda began with stating the "limiting factor" that "Uganda is and must remain for the present at all events primarily a country of peasant agriculture"[14].

Of course it is not difficult to understand the necessity of this limiting factor. The foreign power cannot allow the colonial people to become economically strong, for they will then begin to control the economic life of the country and, in turn, the political life. Therefore the bulk of the Uganda Africans must remain as "traditional" peasants, while the attention of the dominating foreign bourgeoisie could be directed to some other avenues of profit-hunting as well, which they need so much when their position is adversely affected in other parts of their empire.

There is also the point that since the cotton economy is reaching the saturation point, further exploitation of Uganda's resources and labour must be directed concurrently to extracting mineral and other "natural wealth" of the land and conducting a part of African labour to this channel of work, for in the stagnating agrarian economy the African labour force is becoming more and more redundant and so for its full utilisation attention must be given beyond crop production.

Thus, in a symposium on "Development Planning for the Africans", the report of which came out in 1947, Mr. Deverall, Secretary to the Development and Reconstruction Authority, stated[15]:

> "Problems requiring elucidation were the development of secondary industries to reduce pressure on the land, and alteration of the African's attitude to land and stock, a restriction in the number of human beings, and the study of the type of agriculture best suited to these changes."

Non-agricultural production is thus seen to have drawn the attention of the rulers and their counsellors. However, although the extract speaks of "secondary industries", it has actually been restricted to extractive industries, viz. mining, and primary processing of industrial and agricultural raw materials, as it will be explained in the following.

Evidently, when African labour is cheap and would be available in plenty because of the growing agrarian crisis in Uganda, capital investment cannot go unrewarded. And this is particularly true for Uganda, for she is very rich in natural wealth. Of this, the pioneers in Uganda had spoken repeatedly. Lugard, Portal and others spoke in glowing terms of the riches of Uganda[16]. Even Roscoe, whose main task in Uganda was to conduct ethnological studies besides his missionary duties, pointed out that the British enterprisers should interest themselves in the mineral wealth of the country. For instance, with regard to Bunyoro he noted in 1923[17]:

> "There is no doubt that the land is rich in iron, coal, talc, and possibly the precious metals, and the field offered to the investigator is a promising one."

But even though the field was promising, the British monopolists then found it easier to make enormous profits at a cheap rate by exploiting the fertile lands and the labour of the Uganda Africans as peasants. Now, however, with their rapidly weakening position in the capitalist world and the onset of agrarian crisis in the Protectorate, they find it necessary to pay attention also to this avenue of exploiting Uganda. In the pre-war years, the export of minerals accounted for only about 3 per cent of the total export from Uganda. This was to be increased, but at the same time peasant-agriculture must remain the predominant asset of Uganda. The post-war "development plan" had to take both these factors into account. As put quite characteristically by Sir John Hall (the Governor of Uganda) in 1946, the central objective of the plan was to indicate, given the "limiting factor" that Uganda remains a land of peasant-producers, "how the country can best be developed and its natural wealth exploited to full advantage"[18].

Here lies the crux of the spectacular "Development Plan for Uganda" which opened out the colonial perspective for Uganda in the post-war period.

To examine this plan further, to whose "full advantage" would this plan be employed? From the foregoing discussion it is obvious that it would be mainly for that of the British bourgeoisie, along with which the claims of the U.S.A. monopolists also cannot be ignored in the present context. But here again is the snag that private capital is shy of pioneering but eager for safe investment. This applies not only to British but also to U.S. private capital. Palme Dutt noted in 1953[19]:

> "In Africa the immediate prospects during the first few years after the second world war were regarded as unfavourable for any extensive investment of American private capital, pending the necessary unprofitable 'development' expenditure on transport and utilities (*Report of the United States Chamber of Commerce* for 1949). But Marshall Plan funds were found useful for this purpose, and during the more recent period the penetration of American capital has begun to develop at an increasing pace."

Therefore, just as it was at the time of building the Kenya-Uganda Railway, the Government had now to take the responsibility of paving the way for the flow of private capital to exploit the "natural wealth" of Uganda. The author of the plan, Worthington, noted[20]:

> "The plan is mainly concerned with the expenditure of public funds, but the large expansion of public services for which financial provision is made will be of little avail in helping the country towards increased prosperity unless there is a corresponding increase in the activity of private enterprise on the part of Africans as well as of Europeans and Asians. By providing services like improved communications, water supplies and housing schemes, and by furthering the development of hydro-electric power, it is hoped that a number of factors which have limited private enterprise up to date will be removed. Special opportunities may develop in the marketing and distribution as well as in the processing of primary products. Moreover, enterprise will be directly encouraged because large parts of the proposed expenditure are devoted to public works, many of which are suitable for placing out to contract."

It is worthy of note that the above extract refers to "private enterprise on the part of Africans as well as Europeans and Asians", but obviously, because capital in the hands of Africans is very little, the plan can benefit only the foreign interests. The latter will not only be thus encouraged to invest capital in private enterprises, but can also draw substantial profit from the "public works" to be built according to the plan; for, many of these "public works" are "suitable for placing out to contract." Indeed, that this is a part of the general policy of Colonial Development and Welfare was made clear by Lord Trefgarne, the first Chairman of the Colonial Development Corporation, who, while reporting on the activities of the corporation at a Press conference on the 1st of January, 1950, emphasised[21]:

> "We always prefer to go in with a private concern."

Ways and means were thus found for foreign interests to thrive further on Uganda.

Of the foreign interests again, although the plan spoke of European and Asian interests, evidently only the former were in the mind of the planners. This is obvious from the fact that due to world circumstances and the present state of Uganda's economy, Asians (meaning the Indians in particular) have declined in the favour of the British rulers in the post-war years. The Immigration (Control) Ordinance of 1947 was introduced in order to restrict the immigration of Asians in Uganda. According to this Ordinance, not only have the immigrants to fulfil several conditions (including the deposit of a large sum of money with the Immigration Authorities which would be forfeited in case the rulers decide that the immigrant is not behaving properly), but many of the Asian settlers who had already lived in Uganda for several years could, on flimsy grounds, be considered as not permanent residents of the Protectorate. This discriminating character of the Ordinance was strongly emphasised by an unofficial Indian member in the Uganda Legislative Council in 1949[22]. But the Ordinance is in force and several more rules and regulations have come into operation after that to make the position of Asians (Indians) precarious in Uganda. Hence, the plan is no doubt meant to serve the interests of the British monopolists only. This the progress of the plan and allied activities of the colonial government in recent years has made quite clear.

To facilitate the penetration of private capital for a more intensive exploitation of Uganda's mineral wealth it was first necessary to have full control over the land. In those areas where landlordism was created by agreements this created some difficulties, and particularly so in Buganda —the classical area in Uganda for the creation of landed aristocracy. As noted earlier, according to the 1900 Agreement, land rights were vested in the native rulers. Mr. Churchill, on his visit to Uganda in 1907, described the Agreement as a "human document", and assured the chiefs who gathered to listen to his address[23]:

> "(They) have no fear that it will be encroached upon or melted away so long as they themselves and the people of Uganda faithfully adhere to their portion of the contract. Under that agreement all their rights and liberties are guaranteed, and all their lands, possessions and ancient privileges."

But the assurance of an imperial ruler to the colonial bureaucracy is no better than that of a tiger to a lamb. All the pledges were forgotten when the need came to annex land to "develop" Uganda in the post-war period.

In Bunyoro the difficulty was got over earlier by an amendment to the Agreement which was passed in 1933. But the Buganda landlords were more adamant. Whitley noted in 1945[24]:

> "The Buganda Government were asked to agree to amend the Agreement so as to give power to acquire land for any public purpose. It was pointed out to them that such a power was provided for in Article 26 of the Bunyoro Agreement, 1933, so that there would be nothing unusual or revolutionary about such an amendment but the proposed change aroused a great controversy both in and outside the Lukiko. ... The Kabaka, Kulubya [who later became the Prime Minister] and a few chiefs were in favour of the amendment but the great majority, including the Katikiro [the then Prime Minister], were against it, and it was rejected."

The rejection of the amendment, of course, did not stop the underhand or open manoeuvres of the colonial government to annex land. It was alleged that in 1941 mineral rights on Buganda Church lands were secretly passed on to the government by the Bishop and other Church Trustees —which came to light in 1948[25]. The Baganda demanded the resignation of the Bishop, and it may be recalled that in *Buganda Nyafe* it was stated that "the time is near when we shall be deprived of our land and settle on European estates". Whitley found that this move by the foreign rulers to take away land was one of the causes of the "disturbances" of 1945[26]. But the operation did not stop. When the aforesaid Bill was rejected in the Lukiko, the Prime Minister or Katikiro, Wamala, was forced to resign, and the new Prime Minister passed the Bill. He was assasinated after this incident, and, as noted earlier, the ex-Prime Minister Wamala, Prince Suna and many other persons were deported without trial after the murder, and Wamala and Suna died whilst in deportation. Land alienation went on in full swing, so that in Uganda today "a considerable amount of land is being alienated from African ownership for the purpose of mining, development of industries and for European settlement".

Besides seizing Uganda's land for the monopoly interests, it is necessary to develop power for the extractive and primary processing industries. Therefore, the biggest developmental project which was launched under Uganda's "Development Plan" consisted of building a hydro-electric station at Owen Falls on the banks of the Nile near Jinja at a cost of £ 22 million. This is in the Eastern Province of the Protectorate which is noted for mineral resources. In 1953, "the last of the temporary gaps in the Owen Falls dam was sealed in December and the first of the 15,000 kW. generators went on test"[27]. Power from the falls was expected to be available early in 1954 when there would be a gradual change over from thermal to hydro-electric generation. In the spring of 1954, Queen Elizabeth II opened the dam.

With such development of Uganda, private capital has already began to flow for the extraction of Uganda's precious rescources. While up until 1936 some £ 15 million of private capital was invested in Uganda[28], during 1946—50 only the total private investment amounted to £ 8½ million, or roughly £ 1.7 million per year; and it was officially stated in 1954 that "with new developments in copper and other mining it is probable that annual investment is now running at a rather high figure"[29]. In 1951 there were 20 foreign companies registered in Uganda. Since then their number is likely to have increased. Uganda is therefore bustling with activities.

Tororo, an important town in the Eastern Province and not far from Jinja, has become the centre of this new scheme of exploitation in this region. The 1953 Colonial Annual Report on Uganda noted[30]:

> "The Tororo Exploration Company (Monsanto Chemicals Limited, Frobisher Limited and the Uganda Development Corporation) completed its drilling at Sukulu; it would appear that 100,000 tons of apatite and 1 million lb. of pyrochlore, the valuable mineral used in jet engine manufacture, could be required annually."

Also[31]:

> "Mining of apatite at Busumbu continued throughout the year; the apatite is railed to Kenya for making soda phosphate."

Then[32]:

> "The cement factory at Tororo went into production and supplied several thousand tons to the Owen Falls project; the cement is fully up to British standards and can meet Uganda's normal requirements at competitive prices."

Copper mining and extraction of iron at Tororo also form part of the plan for Uganda's development. In addition, in this region prospecting is going on, for "other valuable minerals are present in the soil"[33].

The neighbouring province of Buganda is also not altogether neglected. As stated in the 1953 Report mentioned above[34]:

> "There was some production of amblygonite from private land in Singo County in Mengo District."

Development of Uganda has also consisted of extending the Uganda Railway from Kampala to the western border, up to Kasese —205 miles from Kampala. Thereby the industrial raw materials extracted at Kilembe in the district of Toro, and in the neighbouring regions, can be quickly transported out of Uganda. Kilembe is now bustling with activity. The 1953 Report noted[25]:

> "Agreement was reached between the Colonial Development Corporation, the Uganda Development Corporation and Frobisher Limited, the big Canadian mining concern (in which U.S.A. capital is also said to have an important voice — RKM), on the financing of major activity at Kilembe. The C.D.C. will provide £ 1¾ million, the U.D.C. £ ¾ million and Frobishers, which have already spent £ 2 million, an additional £ 2 million. It is estimated that output will be about 18 million lb. of copper and 900,000 lb. of cobalt

> a year, increasing the value of East Africa's export by £ 2½ million annually. Sites have been selected for the erection of treatment and smelting plants at Kilembe and Jinja."

Jinja, it will be remembered, is where the hydro-electric power station has been built. The 1953 Report also noted that the "present plan is to instal a roast-leach plant at Kilembe and a smelter at Jinja"; while the mine at Kilembe "will have its own hydro-electric power installation"[36]. Evidently, the huge hydro-electric resources of Uganda will not go untapped for the foreigners' profit.

Moreover, besides these large-scale undertakings, foreign interests are ramifying from these two nuclei, at Tororo and Kilembe. The 1953 Report noted[37]:

> "Union Corporation Ltd., of South Africa, put a prospecting party into the field in Ankole while Electorail of Brussels, of which Compagnie Miniere des Grands-Lacs is a subsidiary, had prospectors and geologists working in Kigezi.
>
> There was vigorous prospecting for the high-priced ores of tantalum and niobium, particularly in Ankole. Kagera mines were granted exclusive prospecting licences in the Kakarena area and small workers pegged numerous titles in the Burama Ridge, Kamwezi and Kitofu."

Prospecting for oil and other mines and mineral deposits have also been going on in Uganda for some time.

Then, there is the extraction of lead in Ankole, installation of sawmills in Bunyoro, and, as previously, the extraction of tin, wolfram and gold. But the extraction of tin and wolfram is not being continued with so much vigour because of "fluctuations in world prices during 1953"[38]. The 1953 Report noted[39]:

> "The price of tin declined by a third and that of wolfram by no less than half".

Even so, the "Kagera mines, operating at Mwirassandu in Ankole, continued to produce most of Uganda's tin. In May the company purchased Kikagati Mines"[40].

As regards wolfram, the British Government came to the aid of the monopolists, since "because of the slump in prices the search for wolfram lost its impetus among small workers and the mining companies operated on a reduced scale"[41]. Therefore, the "Wolfram operators were fortunate in having a minimum price guaranteed by the United Kingdom Government", and, in consequence, "despite some difficulties which have not yet been resolved, the wolfram mines continued with the installation of milling and concentrating machinery"[42].

As regards the production of gold, its "production in Ankole continued to decline but considerable success attended opencast mining for gold at Busia. As a result, gold production rose from 161 ounces in 1952 to 439 ounces in 1953"[43].

While almost the entire course of Uganda's post-war "industrial development" is thus concentrated in the extraction of her "natural wealth" and in transporting them out of the Protectorate, the only manufacturing industry which has so far been included in the plan is the building of a textile mill at Jinja as a joint-venture of the Uganda Government and the Calico Printers Association and the Bleachers Association of the United Kingdom. This mill was expected to start work "by the end of 1955"[44]. No scheme is there for the development of heavy industries or even of other secondary industries besides the solitary textile mill. Capital is flowing only for the extraction and primary processing of industrial raw materials of Uganda and for their transportation out of the Protectorate.

Such then are the major economic "developments" of Uganda in the post-war period, which the *Times* of London in its issue of the 18th November 1953 gloriously announced with the words: "Uganda has embarked on its first programme of industrialisation". This programme of industrialisation, however, has not meant the generation of wealth for Uganda and her people. What has actually happened is that in collaboration with the Colonial Government, and taking advantage of the "public works" built by the money furnished by the Uganda Africans from their incessant toils, Monsanto Chemicals and others are making fortunes out of the mineral resources of the land, and the Calico Printers Association and the Bleachers Association are bent on making money from the projected textile mill. The colonial perspective for Uganda which has unfolded itself in recent years is thus seen to be more severe extraction of her resources and labour by the foreign monopolists without developing her economy.

No doubt, such exploitation of Uganda will also hasten the disintegration of British rule in the area. For the Uganda Africans, brought into these concerns as wage-labourers because of their lack of means to subsist in the stagnating peasant-economy, are more rapidly shorn of their "traditional" outlook, and they grow militant. This disintegration of their "tribal" life and the eventual unity of the wage-earners from various "tribes" will inevitably lead to serious political consequences. The working class of Uganda, which has already made itself felt in the political arena, as it has been noted in Chapter II, may take the helm of Uganda's future course of development. Therefore, while this form of exploiting Uganda is dictated by the needs of Britain as a Colonial Power in its phase of decline, they cannot put too much stress on their loudly-proclaimed schemes of "industrialising" Uganda.

There is also the point that, as explained before, Britain's parasitic economy is in a precarious position. So the colonial "developments" can take only "a non-industrial direction". As quoted by Cheddi Jagan (the imprisoned Prime Minister of British Guiana) in his book *Forbidden Freedom*, Mr. John Strachey stated quite bluntly when he was the Labour Minister of Food[45]:

> "Our national position is really too grave to warrant any indulgence in our particular opinion on the methods of overseas development. By one means or another, by hook or by crook, the development of primary production of all sorts in the colonial territories and dependent areas in the Commonwealth in far more abundant quantities than exist today, is, it is hardly too much to say, a matter of life and death for the economy of this country."

Also, the late Sir Stafford Cripps, the Labour Chancellor of the Exchequer, gave the advice in his address to the African Governors' Conference in 1947[46]:

> "You will, I understand, be considering the question of the development of manufactories and industries in the colonies. Though I take the view that such development is highly desirable so long as it is not pushed too far or too quickly, yet it must be obvious that with the present world shortage of capital goods, it is not possible to contemplate much in the way of industrial development of the colonies."

For these reasons, as facts bear out, the possibility of the Uganda Africans to become wage-labourers remains very limited. In 1951, only 77,246 Africans of both sexes (including juveniles) were employed in the industrial groups of manufacturing, mining and quarrying, construction, and transport and communication. Of the total population in Uganda, as given by the census of 1948, this figure works out at 1.6 per cent. Even taking into account all kinds of industrial groups enumerated in official literature, that is, including agriculture, forestry and fishing, wholesale and retail trade, etc., and even employment under the Protectorate or the Local Governments, the figure of employed Africans of both sexes (adults and juveniles) amounted to 199,792 in March 1951; that is, 4 per cent of the total popuation. Of this figure of 199,792, the census of African employees taken on 30th March 1951 recorded 192,966 as the number of adult male employees, and the Colonial Annual Report of the British Government on Uganda for 1951 noted that since 1949 the increment had been of about 20,000[47]. Thus the rate of adult male Africans taking up an employment in the Protectorate during 1949—51 was at the rate of less than 1 per cent of the estimated number of 2.3 millions of adult Africans in Uganda in 1948[48].

While such was the situation after three years of the Ten Year Plan, the progenitor of the plan had made it clear at the beginning[49]:

> "It is assumed that 50 years hence 15% of the population will be engaged in industrial production, commerce, fisheries and forestry."

In other words, agriculture will remain the primary source of exploiting Uganda's land and labour, which also the later speakers for the Government, including Hailey in 1950, have confirmed.

But here again is the problem that while the fertile lands of Uganda cannot go unexploited, the use of better farming methods is now essential to the agrarian economy, and this cannot come from the impoverished African peasants. Hence, the dilemma that is confronting the rulers since the end of the last war is how the Africans can be kept simply as the source of labour-power, while their traditional existence as "peasants" is preserved. An ingenious scheme of maintaining the *form* of primitive agriculture but completely transforming its *content* has thus become the demand of the rulers. Such a scheme has not yet been put into a blue print, but that something is in preparation on these lines is evident from the attention paid by experts to this question and the suggestions they are putting forward. One of the earliest of such suggestions and one which is very significant to the whole of East Africa is the advice given in 1947 by Dr. B. A. Keen, F.R.S., Deputy Director of Rothamstead and Agricultural Adviser to the East African Governors' Conference. The following extracts from the *Proceedings of the Nairobi Scientific and Philosophical Society*, reporting his address on "Primitive Agriculture in the Modern Economic World", delivered at a meeting of the Society on 8th July, 1947, is of interest in this connection[50]:

> "Dr. Keen then mentioned the following big schemes in which a beneficial change in the organization of primitive agriculture has already been accomplished:
>
> The Sudan Plantations Syndicate (Commercial with a government concession),
> The Sudan alternative livelihood scheme (government organisation),
> The Gash Board, Sudan (government scheme run on commercial lines),
> The Jewish Settlements in Palestine (Jewish National Fund),
> The Latifujah Estates in Iraq (private company),
> The Italian Colonies in Cyrenaica and Tripolitania.
>
> These schemes include a variety of aims and purposes, but all possess a common feature of vital importance, i. e., the fact that the land has been bought or leased from its original owners, reorganised into units of economic size, and then leased to peasant cultivators upon conditions which prevent refragmentation. The essential feature is that the grave — perhaps insuperable — difficulty of overthrowing a fundamental feature of the whole social organisation — the system of land inheritance — is avoided. It continues to operate, but harmlessly. The Sudan Government for example, has leased semi-desert land from its owners, whose interest in it is limited to rents. The S.P.S. was then given a concession to irrigate the area for cotton, the original owners having rights of nominating tenants. The outcome was the most profitable peasant farming seen by Dr. Keen throughout the whole of the Middle East. Inheritance may divide up rent allocations but can no longer interfere with farming organisation. The Jewish community has similarly overcome this difficulty by vesting land ownership of settlement areas in the Jewish National Fund.

> Summarising these examples Dr. Keen pointed out that large organisations were able to afford heavy machinery and to employ trained staff for the improvement of crops and stock, and for the efficient marketing of produce. The key to the whole problems of applying modern agricultural science to primitive peasant agriculture was the reorganisation of land into stable units of economic size. In East Africa the alternative appeared to be either the retention of primitive cultures as living museum-pieces, or their reorganisation into economic groupings. The inhabitants themselves were unlikely to accept the first alternative, for as Trevelyan has said: The Social scene grows out of economic conditions to much the same extent that political events in their turn grow out of social conditions! The second alternative is possible when the individual cultivator is assured of the benefits of his own labour. Ownership of the land is not essential, but a secure tenancy of a land unit of economic size must be obtained before the tenant can make effective use of the agricultural scientist's help".

It will be realised from the above that Keen's suggestion falls in line with the post-war general strategy and tactics of Britain to draw maximum profit from all aspects of the colonial economy by following methods suited to particular conditions. The scheme embodies all the main principles of the policy which can be applied to the agrarian economy. These are: (i) giving "concessions" to imported capital for "improvements"; (ii) paving the way for foreign capital by building public works under government responsibility (such as the Owen Falls dam at Jinja will also be used for irrigating vast tracts of land which can now be taken over by some monopoly concern); (iii) preserving as before the landowners' interest in land rent and in their right to nominate tenants, so that the alliance between the foreign rulers and the preserved or created native feudatory elements in the societies remain ever-strong; and (iv) continuing all the time with the ruthless exploitation of the labour of the people, but not as wage-earners, which would have soon become obvious and led them to launch a more vigorous and united struggle as workers against the exploiters, but as "traditional peasants".

Obviously, this method of "applying modern agricultural science to primitive agriculture" can only accelerate the rate of exploitation of the people without leaving any scope for their future development. For, according to the scheme, over and above the prevailing burden of native feudatory and foreign mercantile interests borne by the people would be added the still more oppressive burden of direct interest of finance-capital in the agrarian economy. Nevertheless, on the surface the scheme gives the impression of leading the people to prosperity by raising the productivity of land and thus apparently solving the agrarian crisis.

This has become necessary to the foreign rulers also from the political standpoint, for the colonial administrators are apprehensive of the "social scene" which is sure to grow out of the present economic conditions, and then "political events in their turn" may "grow out of social conditions". Hence, Keen's suggestion was accompanied by paternal advice to the "primitive" Africans to accept such schemes if they wanted to attain prosperity. It is worthy of note that in the discussion following Dr. Keen's address, Mr. Chambers "quoted Sir Frank Engledow's comment that the primitive cultivator has the alternative of freedom of action or freedom of want"[51]. All such high-sounding words, however, cannot obviate the fact that the "peasants" will have no freedom to attain real prosperity so long as the colonial and feudal interests thrive on their labour and these become more intensified through the implementation of such schemes.

No doubt, such schemes are very spectacular, but their true interest cannot be concealed. This is a still more intensive exploitation of the people by the colonial power while keeping the decadent structure of the colonial societies intact. Evidently, the essential purpose of such schemes is to subordinate African small peasant-economy to the domination of British (and the U.S.A.) finance-capital. This is a characteristic venture of colonialism in the last phase of ever-aggravating crisis of the system in post-war years. In order to put temporary checks to its inevitable decay, which History has ordained for it, the brains behind the colonial system come out with such schemes, and the colonial governments are asked to finance them or pave the way for their fruition.

However, such a transformation of Uganda's agrarian life is still only a spectre haunting the Africans. As far as is known, its blue print has not yet been prepared for the Protectorate. But the fact that the Egyptian Government had contributed £ 4.5 million for the construction of the Owen Falls dam for irrigational facilities suggests that such facilities would not be used only by Egypt and Sudan (in Sudan already such schemes are in operation)[52]; naturally, they would also be utilised for British Uganda*.

But, even if such a scheme is implemented in Uganda, there is no doubt that like in all places it will eventually fail in its purpose of averting the inevitable doom of the colonial system. For, however ingenious and indirect may be its operation, in the Uganda societies it cannot but increase the rate of disintegration which started with the beginning of colonial rule. Like all other colonial measures implemented previously, this scheme provides nothing progressive to replace the old structures of the societies; it only adds to the burden of the people.

Evidently, whether such a scheme is operated or not, the ruination of the Uganda peasantry (either as more heavily-taxed tenants on the newly-demarcated plots under the above schemes or, as it is today, depending entirely upon their own paltry means of production) cannot be stopped. Of these impoverished peasants a small number would of course be employed in the so-called industrial undertakings of mining and primary processing of industrial and agricultural raw materials, and their ancillary activities. Worthington made a mention of this possibility in his plan as "special opportunities in the marketing and distribution as well as in the processing of primary products". But, firstly, because the colonial economy, either in the industrial or in the agricultural sector, will not generate wealth for Uganda and her people, the impoverishment of the people will proceed rapidly; and, secondly, even if some of the ruined peasants became wage-earners (which they have already started to do), it would only lead to proletarianisation of one section of the Uganda society. Disintegration of the society cannot therefore be resisted by the post-war colonial policy; on the contrary, it is being accelerated by it.

Moreover, proletarianisation of the Uganda Africans in an industrial economy is not the future dictated by this policy. If it is borne in mind that according to Uganda's "Development Plan", in fifty years' time only 15 per cent of the population will be engaged in non-agricultural occupations, the conclusion becomes irresistible that the perspective now opened before the peoples of Uganda through the post-war policy of Britain means further domination of Uganda by giant monopoly concerns, more efficient plunder of her labour and resources, and continued enslavement of her people without any outlet towards a truly developing economy. This process cannot but lead ultimately to mass pauperisation, total degeneration of the societies, and their final decay.

---

* It is worthy of note that one should not ignore the qualitative difference between the use of irrigational facilities by countries like Egypt and Sudan, which have attained full sovereignty, and that by the colonies like Uganda. As the former countries are governed by the growing *national bourgeoisie* who are keenly interested in *national* progress, the improvements in their agricultural economy by means of irrigational facilities, for instance, benefit all sections in the society in a certain measure. On the other hand, in colonies like Uganda, where the governing aim is not national progress but colonial exploitation, such schemes as noted above essentially mean added exploitation by finance-capitalist interests over and above the existing forms of exploitation.

2. Peoples' Perspective

What then is the other perspective for Uganda? This point should be examined in the light of the three communities of interest prevailing in Uganda today. These are officially labelled as Europeans, Asians and Africans, but strictly speaking they should be termed as British, Indian, and Uganda Africans. The European interest virtually represents the interest of the British, for the other few Europeans living in the Protectorate are under their protective wing and can hardly be regarded as having an independent voice of their own. The Asian interest coincides very largely with the interests of the Indians settled in Uganda, others like the Arabs having insignificant influence in the Protectorate today. The denomination of Indians in Uganda however means those who have immigrated from the Indian sub-continent and who may now owe different allegiance as noted in Chapter II. The Africans are evidently the Uganda Africans, although there is a sprinkling of non-Uganda Africans from Belgian Congo, Ruanda-Urundi, Sudan and such other neighbouring states. These non-Uganda Africans have very little say in the future of Uganda and also comprise a small proportion of the total African population.

In their political outlook all the there communities are fairly homogeneous within themselves and noticeably heterogeneous between them. The ruling community has the same object in view with regard to the course of events in Uganda as propounded by the Colonial Government. No doubt some social and economic differentiation are present among the British between the grades of Government officials and in trading or other business interests, but such differentiations reflect only the truncated top of the class structure of the mother country. There is no class of industrial proletariat within the "European" community. So the ideology of living on the labour of others plays the dominant note throughout the community, although some with a liberal outlook may attempt to tone down the stark reality of the master and servant relation between the Europeans and the Africans. It is true that this ideology has not taken the pathological character of naked racial hatred as is seen in the Union of South Africa, but it is equally true that any attempt on the part of the Europeans to identify themselves with the aspirations of the Africans to become free from colonial bondage is ruthlessly smashed. Of course such occasions do not occur frequently; for, firstly, those who are brought to the colony to run the administration etc. are properly selected in the mother country, and, secondly, by offering even the least important members of the community comparatively higher salaries than they could get at home (in addition to many social and financial amenities) they are readily bought off by the imperialist interests of their mother country. If, in spite of such measures, a "trouble maker" is detected in the colony, he is at once removed. Several cases of such instantaneous deportation from British central and east Africa in recent years have proved how vigilant the foreign rulers are to crush any challenge to their vested interests. Therefore, neglecting minor variations, in basic policy the whole community may be identified with that of the colonial interest in the mother country and its machinery in Uganda, the colonial government.

Like the British, the Indian community also represents only the truncated top of the class structure of the mother society. It is composed mainly of small traders and businessmen and middle-grade civil servants and such employees in the commercial organisations. There are very few Indian monopolists (virtually only one who is the owner of the sugar plantations and factories); and in this community there is practically no industrial proletariat. In all essentials the Indians in Uganda represent the weakness and aspirations of the upper stratum of a dominated country. They are subservient to a colonial power on the favour of which depends their "prosperity" in this colony and which, like in their mother country, they wish to get rid of. In Uganda, of course, the latter course becomes very difficult for them because they have no roots in the society and so their position is unstable and depends on the will of the colonial rulers unless they identify themselves with the anti-colonial aspirations of the Africans. More will be said later of the tragedy of this dilemma of the Indians in Uganda, but here it should be made clear that by themselves the Indians cannot offer any perspective to Uganda's future.

Within the African community there are substantial differences owing to the rapidly growing class structures within the societies and the national differences which are becoming apparent with the formation of national markets in different regions of the Protectorate. These differences within the African community, however, play a minor role in comparison with the uniform outlook of the Africans towards freeing themselves from colonial bondage. Up till today it may not be untrue to state that as the exploiting foreign interests in the Protectorate undoubtedly represent the most important obstacle to the Uganda Africans in their future perspective of freedom, prosperity and progress, the African community represents a homogeneous body in relation to these interests.

For this reason, as opposed to the colonial perspective which can only lead Uganda to final destruction, there is the outlook of the Uganda Africans for regeneration and progress of their motherland. Although this outlook is sometimes deflected from its goal by the desire of some to return to the old life, this is not the general phenomenon today. It is true that some enthusiasts of Uganda's future progress (belonging almost entirely to the upper stratum of the societies) tend to look backwards. They unduly glorify the old order of Uganda's civilisation, ignoring the fact that it represented a low stage of development; and they persevere to persuade the people to eschew the traits of material and cultural development which the modern world can offer today provided the people are free from bondage. However, like those few who still acclaim that British rule has been for the good of the Africans, they are in the minority. The bulk of the people are more and more coming to realise that just as it is true that British rule has meant arresting Uganda's normal course of development and setting a course of retrogression in the African societies, so it cannot be ignored that while serving their own interest the foreign rulers could not but lay down the material premises for the regeneration of Uganda. Therefore, while rightly condemning British rule for leading Uganda into a blind alley of destruction and decay, they do not overlook the fact that Uganda in her future course of development can and should take full advantage of the "new elements of society" which have been scattered within her territory by the British bourgeoisie but which they do not intend to give the Uganda Africans the chance to make use of for the progress and prosperity of their motherland.

In this respect what Marx wrote on *The Future Results of British Rule in India* appears to be relevant also for the other British colonies like Uganda. To quote Marx[53]:

> "I know that the English millocracy intend to endow India with railways with the exclusive view of extracting at diminished expenses the cotton and other raw materials for their manufactures. But when you have once introduced machinery into the locomotion of a country, which possesses iron and coals, you are unable to withhold it from its fabrication. You cannot maintain a net of railways over an immense country without introducing all those industrial processes necessary to meet the immediate and current wants of railway locomotion, and out of which there must grow the application of machinery to those branches of industry not immediately connected with the railways. The railway system will therefore become in India truly the forerunner of modern industry."

Marx, of course, did not acknowledge the colonial rule as a progressive force in India. On the contrary, while visualising the future industrial development of India, he also noted[54]:

> "All the English bourgeoisie may be forced to do will neither emancipate nor materially mend the social condition of the mass of the people, depending not only on the develop-

ment of the productive power, but on their appropriation by the people. But what they will not fail to do is to lay down the material premises for both. Had the bourgeoisie ever done more? Has it ever effected a progress without dragging individuals and people through blood and dirt, through misery and degradation?

The Indians will not reap the fruits of the new elements of society scattered among them by the British bourgeoisie till in Great Britain itself the now ruling classes shall have been supplanted by the industrial proletariat, or till the Hindoos themselves shall have grown strong enough to throw off the English yoke altogether."

This also remains the main issue before the peoples of Uganda, that is, "to throw off the English yoke altogether"; for the progress of a country does not depend merely on the development of the productive power, but also "on their appropriation by the people". As past and present events indicate, this issue will be satisfactorily solved in Uganda at not a too distant date.

To trace now the emergence of this perspective of freedom of the Uganda Africans from colonial bondage, of their progress in socio-economic life, and of their all-round prosperity, this can be conveniently discussed if Uganda's freedom movement is classified under three categories. These are: (1) struggle for economic development, (2) fight for popular governments, and (3) the anti-colonial movement. Needless to say, the three aspects of the peoples' movement are mutually related, and in the dynamic process of Uganda's life today they are inseparable. But for an analytical study of the peoples' point of view, they may be discussed in a sequential order and so they are set forth below.

3. Economic Development

That from the beginning of British rule in Uganda there were local discontents and outbursts, has been noted in the foregoing pages. But in the earliest stage, these were unorganised and sporadic. It will be remembered, for instance, that the women of Busoga struck work and refrained from growing their crops as a protest against the Colonial Government's policy of slave-labour. There were also other instances of local outbursts against the machinations of colonial rulers to deprive the people of their economic freedom. Such cases were probably first organised in Buganda where they also took a political character. This was concerned with the people's deprivation of the land and the creation of landlords by the 1900 Agreement.

As noted before, in a short while after signing the 1900 Agreement the land-question assumed serious proportions in Buganda. There developed accordingly a politico-economic organisation to give vent to the anger of the people at being cheated of their land and to demand their land back. Thomas and Spencer noted about the situation[55]:

> "Clearly the lot of the peasant was not an easy one, and clearly here were all the makings of agrarian unrest; ... Discontent eventually materialised, not by way of a demand for individual ownership of small holdings, but in an agitation for a return to bataka tenure — for the restoration of land to the clans."

It has been mentioned earlier that even during the semi-feudal rule in Buganda by tribute-owning kings and chiefs, tribal democracy persisted in the clan-ownership of land — the primary means of production of the agricultural people. Now, although in a sense looking backward to restore the ancient institution, the people demanded back their democratic rights.

In 1921 an association, generally known as the "Federation of Bataka", was formed, and the clan leaders took their claims into court. In 1925 a law was passed establishing the right of the people to settle on their clan lands with immunity from eviction. But they were subject to the ordinary obligations of a peasant towards the landowner. Although partially victorious, the indignity of a clan head being a peasant on his own land was felt by the entire community. Thomas and Spencer noted[56]:

> "Native opinion was deeply stirred, and from top to bottom Buganda was divided into two political camps. The Federation, whose representations inevitably developed into an attack on the good faith of the Lukiko, was countered by an association supported by the large landowners."

The Federation of Bataka became the rallying ground of all who had grievances against the native government and their masters, the foreign rulers. As a young Makerere College student recently put it[57]:

> "The Bataka are originally a hereditary society of the leaders of the clans, but today it comes to mean any person who has much love for his country."

A national organisation was thus born, and henceforth the tension between the rulers and the ruled began to find its expression through this organisation. Since they had failed to secure a hearing from the local authorities, the Bataka decided in 1947 to send their representative, Semakula Mulumba, to England in order to present their grievances to the Secretary of State for the Colonies and to Members of the British Parliament. Money was collected throughout the country, and Mulumba arrived in England. From there he began to fight for the demands of the Uganda Africans.

The struggle for economic development did not, of course, remain limited to demanding back previous economic rights of the Uganda Africans. With the flowering of the colonial economy the peoples of Uganda began to see how their progress was blocked by foreign interests. Accordingly they organised themselves behind the Uganda Farmers' Union. This organisation demanded a fair price for crops (especially for cotton and coffee), the right to set-up African cotton ginneries, and other measures for the economic progress of the Uganda Africans. Both the Federation of Bataka and the Uganda Farmers' Union rapidly grew in strength, and consequently the Federation of Bataka was banned in May 1948, and the Uganda Farmers' Union on the 26th of April, 1949.

But the aspirations of the people for a better future could not be stifled thereby. Henceforth the African Labour Party and the League of Citizens of Uganda carried forward the movement. How these organisations protested against political and economic discriminations against the Africans has been noted in Chapter II. They also proposed concrete programmes for the economic improvement of Uganda, particularly in the fields of trade and industry, which the bourgeois leadership of these organisations naturally demanded to facilitate the growth of their class. Nevertheless, such improvements demanded by the national bourgeoisie of Uganda would be of advantage to the entire society in developing its economy, and so they were fully supported by the mass of the people.

Before the second wave of national upsurge swept over Uganda, the African Labour Party submitted a scheme, entitled "The Country Aid and Village Development Scheme", to the Ministers of Buganda on the 13th of January 1949. The scheme contained the following main points[58]:

1. To develop industrial enterprise.
2. In every village the members of the Labour Party will guard their own areas at night so as to prevent arson and stealing.
3. A census of employees and employers should be taken in every village, and unemployment must be eradicated by industrial development.

4. Nationalised industries should be started as:
   (a) Workshops, (b) Art Works, (c) Collective Farming, (d) Weaving and Spinning shops, (e) Brickfields and Garden-fields; etc.

Evidently, the African Labour Party expected that if the Buganda African Administration (which is the strongest of all native administrations in Uganda) took up the scheme, there would be no difficulty in persuading the other African governments to implement it. But apparently no heed was paid to the scheme either by the Buganda African Government or by the foreign rulers.

Then, on the 3rd of November, 1949, the African Labour Party proposed another plan, and sent it to the Kabaka of Buganda for consideration. The plan, entitled "A Five Year Plan", contained the following essential points[59].

1. To obtain an Import Licence to import goods into the country.
2. To form a big central store to supply small retailers at reasonable prices and enable them to get supplies on credit as the foreign traders get aid from the Banks.
3. Every man and woman to join this Company, and the Kabaka to order them to pay two or three shillings for the improvement in trade and the establishment of a fund for general purposes in the whole kingdom. The payment should be made obligatory.
4. To establish depots in every county where the goods would be supplied from the Central Depot and the local retailers would get their goods from these branches either in cash or on credit according to his financial position and past conduct.
5. "The Country Reserve Fund", established from the collection made from every man and woman in Buganda, to be used mainly for educating the masses.
6. Minimum wages to be fixed for the employment of workers in fields, shops or industries, etc.

Again, no notice was taken of the proposal. But the agitation and struggle went on, and in this, as noted before, the growing working class of Uganda (still small in number, but not insignificant in its determination and militancy) played a significant role. As a result, the wage-scales had to be revised after 1950 (although the extent of improvement has been much less than desired), and an African cotton ginnery had to be established (even though it is directly under Government control). Also, "during 1953 new legislation in the form of the Coffee Industry Ordinance was passed whereby provision is made for greater participation by Africans in the processing of coffee. Thus the first six curing works to be erected after the start of this Ordinance will be operated by Africans or associations of African growers"[60].

To what extent the foreign rulers will fulfil the above promise has yet to be seen. But there is no doubt that, however limited their achievements until now, the developing movement of the Uganda Africans has shown to them that when they unite solidly and fight for economic emancipation, no force on earth can stop them from winning their just demands.

This is the perspective of economic development of Uganda which has opened up before her peoples, so that in larger numbers they are actively participating in their national and class organisations in order to reach their goal of prosperity and freedom from colonial exploitation.

4. Popular Governments

Along with the growth of the struggle for economic prosperity has gone the demand of the peoples of Uganda for a democratic control of the native governments in place of the rule of "chiefs" appointed by the colonial rulers. In the early years of British rule there was a tendency in some of those societies where the system of Native Administration was forcibly imposed to agitate for the selection of local authorities on the basis of "traditional or hereditary claims" instead of these being "appointed" by the British according to their discretion and convenience. But soon the idea developed of a democratic control of the native governments all over Uganda. Thus, in regard to the Lango people, among whom chieftainship as a political authority had not yet developed in the pre-British period, Hailey noted[61]:

> "In the Lango District the relative weakness of the clan organization no doubt accounts for the fact that in the first instance the British Administration utilized as its agents 'appointed' chiefs, who for the most part had no traditional position. There was, however, a considerable undercurrent of feeling in favour of the recognition of traditional or hereditary claims. Between 1944 and 1947 there was a section which advocated the creation of the post of Paramount (won Nyachi) for the Lango tribe, presumably in the belief that this would give it a stronger position vis-à-vis the Government or other tribes. There was, however, no traditional precedent for such an appointment, and the idea was probably borrowed from the Western Province [where were the kingdoms of Kitara, Ankole and Toro — RKM]. But it is significant that there existed at the same time a movement, which found expression in the activity of the Young Lango Association, for the association of the 'commoner' element with the work of the Native administration, as a counterweight to what was held to be the autocratic use of their powers by the chiefs."

This attitude to declare the appointed authorities under the systems of Indirect Rule and Native Administration as reactionary and agents of the British, and to have the native governments run by democratically elected members of the societies gained considerable ground after the last war. Huxley, while eulogising on the merits of the institution of Indirect Rule as representing the official policy "to offload more and more control to 'the people' ", came across this thorny problem during her visit to Uganda immediately after the last war. She was told by the native "intelligentsia" in no uncertain terms that "the native government does not represent *progressive* people". In support of this statement, she also received "a letter from the 'non-official representative residents in Kyaddondo county', who, they wrote: 'had no opportunity to present our grievances to you on the occasion of your visit, being prevented by our Saza (County) chief from doing so'". The letter contained "grievances" against some actions of the Kabaka of Buganda and of the appointed authorities in the Native Buganda Parliament. Huxley reflected on the letter[62]:

> "Such a letter would suggest that there is (as one might expect) a considerable and probably growing under-current of dissatisfaction with the people's own native government. To relieve this, the malcontents look to the Protectorate Government, but that is gradually shedding its powers; few seem to realise as yet that you cannot simultaneously have more self-government and more direct intervention by the British."

Here, instead of the peoples of Uganda, Huxley herself seems to fail in realising that shedding of powers of the Protectorate Government to the African Administrations (even if it was true) did not really mean self-government of the people. The reactionary character of these African Local Administrations, run entirely by the British-nominated or approved kings and chiefs, cannot be concealed from the mass of the people. Hence their demand for a democratic popular government.

But the way a representative of the foreign ruling class, like Huxley, reacted to such a suggestion is quite characteristic of the mentality of foreign rulers who know that without the aid of local agents they may not be able to exploit a colony for long. To quote Huxley's remarks on the attitude of the native "intelligentsia" who visited her and spoke for democratisation of the African governments[63]:

"The feeling of this group was, certainly and strongly, anti-native Government, anti-Katikiro — and probably anti-Kabaka, although the open expression of such a view would be seditious. They did not appear to be unduly anti-European as yet. This may have been due partly to good manners and partly to a belief that, when the time comes, a small group of Europeans at the centre would be much easier to get rid of than a wide-flung network of native *functionaires* at the periphery."

However, although the foreign rulers desire to keep the African Local Administrations completely under their control by having them run, without any opposition, only by those whom they can trust to toe their line, with the rapid development of the anti-colonial movement in Uganda in the post-war years (which was high-lighted by the two waves of struggle in 1945 and 1949) they had to make some concessions to the democratic aspirations of the people. In the post-war period some sort of election to a part of the membership to the local African councils (from the lowest administrative level to the Provincial) is held practically all over the Protectorate. In some places the lowest orders of chiefs, the village or parish chiefs for instance, are also elected by popular vote. In all such cases, of course, the approval of the representatives of the Protectorate Government is indispensable; and the councils containing the opposition members also have no law-making powers. But, in spite of such disadvantages, it seems that the native authorities are thus acquiring a tinge of democratic character, and it is also noticed that the colonial rulers are sometimes placed in an awkward situation thereby. Thus, under the caption of "More Trouble in Uganda", the correspondent of the *Times* of London wrote on the 21st of August, 1955, from Kampala[64]:

"While the Buganda affair is drawing to a close and a new Bunyoro agreement is about to be signed, the political situation among the Basoga, the second largest tribe (sic.) in Uganda, has become disturbed.

Trouble has arisen over the appointment to a county chieftainship of Mr. W. B. Mwangu, who for the past three years has been secretary-general of the Busoga African local Government. At recent elections for secretary-general he was defeated. Soon afterwards it was suggested that he should become county chief of Bukoli, but after further consideration, since he is a Catholic and the county is strongly Protestant, the idea was dropped. The county chief of Luuka was then moved to Bukoli and it was suggested that Mr. Mwangu should take his place.

The district council's appointments committee, an advisory body, to which the matter was referred, opposed the suggestion by four votes to three, but the non-voting chairman afterwards expressed his support. Mr. Mwangu was appointed to Luuka. There followed some minor but violent disturbances led by some members of the Busoga branch of the Uganda National Congress, who, the administration feels, were sorely piqued at their failure to secure control of any of the local Government posts at the recent elections.

When the district council met subsequently an overwhelming vote of no confidence in Mr. Mwangu was passed and the council had to be indefinitely adjourned when it refused to proceed to any further business until he was removed. Mr. Mwangu remains county chief."

Whether the aspersions made in the above extract against the Uganda African National Congress are true or not (for it should be noted that later when the district council met there was "an overwhelming vote of no confidence in Mr. Mwangu"), the above incident throws light on the fact that the loyal individuals like Mr. Mwangu can boss over the common Africans at the pleasure of the colonial rulers because the voice of the people has not yet been strong enough to put a complete stop to the dictatorial policy pursued from the beginning of British rule in the Protectorate. Such a situation leads more and more people to believe that the problem of Uganda cannot be solved purely through parliamentarism. Nevertheless, they intend to make use of all possible ways to frustrate the reactionary colonial policy, and therefore mustering their strenght in the administrative bodies and in the legislative council of the Protectorate through their elected representatives has become one of the predominant demands of the Uganda Africans in recent times.

Voting rights of the people, however, are very limited in Uganda. Apart from the fact that the key-positions in the African local governments are filled by persons selected by the foreign rulers, even for the elected positions with a purely advisory capacity or with little administrative power a good deal of control is exercised at every stage of the hierarchy, that is, from the parish or sub-county to the district and upwards. Probably for this reason the Uganda African National Congress, which has voiced the demand for immediate self-government of Uganda, has not yet been able to make much headway in these African bodies. But it is of interest to note that where such a control is not strictly exercised and where the people have greater freedom to elect their representatives the leaders of the national movement win the elections with big majority. This is becoming evident from the recent elections for the future composition of the Uganda Legislative Council, about which something will be said in the following pages.

In face of the determined opposition of the Uganda Africans against the autocratic rule in the Protectorate, the foreign rulers had to announce recently a new form of administrative control and a new composition of the Legislative Council for Uganda. This revision in the fields of administration and legislation for the Protectorate has a facade of *Africanisation*, although in essence it is the old wine in a new bottle. In the administrative sphere, instead of the previous *Executive Council* of the Governor, a *Council of Ministers* was announced in Kampala on the 31st of July, 1955. But the names and designations of the first ministers show that very little change has been effected from the past. As before, all the important positions remain in the hand of the British; and the one Indian "Unofficial Minister", one African "Assistant Minister", and the two "African Parliamentary Secretaries" appear to have been very carefully selected.

Thus, the six "Official Ministers" are British. The Chief Secretary, the Finance Secretary, and the Attorney-General are the same persons who were there before. The previous Director of Agriculture is now the Minister for Natural Resources; and, likewise, the previous Secretary for African Affairs is now the Minister for Local Governments and the previous Secretary for Social Services is now the Minister for Social Services. Among the five "Unofficial Ministers" and the "Additional Members of the Executive Council" also three are British. One of them, who is now the Minister for Work and Commerce, was previously an "unofficial" member of the Legislative Council and is, significantly enough, the President of the Chamber of Commerce. And the two "Additional Members" are the Resident of Buganda and the Chairman of the lint marketing, coffee industry, and grain conditioning boards, respectively. Like the other "European" members in the Ministry, the two "Additional Members" also previously belonged to the Executive Council, the former as Ex-officio and the latter as Official Member.

The Indian Minister recently chosen is Mr. A. N. Maini, who appears to have maintained so far a pleasant relation with the foreign rulers. As reported in the *Times* of London of August 1, 1955: "Before coming to Uganda he was for a time acting member of the Kenya Legislative Council. He has served on most of Uganda's statutory boards, and was the first mayor of Kampala for five years until his recent resignation." Yet he has not been given any portfolio, and will sit in the Council as an "unofficial minister".

The selection of the African "minister" and the "parliamentary secretaries" is even more instructive. Mr. Z. C. K. Mugonya, who has been chosen as the "Assistant Minister for Social Services", was previously the Chief Minister of the Kingdom of Ankole from 1948. Mr.

G. Oda, who has been selected as the "Parliamentary Secretary for Natural Resources", was first a Divisional Chief, then the Secretary of the West Nile African Local Government, and lastly a member of the Legislative Council for West Nile and Madi in Northern Province. His successive promotions at the will of the colonial rulers suggest his proficiency in maintaining "law and order" in the Protectorate. And Mr. D. J. K. Nabeta, who has been selected as the "Parliamentary Secretary for Local Governments", was a member of the Legislative Council for Busoga in Eastern Province and is a "prosperous trader" according to the *Times* of August 1, 1955. Two more African Ministers (Minister for Rural Development and Minister without portfolio) were to be appointed, and it was rumoured at the time of the first announcement of the Council of Ministers (31st July, 1955) that they would be selected from the Baganda.

Referring to the above mentioned changes in Uganda's Central Government, the London *Times* of August 1, 1955, commented: "This would seem fair distribution of posts among those with a claim to them". But there can be little doubt that this form of *Africanisation* of the governmental apparatus hardly at all caters to the demands of the peoples of Uganda. Evidently, not only the few Africans who have been and are to be taken into the *Council of Ministers* are selected essentially because of their loyalty to the foreign rulers, but what is of greater importance to note is that the key positions in running the colony in favour of Britain's colonial interests are held by the British and by those particular individuals who were ruling before.

Thus, the position of the Chief Secretary, "the head of the Protectorate Civil Service", through which the Governor "can exercise general control over the direction and conduct of public affairs", is reserved for a British. The positions of the Finance Secretary and the Attorney-General, so vital to keeping Uganda amenable to colonial exploitation and to the maintenance of colonial law and order, are also reserved for the foreign ruling class. Then the Ministry for Natural Resources, which is so very important for a successful execution of Britain's post-war policy in Uganda, as outlined above, is also reserved for a Briton. Also, the African Local Governments are controlled by the British Minister for Local Governments; and the Ministry for Social Services, which is primarily concerned with publicising Britain's noble role in Uganda and with "sensing public opinion" against colonial rule (as has been described earlier), is in the hand of a member of the ruling class. Moreover, the British Minister for Work and Commerce is also the President of the Chamber of Commerce, which has a good deal of say in the matter of *colonial trade*; and, of the two British "Additional Members", as the Chairman of the lint marketing, coffee industry, and grain conditioning boards, one can look after the systematic exploitation of the Uganda Africans through the production of cotton and coffee, while the other is entrused, as the Resident of Buganda, with the task of keeping the Baganda (the most politically conscious people in the Protectorate today) under control.

On the other hand, out of the few non-British members in the Government, the Indian Minister is so far a non-entity, and the African Assistant Minister and the two Parliamentary Secretaries have the responsibility to help the foreign rulers in running the strategic departments of "Natural Resources", "Local Governments", and "Social Service". Thus it is seen that the controlling power of the colonial rulers remains as strong as before; only a new sign-board of *Council of Ministers* has been put up in place of the previous *Executive Council* of the Governor.

This form of *Africanisation* of Ugands's Government is similar to the *Indianisation* of the Government of British India according to the Constitution of 1935[65]. There, faced with a formidable opposition from the Indian people against the autocratic rule of the foreign rulers and being unable to ignore completely the ever-growing demand of the Indian movement for independence, the British rulers were forced to announce several constitutional changes one after another. But every time they were obliged to bring in a new Constitution, they announced to the world that with the progress of their "civilising measures" more and more control were voluntarily given over to the Indians, although, like the two previous Constitutions (considering the Morley-Minto Reforms as the first), the Constitution of 1935 not only did not concede what was demanded by the Indian people, but it made a mockery of democratisation of the Indian government apparatus. Mahatma Gandhi wrote in an article in his journal *Harijan* of August 1938:

> "Democratic Britain has set up an ingenious system in India which, when you look at in its nakedness, is nothing but a highly organised military control. It is not less so under the present Government of India Act. The Ministers are mere puppets so far as the real control is concerned. The Collectors and Police may at a mere command from the Governors unseat the Ministers, arrest them and put them in a lock-up."

And Professor A. B. Keith, "the leading constitutional authority in Britain", frankly stated while commenting on this Constitution[66]:

> "It is difficult to resist the impression that either responsible government should have been frankly declared impossible or the reality conceded; . . . it is too obvious that on the British side the scheme is favoured in order to provide an element of pure conservatism in order to combat any dangerous elements of democracy contributed by British India.

The recent changes in Uganda's governmental apparatus present a striking similarity to what the same Colonial Power attempted in another colony about two decades earlier in order to stem the tide of the freedom movement of the Indian people, and failed in its attempt as is inevitable with all measures taken to keep a people permanently in bondage. Now it is the turn of the Uganda Africans to see through this fraud.

In the future composition of the Uganda Legislative Council, however, a slight change has been proposed, although here also the decisive control remains in the hands of the foreign rulers. Thus the Council has been enlarged to 60 members and the Africans have been allowed to form half the total number. But 12 out of the 30 African members are to be the representatives of the Governor and the Governor himself will have the right to vote. The majority for the British Government is thus secured. Even so, the recent elections of African members to the Council indicate which way the wind is blowing. In spite of the fact that these African members are not allowed to be elected by universal franchise ("At first they are to be chosen by an electoral college comprising representatives of each Buganda county"[67], and in other regions of the Protectorate the choice is even more restricted), the correspondent of the *Times* of London reported from Kampala on the 29th of August, 1955[68]:

> "At the 20 saza (county) head-quarters throughout Buganda village representatives elected members of the Legislative Council; and the Great Lukiko of Buganda this morning nominated candidates for the list from which Mr. D. M. Kintu, the new Katikiro, will chose his five colleagues. Results in the county elections were this evening available only from five places, the sazas of Kyadondo, Kyagwe, Busiro, Mawokota, and Singo. These reveal that in each instance a prominent member of the Uganda National Congress has been returned as one of the three representatives from each county."

The final results of the election of the five Baganda members to the Legislative Council is even more striking. Out of the five, four belong to the Uganda National Congress. On the 5th of September, 1955, out of a possible total of 60 votes, they polled 48, 47, 33, and 22, respectively. The only other non-Congress candidate who was successful in the election was a Roman Catholic priest. He polled 22. The other three candidates, Mr. E. M. K. Mulira

(President of the Progressive Party — an organisation favoured by the regime), Mr. J. W. Kiwanuka, and Mr. Y. K. Lule, were defeated. The correspondent of the *Times* commented on this election from Kampala[69]:

> "As it is, only opposition to Congress has come from some Roman Catholics and some politically discredited members of last year's Hancock committee."

More will be said later about the Hancock Committee, but here it is worth recalling that the main demands of the Uganda National Congress are for the democratisation of the state apparatus and for immediate self-government.

In this way, however limited is its scope even today, the will of the people is being felt in the Legislative affairs of the Protectorate. And probably because of this, lamenting over the impotence of the Buganda Lukiko (where the colonial rulers have still some loyal elements), the correspondent of the *Times* of London noted in his desptach from Kampala[70]:

> "This is the second time this year that because of the inactivity of Lukiko leaders Congress have found the ball at their feet. On the first occasion they promptly kicked it out of touch by their demand for immediate self-government and their threat to declare it if it was not granted, against which there was a sharp Baganda reaction followed by a surreptitious retraction by Congress. It remains to be seen whether they have learnt better ball control.
>
> Their relations with the new Buganda Government, particularly over any claim to represent the authentic opinion of the Baganda, will be as interesting as the stir they will assuredly make in the Legislative Council. By agreeing to Baganda participation in the Legislative Council, and more particularly by accepting the recommendation that the Lukiko should not itself elect the Baganda members, the Lukiko has embarked on what in view of its preferences for self-sufficiency is a slippery slope."

No doubt such a development of the fight for popular governments, as is evident today, has a significant bearing on the future perspective of prosperity for Uganda, for it makes the colonial regime even more unstable. However, this fight cannot be successful unless there is a broad-based unity of the Uganda Africans for this purpose. Therefore, as alleged in the above extract, to what extent the Congress demand for immediate self-government has precipitated "a sharp Baganda reaction" will be discussed in the next pages which will trace the development of the struggle of the Uganda Africans for their goal of independence from colonial domination, as this is indeed the fundamental issue before creating a prosperous Uganda.

5. National Independence

Side by side with the economic struggle of the Uganda Africans and their fight for popular government developed their political movement in order to control their own destiny. And here also in the beginning the struggle was spontaneous and sporadic, except the brave resistance given by Kabarega, the King of Bunyoro. Moreover, the superior strength of the British, their assurances of being benefactors to the people, their ingenious moves such as splitting the people into factions, and finally their clever scheme of letting the Africans rule the Africans through the systems of Indirect Rule and Native Administration, could easily put a stop to recrudescence of such "troubles" and consolidate their regime all over what is now the Uganda Protectorate. But by thus bringing all these societies under one system of government and very soon spreading the economy of cotton production all over the Protectorate, and thus making the Uganda Protectorate a uniform political and economic area, the British also united the Uganda Africans to fight against further colonial manoeuvres and against the root of all evils — the colonial bondage.

In this anti-colonial movement, the Baganda have taken the leadership for the whole of the Protectorate area. The reason is obvious. These were the people who were most developed in this part of Africa when the British appeared. Also because the colonial regime was most consolidated in Buganda, the Baganda people "progressed" farther than other Uganda Africans in the early years of British rule. From a survey conducted before the last war it is seen that Baganda as wage-labourers are mostly employed in skilled or semi-skilled jobs, while most of the unskilled labourers required in the Protectorate were supplied by the less-developed peoples, like the Bagishu, Bakedi, and those living in the West Nile District in the Northern Province. This can be seen from the following table obtained from a government report[71].

Table V.1

People	Percentage of total number employed in industry				
	Cotton	Mining	Sugar, Sisal, Timber, Fuel	Plantation	Total
(1)	(2)	(3)	(4)	(5)	(6)
Natives of the Uganda Protectorate					
Baganda	2.4	0.8	0.6	1.4	1.2
Basoga	2.6	—	1.2	—	1.05
Bakedi	10.7	—	1.8	—	2.7
Bagishu	3.3	—	9.7	23.3	9.8
Teso	10.2	—	—	—	1.9
Acholi	5.6	—	6.8	—	4.2
Lango	13.1	—	6.8	—	5.6
Banyoro	1.5	—	1.7	—	1.1
Banyankole	—	27.8	—	—	4.1
Bakiga	—	13.6	—	—	2.0
Bakonjo	—	15.5	—	—	2.3
West Nile	19.3	—	42.5	44.9	32.4
Karamajong	0.2	—	—	—	0.05
Total Uganda peoples	68.9	57.7	71.1	69.6	68.4
Immigrants to Uganda Protectorate					
Banyaruanda	15.4	39.7	20.4	23.7	23.0
Congo	—	2.6	—	—	0.4
Congo (kakwa)	3.0	—	4.8	—	2.8
Sudan	1.7	—	—	—	0.3
Kavirondo & Jaluo	11.0	—	3.7	6.7	5.1
Total immigrants	31.1	42.3	28.9	30.4	31.6
Grand Total	100.0	100.0	100.0	100.0	100.0
Sample	4105	3300	10315	4500	22220

Since the last war a definite class of industrial proletariat has emerged among the Baganda, although in bulk it is yet small. The same is also observed among the Basoga, the Batoro, and other peoples like the Banyoro, Iteso, and Bagishu; but their "development" in this respect is less than that of the Baganda.

Moreover, Buganda received the greatest "benefit" from the cotton economy which was first started in Buganda in the first landlord-ridden African society in the Protectorate, and it took nearly ten years to extend the economy over the whole of the Protectorate[72]. Also, during the early period of British rule, the price of cotton paid to the growers was always higher in Buganda than elsewhere, such as during 1921—27[73]. These "benefits" aligned the Baganda more quickly against the foreign rulers, for the peasants and workers soon found how ruthlessly they were exploited and the growing bourgeoisie also realised that their future was blocked by foreign interests. Anti-British feeling grew up in Buganda in such a short time that, during her anthropological studies among the Baganda in the early 'thirties, Lucy Mair found[74]:

> "The native recognition of the superiority of British justice is by no means as universal as it is sometimes assumed; I have heard natives, quite without rancour, explain unexpected decisions by assuming that the British officer was bribed."

Huxley, after the last war, found the situation much more acute. Describing the activities of the "Combined Welfare and Propaganda Teams" of the Public Relations and Social Welfare Department in the Western Province "on the virtues of soya beans", she noted[75]:

> "The last demonstration ends with the serving out of savoury dishes to the audience, a highly popular move; but when one of the Buganda team saw the performance he shook his head saying: — That sort of thing's all very well for backward places like Ankole and Kigezi, but it would never do for Buganda! ... We Baganda are too advanced ... The Baganda would expect the dishes to be poisoned."

Whether the above report was exaggerated or not, the fact is that anti-colonial feeling is most strong among the Baganda people. Naturally, therefore, they took up the leadership of the freedom movement in Uganda. And for this reason the spokesmen of British rulers today often speak of the Baganda as "trouble makers" in the Protectorate. This, of course, cannot hide the reality that everywhere throughout Uganda the mass of the people nurture the same sentiment as the common Baganda, although previously they may not have always played an active or a militant role in the movement. Moreover, the upsurges of 1945 and 1949, and the later developments in the Protectorate after the deportation of the Kabaka of Buganda in 1953, leave no shadow of doubt that the freedom movement against colonial bondage has permeated throughout the Protectorate.

In its beginning, this all-Uganda outlook against the schemes of the Colonial Power was probably first evident in 1921 when a proposal was in the offering for the amalgamation of the Uganda Protectorate with the other states in British East Africa for the eventual emergence of what is known today as the East Africa Federation. By this Uganda would have been turned into a White Man's Country, where the "white" minority was to rule continually over the "black" majority, as has happened in the Union of South Africa, Rhodesias and Nyassaland. So this was the beginning of the united resistance of the Uganda Africans against further schemes of colonial bondage, and in this the Baganda played the most prominent role.

In successive periods, along with successive steps taken by the British Government to bring in the Federation, the opposition of the Uganda Africans gathered momentum. In 1924 the Ormsby-Gore Commission was set up with the same purpose in view as in 1921, and the Government faced stronger African opposition. Next there was the Hilton Young Commission, and in 1929 came the British Government White Paper on Closer Union of the Dependencies in East and Central Africa[76]. But the African opposition had also stiffened further. This opposition had by now been so organised that in 1931 a deputation from Buganda went to the House of Commons in London to protest against the scheme of Federation.

The scheme was, of course, not given up. In 1947 came the White Paper No. 210, "Inter-Territorial Organisation in East Africa (revised proposals)", and this revised version of the first step towards the fruition of the scheme was imposed against a strong protest from the African and Asian population of the whole of East Africa. On behalf of his countrymen, Semakula Mulumba sent a strong protest from England to the United Nations General Council, but the British representative in the Council affirmed that the proposal was within the purview of domestic matters of his government, and so the Council need not discuss it at all[77].

The next step was taken not long afterwards. In July 1953, the Secretary of State for the Colonies in the British Government made a statement re-propounding the government policy of East Africa Federation. But this time the opposition flared up with so much vigour that even the African ruling class could not but give at least a partial expression to it. The upshot was that the Kabaka, the traditional ruler of Buganda, was deposed and deported to England without much ceremony.

The anti-colonial movement of the Africans is of course not limited only to their resistance to the East Africa Federation. It also expresses itself in regard to the British scheme of further encroachment on the rights of the Africans. In such cases, even if those who are immediately affected belong to the upper strata of the societies, the people as a whole fight against such schemes. Notable among them is the attempt of the colonial government in the last few years to take away land from the landowners of the "developed societies" for the exploitation of Uganda's natural resources by the monopoly concerns or for white settlement.

How such a scheme encountered strong resistance from the Baganda Africans, and how it has given a filip to the rapid development of the freedom movement of the whole of Uganda with the slogan of *Uganda for Uganda Africans* has been noted before. As a result, while the British have so far been able to carry on their designs to a certain extent even in the face of such vigorous opposition, they have become very apprehensive of their future in Uganda. Hence they have resorted to: —

(1) giving some concession to the growing African bourgeoisie in the form of opening an African cotton ginnery and instituting the Concillar System in Native Governments (of which more later);

(2) ingeniously putting-up the Indians as the hindrance to the aspirations of the African bourgeoisie;

(3) endeavouring to keep the African intelligentsia within their fold by organising social clubs among them under the auspices of missionary societies or the Government Welfare Department, where politics are of course rigidly banned;

(4) setting-up a Special Branch of the Criminal Investigation Department in order to sense public opinion and inform the authorities in good time so that they may crush any political grouping; and, finally,

(5) trying to smash ruthlessly any attempt of the mass of the peasantry and workers for economic improvement or political rights, for which the police force has been largely increased and para-military units formed in recent times.

Needless to say, as history has shown in other colonies, such measures cannot check the mounting anger of the people and stop them from achieving their goal of freedom, prosperity and progress. Therefore it is seen that in spite of the above-mentioned measures and anti-popular legislation (such as the anti-strike laws passed by the Protectorate Government in the Uganda Legislative Council with its picked-up members of 20 "Europeans" out of the total of 32 in 1950—52, with only 8 selected African members), the peoples' movement after passing through two major waves of struggle in 1945 and 1949 reached such a mark that the Uganda Africans began to put forward resolutely three primary political demands to forestall further oppressive bondage and to make way for their independence from colonial rule. As noted at the beginning of this study, these three demands are:

(1) The British Government should not push Uganda into the East Africa Federation;

(2) Uganda being a Protectorate and not a Colony, it should be transferred to the Foreign Office instead of being under the Colonial Office of the British Government; thereby, at least partially, acknowledging the political aspirations of the Uganda Africans; and

(3) a time schedule should be announced for the independence of Buganda, the largest and the most developed province of Uganda both economically and politically.

These demands were most emphatically voiced in Buganda, for here the freedom movement had so extended its influence that even the King of Buganda (the Kabaka) and the Buganda African Parliament (Lukiko) publicly went against the dictates of the colonial rulers. This resulted in November 1953 in the deportation of the Kabaka after, as announced by the British Government, he refused to "co-operate loyally" with the foreign rulers in forcing the Buganda Parliament to withdraw the resolution incorporating the above three demands. But the deportation of Mutesa II, the deposed Kabaka, did not put a stop to the growing anti-colonial movement. On the contrary, it further united the peoples of Uganda, to all of whom reinstallment of the Kabaka became a symbol of establishing their right to independence.

This movement attained such strenght that even a large section of the native rulers began to give active support to the popular demand for the reinstallment of the Kabaka and the fulfillment of the above demands. Deporting the Kabaka and declaring a *state of emergency* on the 31st of November, 1953, the colonial government thought that the people would be brow-bitten or at least the native rulers would renounce the demands of the Buganda Parliament. On the contrary, on the 2nd of December 1953, the Buganda Lukiko rejected the British Government's charge of "disloyalty" of the Kabaka and passed a resolution calling for his return. Also, instead of withdrawing their previous demands, the Lukiko pointed out to the foreign rulers that the demands put forward by the African Parliament, and not individually by the Kabaka, will remain in force.

Failing to isolate the native authority from the people, the colonial government made several attempts to engineer "palace revolution". The aim was to find a docile person from the Buganda Royal House, sit him as the head of the Lukiko in place of Mutesa II, and thereby endeavour on the one hand to hoodwink and pacify the public and on the other to pursue the colonial policy. But all such attempts were in vain. Even if anyone from the Royal House was willing to oblige the foreign rulers, none dared to go against the wish of the solidly united people.

Indeed, the people's anger was so roused by the unjust deportation of the Kabaka and their movement was so well organised and strong that, however embarrassing it was to the colonial officials, during the vist of Queen Elizabeth II to East Africa in 1954 they had to change the venue for her reception from Kampala (at the outskirts of which the Buganda Native Government has its headquarters and where a large number of Baganda live) to the Government capital Entebbe (where live very few Africans other than the small number of government employees).

Moreover, subscribing to the demands of the people, the Buganda Parliament now asked the Colonial Government "that the Queen's visit to be postponed, as the country would be in a state of mourning for the deposed Kabaka"; and on March 3, 1954, the Governor of Uganda had to announce from Kampala that the Queen's visit to the Lukiko was cancelled!

While the colonial rulers had thus to lose face once and again, the movement of the people went a step farther. With the Uganda African National Congress at their leadership, the people declared a boycott on all buying, except of necessities, until the Kabaka was reinstated. The boycott was evidently so successful that the British Colonial Secretary, Mr. Lyttelton, declared in the House of Commons in London that this was the main reason for reimposing the state of emergency which, expecting a "change of heart" of the Africans, had been lifted on the 31st of March and which was now reimposed from 31st May, 1954.

In spite of this and other colonial measures introduced to intimidate the people (including an open reign of repression and terror which was unleashed with the announcement of the state of emergency), and contrary to the wishes and expectations of the foreign rulers, vigorous and disciplined protests and struggle went on from the African side to vindicate their cause. It developed to such an extent that even the Uganda women challenged the British Colonial Secretary. On behalf of the Baganda Women's League, Mr. Musa K. Parma-Ntanda was authorised by the Nabagareka, the Queen of Buganda, to make a challenge to Mr. Lyttelton on the issue of the deposition of Mutesa II. After he came over to the United Kingdom, he was authorised to speak also in the name of the women of the whole Protectorate of Uganda, thus indicating how the peoples of the entire Protectorate had taken a united stand against the dictatorial policy of the British Government. In a statement issued in London, this representative of the Uganda women declared:

> "Anyone who studies the White Paper, the Colonial Office statement of the 22nd December 1953, and the Lukiko delegation's replies, will undoubtedly reach the conclusion that the Secretary of State finally based his decision not on the constitutional issues raised in these documents, but on his and the Governor's assessments of the personal character of the Kabaka, which are irrelevant and unsubstantiated."

He also added that the plan of the Governor of Uganda was to get rid of the Kabaka in order to replace him by a docile "king", — a manoeuvre which the Governor thought could be accomplished very easily. But he found himself in a very difficult position, for the Baganda were not inclined to oblige him.

In face of the unity and determination of the Uganda Africans (and specially of the Baganda) to foil further British manoeuvres and for the return of the Kabaka, the colonial rulers then decided upon a new plan. This was of course nothing but another version of their old policy of *divide et impera*. Attempts began to be made to launch a political party for the Africans which would disrupt the unity and support of the Uganda African National Congress. A small political organisation was formed, styling itself the United Uganda Association, which claimed to comprise peoples from all communities in Uganda today and which expressed its aim to establish a new all-African political party. In September 1954 the organisation consisted of only 25 people, and had to work secretly with the unofficial blessing of the Protectorate Government. Although official rules in Uganda do not allow civil servants to belong to political organisations, no attempt was made by the colonial government to prevent loyal civil servants from joining this United Uganda Association.

This organisation began to work to establish a body which would, in their opinion, be capable of undertaking the government of the country when it will be handed over to the "people" by the foreign rulers. But popular feelings were so strong against such moves that the organisation was forced to work underground to save its members from the wrath of the freedom movement led by the Uganda African National Congress.

As opposed to this organisation, the Uganda African National Congress reaffirmed in a statement its stand against "any Lyttelton multi-racial form of Government as introduced in Kenya", the very bitter taste of which the Kenya Africans are now feeling through the Mau Mau terror of the White Settlers and the Colonial Government of Kenya. The Uganda African National Congress again repeated that "Uganda has been and will remain entirely an African state".

Such a bold and definite attitude of the Uganda Africans put the foreign rulers into a quandary. To pacify the people a Commission was previously announced to inquire how to come to terms with the people, and on the 21st of June, 1954, the Hancock Mission began its work. At the same time the Governor of Uganda made it clear that the deportation of the Kabaka was final. Edicts were sent out to all chiefs to inform the people that the Kabaka would never return; political leaders were exiled; newspapers were banned; and, as it was complained, internal affairs of the Royal Palace were also interferred with. Yet the situation became more and more embarrassing for the Government. On September 27, 1954, a civil action began in the High Court in Kampala to test the legality of the British Government's withdrawal of recognition from the Kabaka. The case was brought by three of the Kabaka's nominees in the Lukiko against four "unofficial" members of the Lukiko and the Attorney-General. The supporters of the Kabaka claimed that the four "unofficial" members of the Lukiko were not entitled to sit in Buganda Parliament as their appointments were not confirmed by the Kabaka!

Indeed, the situation was now found to be so difficult that even the Conservative weekly, the *Spectator* of London, had to advise the Colonial Secretary in relation to the Uganda crisis that there are times when one can save one's face only by losing it! Another English newspaper, the *Manchester Guardian*, also sounded the warning note:

> "What happens in Uganda is of course crucial for the whole of eastern and even central Africa. Not the Buganda alone but all the peoples in the Protectorate collectively, can be either our most valuable friends or our most dangerous enemies. Conditions there will set the pace either of co-operation or resistance."

Such is the spectre which began to haunt the British rulers in Uganda. Evidently, as it is implicit in the above warning, the anti-imperialist movement has permeated all over the Protectorate. The wise men had therefore to take a note of it.

6. Colonial Manoeuvres

But did the determination of the Uganda Africans to end with the colonial autocracy or the advice and warning from the British newspapers to adopt a "wise" policy changed the colonial manoeuvres in Uganda? On the contrary, with their characteristic desire of holding on to a colony up to the last moment, the foreign rulers are resorting to several ways of obstructing the eventual victory of the anti-colonial movement of the Uganda Africans. Although these manoeuvres are often inter-linked and may work simultaneously, for a proper elucidation they may be classified under four categories. These are: (a) repression and a reign of terror when all other weapons fail, (b) conciliation and compromise with the vested interests in the societies in order to isolate a section from the mass of the people, (c) fomenting national disruption to weaken and dissipate the peoples' movement, and (d) utilising a scape-goat in order to deflect the just anger of the people from the real and the main culprit. Since all these attempts have a significant bearing on the eventual outcome of the freedom movement of the Uganda Africans and the fulfillment of their perspective for a free, prosperous and progressive Uganda, on the following pages they will be discussed one by one.

6(a). Repression and Terror

It is well-known that with the aggravating crisis in the British colonial system in the post-war world, the foreign rulers are more often taking to a reign of terror and repression in the colonies than they had to in the pre-war years. This has become particularly marked by its brutality in colonies like Malaya and Kenya. And now it came to be the turn of Uganda too, for her peoples also raised their heads to achieve their birth-right of freedom and democracy. Therefore, as during the first two waves of the peoples' struggle in 1945 and 1949, from the beginning of the present crisis in Uganda the colonial rulers resorted to their policy of force and repression. Quite candidly, in its issue of the 30th of August, 1955, the *Times* of London reminded the Baganda people of "the troops who overawed them as soon as the Kabaka was removed".

But when the people refused to be "overawed" by such naked show of brutal strenght, there were more to follow. At the end of May 1954, according to a report in the *Manchester Guardian*, several members of the Uganda Legislative Council (56 members including 20 Africans —none elected, and 28 "Europeans" and 8 "Asians") were pressing for the resumption of the state of emergency in Buganda and the deportation of all Uganda African Congress leaders. Remembering Kenya in 1952, this demand from white "unofficials" has an unpleasantly familiar ring. With the Jinja Hydro-electric power plant now open, the Kilembe copper-cobalt mines (with the new railway to their smelter at Jinja) opening in 1956, the Uganda Development Corporation's cement factories being developed to supply all East Africa, and with the increasing importance of Uganda's coffee and cotton in the world market, the country is growing more and more attractive to the foreign monopolists as a field both for investment and local production. But the price for investments is a regime subservient to their interests as well as plenty of cheap and docile labour. Hence the white "unofficials" could not remain quite in face of the resolute opposition and boycott of the Uganda Africans. The national aspirations of the peoples of Uganda must be crushed at any cost. That was their demand to Mr. Lyttelton.

Mr. Lyttelton was apparently only too ready to serve the white "unofficials", and as a result the state of emergency was reintroduced on 31st May, 1954. Moreover, according to the *Manchester Guardian* of the 1st of June, 1954, three vernacular newspapers, namely, the *Uganda Post*, the *Express*, and *Eyogers*, were suppressed under emergency powers; and the Fourth (Uganda) Battalion of the King's African Rifles (stationed in Jinja) was made ready to move on Kampala at short notice. Further, even though the Uganda African National Congress had publicly forbidden all forms of intimidation, "the Special Constabulary was being called up, and the Commissioner of Police was appealing for volunteers of all races to undertake police patrols in and around Kampala so as to release the regular police for duty in the rural areas".

These measures cannot but lead one to compare the situation in Uganda with that in Kenya. Here also the African National Congress had publicly forbidden all forms of intimidation, and had asked the people to organise a movement for independence in a law-abiding but determined manner. Yet the colonial regime there went on with meeting the situation by preparing for extreme violence. And, working in conjunction with their neighbouring administration, the Uganda Government passed the *Immigration Control (Amendment) Ordinance 1953* which, while purporting to prevent the entry of undesirable Africans into Uganda, was declared in an official report prepared by the Uganda Government as "only to be applied in practice to Kikuyu", that is, the most militant people in the freedom-movement of Kenya[78].

In Uganda also the colonial government prepared for on even more open reign of terror. A report from Kampala in the *Manchester Guardian* stated on the 16th June, 1954: "The highest priority is to be given to the strengthening of the Uganda Police Force". The report also stated that the Uganda Police was to be reinforced with men from the Malaya Police. The Government "recommended new security measures, including the searching of selected areas and interrogation of all suspicious persons found there". In the light of what inhuman torture went on in Kenya in the name of interrogation[79] one can well imagine what the colonial rulers of Uganda were up to. One should also note the significance of bringing in those from the Malayan Police who had several years practice in Malaya in suppressing the movement for national liberation and had even gone to the extent of being publicly photographed with the blood-dripping heads cut-off from the body of Malayan patriots. When this practice was exposed in the London newspapers in 1952 it created much consternation.

The prospect of a mass massacre thus opened before Uganda, as it was being put into practice in other colonies within the British Empire, like Malaya and Kenya.

6(b). Conciliation and Compromise

But even in the face of Uganda beeing drowned in a bloodbath of her patriots for the monopolists' money bags, the people went on fighting as resolutely as before. So the rulers were forced to come down a little from their previous uncompromising position. As the *Times* of London noted on the 17th of November, 1954, in its summary of the Hancock report on the constitutional changes for Uganda, the British Government "will not raise the issue of East African Federation while public opinion remains against it".

This was indeed a victory for the peoples of Uganda. But their anger would not subside until their other demands were met and the Kabaka was reinstated. So when in November, 1954, the Governor of Uganda, Sir Andrew Cohen, tried to explain the British proposals in the Lukiko, he was shouted down amid cries of "Stop reading that" and "Stop that nonsense". The Governor was forced to stop speaking, and to leave the Lukiko without his speech having been made. The *Times* of London reported that the 10,000 strong demonstration formed in Kampala on this occasion was the greatest "in living memory".

Of this the rulers had to take a very serious note. Therefore, while they had previously announced that the exile of the Kabaka was final, now they had to agree to his return to Buganda. But this surrender on the part of the British Government was made conditional on the acceptance by the Lukiko of a series of checks proposed by the former in order to try and hinder the Kabaka and the Baganda from moving towards their goal of independence. Simultaneously, while the British Government proposed some administrative "reforms" for the whole of the Protectorate, it made it clear to the Uganda Africans that self-government is not on the agenda. The central objective in the proposal of the British government was to establish fully that Uganda must remain a colony as before and that Buganda must henceforth unequivocally accept its position as a part of the colony. Thus it was declared that Buganda will be subjected to the same conditions as in other parts of Uganda (to the extent that the Lukiko must accept the "advice" of the Governor and can be dismissed if they do not take it), that the Governor will appoint a Resident in Buganda and assistants in other parts of Uganda, and that the Uganda police under the Governor's control can be stationed in all parts of the Protectorate. In return to these conditions, it was proposed, the Central Government and the Legislative Council would be *Africanised*, and the Kabaka of Buganda would be allowed to return from exile provided the "reforms" are accepted by the peoples of Uganda and the Governor finds them working to his "satisfaction".

This gesture of conciliation and compromise did not go unrewarded. The Kabaka stated in an interview in London that he would abide by the decision of his Lukiko, and after a long deliberation, on the 12th of August 1955, the members of the Buganda Lukiko "voted unanimously for acceptance of the new constitution which will open the way for the return of the Kabaka"[80]. On the 17th of October, 1955, the Kabaka returned to Kampala from his exile in London and on the next day both he and the Governor of Uganda signed the New Agreement.

According to the terms of this agreement, the Kabaka will henceforth remain as a mere figure-head, — a constitutional monarch, as it is generally known. Under the caption of "New Régime for Buganda", the *Times* of London reported on the 13th of August, 1955:

> "The new constitution contains a 'solemn undertaking' to be given by the Kabaka before he assumes the functions of his office as constitutional monarch. This reads:
>
> 'I do hereby undertake that I will be loyal to Queen Elizabeth II, whose protection Buganda enjoys, her heirs and successors, and will well and truly govern Buganda according to law, and will abide by the terms of the agreements made with her Majesty and by the constitution of Buganda; and that I will uphold the peace, order and good government of the Uganda Protectorate, and will do right to all manner of people in accordance with the said agreements, the laws and customs of Buganda, and the laws of the Uganda Protectorate, without fear or favour, affection, or ill-will'."

Besides neutralising the Kabaka of Buganda, a similar agreement was signed by the Governor with the "king" of Bunyoro-Kitara on the 4th of September, 1955. Under the caption of "Democracy for Bunyoro", the *Times* reported on the 5th of September, 1955:

> "The agreement replaces the Bunyoro agreement of 1933, which has hitherto regulated the relations between Britain and the Omukama and people of this attractive small kingdom situated atop a steep escarpment on the eastern shores of Lake Albert, in western Uganda.
>
> Eighteen months ago, the Omukama joined with the two other rulers of kingdoms in the Western Province to express to the Governor of Uganda apprehension that they were placed between the Protectorate Government and their own people, much in the position which the Kabaka of Buganda had felt to be so difficult a few months before. The rulers were told in reply that Professor Sir Keith Hancock would visit each of their kingdoms during an interlude in his constitutional mission to Buganda.
>
> In the event, Sir Keith Hancock could visit Bunyoro only for a day or two, but he continued to take a personal interest in the work of the Bunyoro constitutional committee which, with the assistance of the district commissioner, Mr. Kenyon Gowers — to whom a number of sincere tributes were paid at the ceremony — drew up the new agreement which has been accepted by the Omukama and Rukarato (council) of Bunyoro and by her Majesty's Government."

Probably such agreements will also be signed with the "kings" of Ankole and Toro, as the above extract tends to indicate. But, whether this legal measure is further applied or not, it appears from the recent happenings that the situation in Uganda will soon return to "normal". But does it mean that the freedom movement of Uganda will also end thereby? On the contrary, an analysis of the 1953—55 wave of struggle suggests that it will now take a new course under the present circumstances.

Assessing the outcome of the recent peoples' upsurge in Uganda it is found that while the British rulers have been able to maintain their hold on the colony by means of conciliation and compromise, the peoples of Uganda also have achieved some positive results. Of them, three appear to be of primary importance.

Firstly, the anti-colonial outlook of the Uganda Africans has now permeated all over the Protectorate. This is evident from several incidents mentioned in the course of describing the development of the upsurge of 1953—55. It has also been mentioned that this was implicit in the previously quoted comment of the *Manchester Guardian* on the *depth* of the so-called Buganda crisis.

Secondly, because of the tremendous force generated by the recent anti-colonial upsurge, for the time being at least the British Government has been compelled to give up the scheme of East Africa Federation. Referring to the "new constitution" for Buganda, the *Times* of London wrote on the 13th of August, 1955:

> "A note records Britain's assurance that she had 'no intention whatsoever of raising the issue of an East African federation while local opinion in Buganda remains against it'."

This is indeed a substantial victory for the peoples of Uganda, as they could prevent by their heroic struggle the more ruthless form of colonial domination over their land as it is now experienced in the countries belonging to the Central Africa Federation.

Thirdly, although it has been offered as a sop to the people, the limited but the beginning of the expression of the voice of the Uganda Africans in the legislative and administrative affairs of the Protectorate cannot but have a far-reaching effect. Summarising the "new constitution" the *Times* noted in the above-mentioned issue:

> "Under the new agreement Britain undertakes to arrange, during 1957, for a review of the methods by which representatives to the Uganda Legislative Council are chosen. . . . They [the reforms] include provision for direct election of Buganda's representatives to the Uganda Legislative Council (Lower House) by 1961 at the latest."

Whether the British Government will fulfill its promise or not (and this again will depend on the strength of Uganda's struggle for freedom), it appears that parlimentarism will now become an important weapon for the Uganda Africans in the ensuing phase of their anti-colonial movement. And, in this line of struggle also, although this alone may not bring them independence and prosperity as the recent happenings in British Guiana so clearly illustrate[81], the peoples of Uganda can attain a good measure of success. This the recent happenings in Uganda tend to indicate.

Thus it has been noted in the foregoing pages that, although the process of *Africanisation* of the administrative and legislative bodies according to the new "reform" measures is a mockery of democracy, wherever there are opportunities for the people to express their opinion they are electing their representatives mainly from the Uganda National Congress presumably because this organisation has taken up the leadership of Uganda's freedom movement. Moreover, from whatever stray information are available from the press, it is seen that the loyalist elements in the Uganda societies are having a very rough time in spite of the full support given to them by the colonial regime. The incident of Mr. Muwangu in Busoga has been mentioned earlier. In Buganda, such an alignment of forces is even more marked. For instance, in the recent election for the post of Prime Minister of Buganda, the outgoing Prime Minister, Mr. Paulo Kavuma, suffered a miserable defeat by polling only 4 votes when the successful candidate, Mr. Mikaeri Kintu, polled 40. Explaining this remarkable defeat of the previous Prime Minister as well as of the outgoing Chief Justice, Mr. Matayo Mugwanya (who polled 37 votes), the correspondent of the *Times* of London wrote in the issue of the 25th of August, 1955, of the paper:

> "Earlier this week a Luganda newspaper edited by a member of Uganda National Congress denounced Mr. Mugwanya and Mr. Kavuma as enemies of the Kabaka. That Mr. Mugwanya [a Roman Catholic] came so near to success in the poll had been expected in view of efforts of the Roman Catholics — with some support from other quarters — to secure the post for the first time for over 60 years."

These incidents suggest that, as the champion of the freedom-demand of the peoples of Uganda, the Uganda National Congress will in future receive extensive support from the Uganda Africans to represent their interests in the legislative and administrative matters of the Protectorate. This development cannot but have a significant effect on Uganda's political life. Also it is not totally impossible that, with the growing strength of these representatives of the people in the legislative and administrative bodies of Uganda, they will also begin to influence their non-Congress African colleagues. In that case the situation will surely become very serious for the colonial regime. That there is such a possibility could not be overlooked by the stalwarts of British colonial rule even at this stage of jubilation over the "solution" of the crisis in Uganda. Thus, in his article entitled "A Solution in Buganda", the Kampala correspondent of the London *Times* wrote in the issue of the paper dated August 30, 1955:

> "The signs are that by the time Africans come to dominate the central ministry, there will be something like an alignment between ministers and politicians who are African, and administrators, who, in spite of the efforts being made (!), can mostly be European. If anything occurs — or persists — to alienate them from each other, a serious breakdown could very easily follow. Uganda, as those who knew the Sudan or the West Coast will aver, is on this matter most delicately placed."

No doubt, the colonial rulers do not want to place Uganda so "delicately" as Sudan or Gold Coast are placed today with their African governments. Therefore, colonial manoeuvres to maintain its domination over Uganda are taking new forms. These will be discussed in the following pages.

6(c). National Disruption

There is no doubt that in spite of all forms of repression, terror, and subtle compromise, the movement of the people for their freedom and progress cannot but be victorious. As happened in India for instance, through successive waves of struggle the peoples of Uganda will also eventually win their birth-right to become masters in their own country, to eliminate the colonial and other forms of exploitation, and to embark on a new life of peace, prosperity and progress. But in this march towards freedom and normal course of development the peoples of Uganda are likely to encounter several obstacles besides ever-increasing repression and a reign of terror which will, no doubt, mount in virulence along with the growing intensity and extension of their anti-colonial upsurge. These obstacles are likely to be encountered because of the ruler's manoeuvres of (1) sowing dissension within the African community and (2) letting the Africans against the Indians in order to save their own skin.

To consider here the first of these two manoeuvres, it is obvious from the present course of British policy in Uganda that following their traditional principle of *divide et impera* the foreign rulers are trying to foment tension within the African community on a class or a national basis. There is no doubt that the acquiescence of the foreign rulers to the Concillar System of native rule in the post-war period (a facade of democratic administration by elected representatives of the people who are not given any real control), the grant of a licence to open an African cotton ginnery (government-controlled of course) the passing of the 1953 Coffee Industry Ordinance "for greater participation by Africans in the processing of coffee", and the recent announcement of administrative and legislative "reforms" are evidences of partial victories won by the people. But there are reasons to believe that while granting these "concessions" the colonial rulers had the intention of buying-off a section of the upper strata of the African societies by giving them some privileges and facilities, and as a last resort intimidating them to submission with the threat of withdrawing the privileges given to them previously.

It is interesting to note here that an Acholi told the author in 1950 that the few scholarships offered in Uganda by the British Council (a semi-Government organisation) for higher education abroad go almost entirely to the Baganda. For, as he characteristically put it: "The Baganda *cry* so much that the government is compelled to listen to their demands". This is no doubt a compliment to the strongest freedom movement in Buganda (for the Acholi youth referred to the upsurges of 1945 and 1949 which were stronger in Buganda than elsewhere in the Protectorate), but there is also the point that by giving "suitable" training abroad to some Baganda the Government intends to further its own cause.

That such a policy is likely to be followed more rigorously and particularly for the colonial administration of Uganda is suggested by the recent utterances on how the "African ministers" under the "new constitution" should be developed. Thus, after pointing out the future possibility in the newly organised central government and the legislative council of the African ministers and "politicians" to line up against the dictates of the "European" administrators (quoted earlier), the correspondent of the London *Times* gave the advice:

> "A bridge can be provided by the new African ministers in the central government, the majority of whom are unlikely for the time being to be politically affiliated. They will need first to catch the public eye. It will then be desirable to help them to continue to sit in the council of ministers once the politicians have reached so far, so that their experience, of which the politicians may be glad, shall not be wasted. Alternatively, it might be possible to school them for the post of permanent secretary rather than minister. None of this will be easy, but some such object needs to be tacitly defined."

In other words, the policy of isolating the native rulers from the people will be maintained as before, and since these persons will be chosen by the Governor they are "unlikely for the time being to be politically affiliated". Moreover, there is the possibility, as is known from other colonies, that the foreign rulers may be able to buy-up some "politicians" as well. Because of the low stage of *social* development of the Uganda Africans as a result of their colonial existence, those who are at the helm of their affairs at the present time have generally come from the upper strata of the African societies. Representing the vested interests in most cases, they may be less consistent in their attitude and activities towards national emancipation than are the common people. Therefore, some of them may be vulnerable to the shrewd colonial policy of isolating them from the masses with the offer of a "good prospect". But, firstly, as evident from the results in some other colonies, this policy may act as a boomerang; and, secondly, whether the foreign rulers can succed in buying up a section of the upper strata of the African societies will evidently depend on the strength of the African freedom struggle, that is, on whether it can control and rise above sectional interests.

So far it appears that the foreign rulers have not achieved a great success in this respect. A definite section of compromising African bourgeosie is not perceptible in Uganda. So the anti-colonial movement can embrace all sections of the people. But there can hardly be any doubt that in the future perspective of Uganda's development the foreign rulers will certainly not give up this policy which, for a time at least, they have found to have worked satisfactorily elsewhere. The peoples of Uganda have therefore to be on their guard.

The Uganda Africans have also to take note of the attempts the foreign rulers are making to split the Uganda Africans on the basis of their national differences. It has been explained in Chapter III that before the advent of the British in what is known as the Uganda Protectorate today most of the African peoples had passed beyond the tribal stage; many of them had developed into nationalities, and the time was ripe, especially for the Baganda, to develop into nations with the beginning of the formation of national markets.* Today,

* It will be noticed that the concept of a *nation* as accepted by the author in this study is not the same given by several authorities. It is, however, accepted by many reputed anthropologists that *nationality* is the next stage of social development after the *tribal* stage, and that a *nationality* is a historically-evolved stable community of people which speaks one language and not a dialect, inhabits a particular territory, and possesses a distinct socio-psychological life of its own as is manifested by its specific culture. Professor Kroeber, for instance, stated (cf. *Anthropology*, Harcourt, Brace and Company, New York, 1948, p. 281):

"Where 'tribes' of fifty or a hundred thousand or a million persons are spoken of, they are either organized into a supertribal state, or they are really no longer tribes, but nationalities. Nationalities possess a culture and a speech that they recognize as common in essentials, but they may constitute many societies. This means

although under colonial conditions the development of the national market has been seriously hampered, nevertheless it has developed among some of these peoples and among others signs are apparent for its emergence. For reasons mentioned earlier, of all these peoples the national formation is most visible among the Baganda, and because of the Baganda's most militant attitude to British rule in Uganda the foreign rulers are endeavouring by subtle propaganda and suitable actions to isolate the other African peoples from them by suggesting that the Baganda want to extend their influence all over Uganda just as they intended to do before the British came.

This policy has not come out in the open yet, but one cannot ignore the attempts being made by the British to speak of the "distrustful nature" of the Baganda, as mentioned before; to speak of the "queer channels" in which the Bantu mind works (as mentioned in Chapter II), evidently meaning thereby the attitude of the Baganda — the foremost of the Bantus in Uganda today — to the colonial regime; and simultaneously applauding the straightforward nature of the simple Nilotes who are as yet less militant in the anti-colonial struggle than the Bantus. There can hardly be any doubt that such attempts are designed to drive a wedge into the solid block of African resistance to colonial exploitation.

The peoples of Uganda have to be particularly watchful to prevent any disruption of their freedom movement from this aspect, for the differential rates of "development" of the regions of Uganda can provide the material basis for such a manoeuvre. Thus, because of

---

that, in the case of a nationality, ethnically and culturally we are dealing with a single people, but politically often with multiple independent societies, each one autonomous and owning its own distinct tract of land."

To what extent an internal *political* differentiation can be drawn within a nationality is a matter for discussion; but, apart from that, as regards the concept of a *nation* Kroeber is seen to identify it with a *state* (*ibid.*, p. 226). And this viewpoint is also found explicitly stated in the British *Notes and Queries on Anthropology* as follow (cf. the Sixth Edition, Routledge & Kegan Paul, London, 1954, p. 134):

"When there is a well-developed central authority, exercising final legal, administrative and military power over a group of people occupying a clearly defined territory whether or not they are of uniform culture and of homogeneous ethnic origin we may speak of a state and of the people as a nation."

But in the case of Uganda, for instance, this definition will naturally lead one to the conclusion that if for the reasons stated in Chapter III the Uganda Africans are not to be classified as "tribals", all the peoples together must form a "nation", for they are all under one state power, viz. the Uganda Protectorate Government. And how fallacious such a conclusion will be is obvious from the present discussion, for it will fail to explain the growth of tension between the different African nations and nationalities inhabiting the Uganda Protectorate today.

Kroeber has steered himself out of such difficulties by stating that it "is certainly clearest always to use the term 'nationality' when the reference is ethnic, and to add the epithet 'political' to 'nation', or to substitute 'government'". This he suggested evidently because he finds it impossible to distinguish the "dictionary definition" of a nation as "a people of common origin, tradition, and language" from the definition of a nationality. (cf. *Anthropology*, p. 227).

The author, however, is inclined to the view that the difference between a *nationality* and a *nation* lies in a further stage of social development, viz. in the development of the *national market* which comes into existence with the development of the capitalist mode of production and the emergence of a class of *national bourgeoisie*. Unless such a distinction is drawn, it becomes difficult to explain how at a certain stage of social development *national movements* emerge under the leadership of the national bourgeoisie, as they have been found in the colonial countries as well as in the multi-national sovereign states, such as in Eastern Europe, each of which is or was held by one political power.

Incidentally, the distinction between a nationality and a nation, as drawn above, also explains how in those areas of Europe, where capitalism developed at a rapid rate and feudalism disappeared quickly, the formation of nations also coincided with the establishment of national states (such as it was with the British, French, and others in Western Europe); while in Eastern Europe, where feudalism still lingered on and the rate of capitalist development was slow, multi-national states came into existence (such as it was in Austria-Hungary and Russia), in which the nationalities which had not yet been able to "consolidate themselves economically into integral nations" were pushed into the background and others which had developed into nations (for instance, the Great Russians in Russia) began to dominate over the rest.

its historical antecedence and its close contact with the British from the onset of British rule in Uganda, the Buganda society has undergone the colonial development at a more rapid rate than other societies in the Protectorate. Just as a class of proletariat has emerged in the Buganda society today, so has a class of national bourgeoisie and the middle class; and, in general, among the non-Baganda a *feeling* has developed that the Baganda show off their sophistication. For instance, in 1949, an Acholi student at Makerere College told the writer that the Baganda make fun of the Acholi as primitive people, but they cannot say anything to him because he knows how to deal with them. Also, as has been mentioned earlier, many non-Baganda think that the Baganda are more favourably treated by the Protectorate Government than they. No doubt, this unequal position of the Baganda today is not merely due to the favouritism shown to them by the British in the early years of colonial rule when "Buganda was doubled in size as a reward for its alliance with the British"[82]. It is more due to the strength of their present-day movement for social, economic, and political improvements. This has resulted in a number of changes in the life in Buganda; as for example, the first African cotton ginnery has been opened in Buganda. Moreover, some more changes are anticipated in the future which may accelerate the differential rates of "development" of different regions of the Protectorate. Thus the Baganda as well as those living in the Eastern and Western Provinces round about the industrial centres of Jinja, Kilembe, etc. will be affected more directly by the post-war "colonial developments" than, say, those living in the Northern Province. So, from an objective point of view, it appears that there will be plenty of scope for the colonial regime to foment tension between the Baganda and the Bunyoro, Basoga, etc., and between those living in the areas undergoing "development" and those living elsewhere.

That such a policy may be pursued in the future is suggested by the pregnant remarks of the correspondent of the London *Times* from Kampala who is certainly in the know of things in Uganda's colonial government. Referring to the solution of the Uganda crisis and after remarking on the "stir" the Baganda members belonging to the Uganda National Congress "will assuredly make in the Legislative Council", the correspondent wrote in the *Times*' issue of the 6th September 1955:

> "Relations between Buganda and the protectorate government would seem to have been successfully solved. Relations of Buganda and, indeed, other component parts of the country, with a future self-governing central Government of Uganda are probably not altogether the same thing. That chapter may well have opened today."

Whether this "not altogether same thing "will finally lead to the enactment of another tragedy like the division of the sub-continent of India into the Indian Union and Pakistan has yet to be seen[83]. But those who want to solve the problem of Uganda must be on their guard against such a manoeuvre. There is no doubt that this can be forestalled by the Uganda Africans provided even more solidly they forge the unity of all peoples of Uganda behind their common cause of freedom from colonial bondage and at the same time take note of the aspirations of the different nations and nationalities living in Uganda today. From the way the peoples' movement is seen to be conducted in Uganda today it is expected that such will be its course in the future perspective.

6(d). Scapegoat

There is, however, one policy in which for a time the foreign rulers may attain some measure of success. This is in partly diverting the peoples' anger from their direction to the foreign intermediaries in Uganda today. These intermediaries who may thus partly serve the cause of the foreign rulers are the Indians who have settled in Uganda. Their role in the colony has not been altogether a good one, and since they occupy a significant position in Uganda today their role in the future perspective of Uganda needs a deeper understanding. In the following passage therefore a brief account is given of the past and the present role of the Indians in Uganda.

It has been explained in the last chapter that in the cotton economy of the Protectorate, which is the backbone of Uganda's life today, the Indian community plays the role of intermediary between the British and African interests. This intermediate position of the Indian community between the Africans and the British is not accidental. The Indians (the term being used to include all people living within the geographical sub-continent of India) have an association of long standing with East Africa. As today, this association was mainly in the sphere of trade and commerce. But it is worthy of note that, although they have traded on the east coast of Africa since Roman times, they never represented the ruling class or the highest stratum among the foreigners. In the past they were the "middlemen" between the wealthy landowning Arab proprietors and the African peoples. Some of them might have been petty bankers and also ship-owners, but even in such capacities they were hardly anything more than middlemen between Arab merchants and African peoples. Coupland wrote about those times[84]:

> "Much of the ocean-going shipping was Indian-owned and Indian-manned; and since the Arabs in general seem never to have shown much aptitude for the technique of business, it is probable that the Indians were from the earliest days what they still are in East Africa, the masters of finance, the bankers and the money-changers and money-lenders. Probably too, much of the actual trade was either directly in their hands or managed by them for wealthy landowning Arab proprietors, and probably then as now they monopolised the lesser retail trade."

Lugard also mentions in his memoirs on the rise of British East African Empire how during his inland journeys he came across Indian assistants of Arab masters driving a band of African slaves[85].

In the political sphere also the Indians were intermediaries in the past. They were the cause of the first British relations with Zanzibar, which as will be remembered was the prelude to the conquest of East Africa by Britain. There a British Consulate was opened at the Sultan's Court to protect the interests of the Indian subjects of the "Maharao" of Cutch. This later resulted in establishing the British East African Empire. When the Royal Charter was conferred upon the Imperial British East Africa Company by Queen Victoria in 1888, it was stated in the preamble that the possession by a British Company of the coast-line which included the Port of Mombasa would be advantageous

> "to the commercial and other interests of our subjects in the Indian Ocean who otherwise become compelled to reside and trade under the Government and protection of alien powers."

The Indians also directly helped the British to conquer Uganda. Indian soldiers took part under British Command in occupying Uganda[86].

Maintaining their historical role, the Indians began to migrate inland after the British rulers and European settlers had moved into Kenya and Uganda. Indians were recruited for the railways at first on a system of indentured labour which was abolished in 1917[87]. But African labour from an early period was ousting the Indians from this basic occupation in the society. "Great numbers of (native) labourers at Entebbe" were mentioned by Rev. W. Chadwick in 1905 who also recalled "the odours in the labour camps of the Indian coolies at work on the railway" and compared the situation with "the comparative sweetness of this place"[88]. Some merchants and professional men also came during the same period of railway construction[89], and later free emigration followed and merchants and professionals arrived as opportunities for their services expanded. From the beginning they took charge of the retail trade and spread it into the remote corners of the country. The Indians, especially those from Goa, also joined the Government departments as clerks and subordinate officers from the beginning of British rule[90]. Their position in the life of Uganda as "middlemen" between the British rulers and the subjugated Africans was thus finally determined.

It is worthy of note that their stability as intermediaries in Uganda was possible only under a colonial power. Although the Indians had worked with the Arabs for a much longer period, they could not have thus established their position. But with the advent of the British, they penetrated into the remote parts of the country and settled down to exploit the Africans as intermediaries. The following account from Johnston's memoirs, published in 1902, shows how they gradually established themselves in the economic life of the country[91]:

> "At the time of writing, British trade — that is to say, commerce conducted by natives of the United Kingdom — is by no means to the fore in the Uganda Protectorate. ... [The] British firms, in fact, have decided to hold their hands and restrain their enterprise until the completion of the railway and the placing of capacious steamers on the Victoria Nyanza.... Meanwhile two German firms are pushing ahead in all directions. ... Almost-simultaneously with the German came the British Indian. He rendered further benefits to the European community by bringing down the German prices, and by opening stores at places whither even a German hesitate to penetrate. Within the term of my Special Commission, Indian traders advanced their posts from Kampala (Mengo) to Toro and the vicinity of the Congo Free State, to five places in Unyoro, and all the posts at which European or native soldiers were established in the Nile Province [much of the Northern Province of Uganda today], besides opening bazars at all the stations in the eastern half of the Protectorate. This commercial enterprise of the Germans and British Indians, of course, added largely to our local revenues, and did a great deal to encourage the natives to embark in trade in the products of their country. To the British Indians I can only wish unlimited success, since they trade under the British, and create trade, first in a small way and then in a large way, where no trade has hitherto existed. ... A native of Kach in British India is erecting hotels for British visitors at the principal stations on the Uganda Railway, and elsewhere in the Protectorate. ... In the case of British firms, I am bound to admit that they are somewhat apt to ask for guarantees, for assured contracts, for concessions of land or monopolies of production, before they are able to venture their capital and enterprise."

This role of the Indians in the life of the Protectorate was fully revealed when the colonial power began the determined exploitation of the resources and labour of the country through the cotton economy. The Indians participated in this economy first as buyers of cotton from the African growers and as carriers of the seed cotton to the ginneries owned by the British. It may be of interest to note that the occupation of cotton-buying was previously the occupation of the Africans. Carter noted[92]:

> "In Buganda, for about ten years thereafter [1906], the bulk of the cotton was purchased by wandering native buyers, ... the number of Indian traders settled in the out-districts of the province [of Buganda] at that time [1915] was small, though, no doubt, they took some part in the purchase and collection of cotton. The industry developed later in the Eastern Province. ... Cotton was bought and collected mainly by Indian traders. ... In the early part of the year [1917—18] the middleman was able to buy at low prices from the native, and to sell, often within an hour or two, at very high prices to the ginners. It was in that year that several fortunes were made among Indian cotton-dealers, and

many were attracted to the trade of middlemen, which had now become established as one in which profits could be made quickly, with little or no capital."

The Indians thus first displaced the Africans from the operation of the cotton economy other than the production of the crop, and later they also took over almost completely the ginning industry from the Europeans. Thus they have fulfilled their role as intermediaries between the Africans and the British in the main economic organisation of the country as well as in the trading business.

It is seen from the above that throughout Uganda's history the Indians' role has been subservient to the British rulers and their interests in this colony. Throughout the period they have been used as pawns in the game of conquering and stabilising the British interest in Uganda. With the advent of the British in Uganda, military service was necessary to subjugate the country and the people. Lugard, at that time, praised the Indian soldiers who, under British command, were helping to establish British domination over Uganda. Then came the time when labour became necessary to build railways to take Uganda's resources out of the country. Indian laboureres were imported. But they came as indentured labour, so that when the Africans were able to take up this lowly occupation and there was no further need to use Indians for this basic colonial task, they were relieved. Next it was necessary that some individuals should bring trade into the remotest parts of the country and thus pave the way for the introduction of a cash economy. The Indians braved the difficulties. Johnston blessed their pursuits because they traded under the British flag! Also, reliable man-power was needed to assist the British officials to run the colonial administration, and so from the beginning the Indians (either from British India or from Portuguese Goa) took up subordinate and clerical jobs under the Protectorate Government because the remuneration was comparatively higher than in British India or Goa. In this way, from the beginning of British rule in Uganda, the Indians have served the interests of the foreign rulers; in return, they have received a share in the profits from Uganda as cotton ginners, merchants, etc., but politically and for the sake of economic gain they have all along remained at the mercy of the British rulers. So far they have always been completely divorced from the African peoples in regard to the aspirations of the latter for a better and progressive life. Regrettable as it is, up till today the Indians and Africans in Uganda have nothing in common; the prosperity of the former has always been and still is at the expense of the latter.

The foreign rulers have been fully aware for a long time of this dual position of the Indians in Uganda, and so, to serve their own ends, they have taken full advantage of the Indians' profit-seeking but politically insecure position and the growing bitterness of the Africans towards them as the immediate exploiters.

In old times while the British "pioneers" praised the role of Indians who served their interests in Uganda, they saw to it that the latter did not become insolent and claim equality with them instead of remaining servile and respectful. Thus, Jackson, the Lieutenant-Governor for the East Africa Protectorate (1907—1911) and the Governor and Commander-in-Chief of Uganda (1911—1917), describes an incident in Uganda in his memoirs entitled *Early Days in East Africa* as to how his servant took an Indian trader severely to task for not properly *salaaming* (saluting) him when he met that trader on the road during his journey between the Molo River and Nakuru. Referring to this incident he noted[93]:

> "In my experience such and similar acts of deliberate insolence were entirely confined to ex-coolies and their low-caste friends and relatives who came to the country at their instigation; and not to the genuine mercantile class of the coast, whose business passed from father to son."

This "genuine mercantile class" of Indians in East Africa was no doubt very respectful to the British rulers, unlike the Indian workers who came to build the railways and many of whom later stayed to earn their livelihood as retail traders. Moreover, the former class was also useful in persuading the other Indians to remain servile to the British masters. Thus, further commenting on the incident mentioned above, Jackson noted[94]:

> "Where, I ask, could you meet in India, or out of it, a more charming little old gentleman than Alidini Visram of Uganda, respected by every one in the country, high or low, white or black? Or his representative at Entebbe, Rashid? Those two between them did everything they possibly could by persuasion and example to check and counteract the movement, when it began to take root; but they were powerless, and the disciples of Ganghi have since carried the day."

To the "Colonials", like Jackson, Mahatma Gandhi symbolised the anti-colonial movement of the Indian people. This movement, however, never took a positive root among the Indians in Uganda. Here, while probably all the time resenting the master-like attitude of the British, the Indians went on fulfilling their role of intermediary, and thus being utilised by the British colonial interests.

It may be of interest to note how, even in recent years, the Indians have been used against the Africans, and have also been condemned by the same person when the occasion demanded a "concession" to the Africans in order to direct their bitterness against exploitation into a safe channel. Thus, Whitley, as Commissioner of the Inquiry into the 1945 "Disturbances", recommended that *reliable* Indians should be given licences to keep firearms to protect their shops — which were spread all over the country — from being looted by African "hooligans". In his report he also mentioned how, during the "riot", some Africans were "accidentally" shot by the foreigners[95]. At that time, Indians were needed as an ally of the reactionary forces, and unfortunately it is true that the Indians welcomed the suggestion: in fact, they themselves made the request for firearms to Whitley, so servile had they become to the foreign interests in order to make money in Uganda. But the same Whitley, who in 1945 supported the Indians in their demands for firearms to be used against the Africans, in 1948 as the Chairman of the Uganda Cotton Industry Commission, charged them with unscrupulous and dishonest practices, and thus appeared as the champion of the just cause of the Africans! Of course, he did not mention a word in his report about the *real* state of the cotton-growers and whose unscrupulous and dishonest practices have brought disaster to Uganda Africans. In this report there is a thinly veiled attempt to declare that the Indian businessmen are corrupt and that they always engage in exploiting the Africans illegally. Such statements, of course, had to steer round legal difficulties. For example, the report stated[96]:

> "It has been suggested to us that, if only African buyers were allowed and no Indian buyers, that might stop cheating but most of the growers who appeared before us were emphatic that African buyers also cheat. The only advantages, according to one shrewd cotton grower in Lango, would be: —
> (1) That whereas the Indian buyer cheats systematically and cleverly, the African would be much less expert and more easily detected; and
> (2) That if an African buyer cheats the money thus dishonestly obtained stays in the country, whilst in the case of the Indian it 'goes to Bombay'."

The above comment of the Lango in a report of only 44 pages appears not to serve any useful purpose except to disparage the Indians without falling into legal difficulties, by pointing out their presumed tradition and efficiency in cheating. It is worthy of note that the Protectorate Government was in a rather difficult position because of this direct reference to the Indians in the report, and tried to make ammends for it in a preface to the report, entitled "Notes on the Report of the Cotton Industry Commission, 1948". In this Note it was stated:

> "In paragraphs 57—66 of the Report the Commission has drawn particular attention to the prevalence of dishonest practice in the buying and selling of seed cotton. The Commission was naturally impressed by the evidence of malpractices which it received, but it must be remembered that the Commission's enquiries related solely to the cotton industry. If the Commission's enquiries had extended beyond that industry to other commercial operations it would doubtless have found similar evidence of malpractice, for it is a deplorable fact that dishonest practices and fraudulent transactions are prevalent throughout Uganda today; and, as the records of convictions for such offences in the Kampala and other Courts show, the offenders are by no means confined to one community or one race only."

One cannot, however, ignore the fact that for many years the Chairman of the Commission was in the Judicial Department of the Protectorate Government, and was thus in possession of the knowledge "of convictions for such offences" mentioned in the Government Note. The other members of the Commission were also connected with Government Administration for many years. Therefore, it would be naive to assume that they did not know that, as stated in the Government Note, the lack of moral scruples in the Protectorate was not confined to "one community or one race only". Indeed, from the preceding discussions it should be evident that the roots of this widespread corrpution are much deeper than generally stated, and the blame for such a state of affairs in Uganda today should be laid primarily on the ruling community. Colonial exploitation and the hunt for easy money under government protection naturally affect the "moral" fibre of the exploiters, and then this can be easily emulated by the intermediaries. But, instead of going into an impartial discussion on this particular problem of Uganda (one of the many serious ones), the above observations in the report seem to be more political than taking up a moral stand against the exploiters. The main exploiters remain behind the scene.

Such a policy of divide and rule is, of course, not unprecedented in Britain's colonial history. The fire is now directed against the Indians; and their insecure position and "a-moral" existence in Uganda, as a party to the exploitation of the country and the peoples, is recoiling on them without calling forth the least sympathy from the other two communities.

But why was it necessary for the colonial masters to become so cruel to the Indians in Uganda, who have all along served them loyally? Obviously, there is no moral principle involved in it. The justification is simply that of the survival of colonialism in the most delicate situation the British now face in Uganda.

The previous need for using the Indians as intermediaries will cease with the full implementation of Britain's post-war "Colonial Development" schemes for Uganda. Moreover, it is not unlikely that if the situation is left as it is, in course of time the Indian capitalists will stand as a rival body to the British capitalists in "developing" Uganda. Almost all the ginneries are now in their hands; the two sugar estates and their factories, the largest industrial concerns in Uganda until recent times (which in 1953 employed some 22,000 workers out of the total of 250,000 Africans in Uganda[97] in paid employment at any one time in the year*), are also run by Indian capital. And now that Britain has no direct influence on the Indian sub-continent, the Indian capitalists in Uganda may not be so easily controlled as hitherto. Hence, in the post-war years an immediate need developed to restrict further control of Uganda's life by the Indians.

* The 1953 Colonial Annual Report on Uganda wrote at the same time: "... the numbers permanently employed are very much smaller" (cf. *loc. cit.* 97, p. 17).

Furthermore, with the inevitable growth of some educational facilities among the Africans and the formation of a class of nascent bourgeoisie among them, the upper class of Africans are demanding trading facilities and government jobs. For a peaceful introduction of Britain's postwar policy of colonial exploitation, they have to be appeased as far as possible. Moreover, in their elemental anger, the Africans in general regard the Indians as the direct party of exploiters. The tension between the rulers and the ruled can, therefore, be obviated at the moment by canalising it on to the scapegoat — the Indians. British policy towards the Indians is consequently, (1) to restrict further immigration into the country, (2) gradually to curtail their present influence in the economic life of Uganda by action and propaganda, and (3) with further control by Anglo-American monopolies of Uganda's life and labour, and with some inevitable "progress" of the Africans in business and education, to liquidate them in due course of time.

There are several indications to show that this policy is being put into effect in the post-war years. The purpose of the *Immigration (Control) Ordinance of 1947* has been mentioned before. Further restrictions were imposed later. The "Pecuniary Qualifications" of the Immigration Control Regulations of 1949 were meant further to restrict the Indians' control of Uganda's economy as retail traders and technicians. Thus, while debating the *Immigration (Control) Bill of 1948*, the Unofficial Member of the Uganda Legislative Council, A. N. Maini, noted[98]:

> "As one looks around East Africa . . . the lack of provision of a big initial amount of capital has not hindered many of these immigrants from being useful to the development and prosperity of these lands. The fixation and increase of these amounts is in my humble opinion definitely wrong and harmful. In practice it will compel many useful craftsmen and trained people like electricians, mechanics, foundrymen and others not to come to these territories at all or to seek jobs with existing firms on their terms. . . . If £ 2,000 is going to be the necessary pre-requisite I doubt if we shall get many of these people to come to the country."

Indeed, that is what the British Government is up to; to restrict the immigration to Uganda of technically qualified and other Indians, so that the Indian community's influence over the Protectorate may not increase any further.

Furthermore there are attempts which, while being of some advantage to the Uganda Africans in certain ways, bound to react adversely on the present source of living and profit-making of the Indians. Thereby there is the attempt to oust the Indians from the important branches of economic life of the country. Thus, for instance, since 1946 co-operative societies among the Africans have been encouraged by the formation of a Co-operative Department, with a view to taking over the control of retail trade from the Indians in course of time. Also, the present policy of the Protectorate Government is to put Africans, wherever possible, into posts previously filled by Indians. In the post-war period, posts such as those of Assistant Medical Officer and Sub-Inspector of Police have been approved for Africans. Local Civil Service posts are also given to Africans, and Asians (Indians) are not encouraged. In continuation of this policy the Indians will, in the future, be ousted from all government departments, although discrimination against the Africans with regard to the posts monopolised by the British would not be eliminated thereby.

It is, however, difficult to oust the Indians so readily from their control of the cotton economy in its intermediate process. Attempts are, of course, being made, and in the course of time Indians' influence in the cotton economy will be very much curtailed, if not completely abolished. The 1948 Cotton Industry Commission had recommended that the Africans should be encouraged as cotton buyers, African employees in the ginnery must do the actual weighing of growers' cotton, the ginning profit should be curtailed, and the ginneries cut down in number. As previously noted, the Commission also made indirect propaganda to discredit Indians as honest businessmen. Since then the ginners' profit has been curtailed — not to enrich the African growers but the British interest on top of the economy, as explained before; and the Protectorate Government is exercising greater control over the ginning industry, and many ginneries have been closed or taken over by the Government.

The future of the Indians in Uganda is thus very bleak if they continue with their present mode of living on a share of the exploited wealth of the colony.

But will this policy give rise to any serious tension in Uganda's life? How do the Indians react to the situation? Do they align themselves with the freedom-urge of the Africans, and unitedly stand against colonial manoeuvres in Uganda?

The tragedy of the Indians in Uganda is that throughout their long connection with the country they have never played a progressive role in its life. The upshot is that, as a community, they have no power to retaliate by themselves alone. They never had any support from the Africans, and now they are no longer wanted by their British masters.

Within the Indian community in Uganda tension could have arisen because of its social and economic differentiations, and some differentiations have indeed taken place during their domicile in Uganda. Some Indians have become fairly rich in Uganda, the most notable of whom is the person who owns the sugar estates and factories. But the bulk of them have remained in the same class position they occupied when they came, namely, the middle class of retail traders, office employees, technicians and intelligentsia. This apparently stable position of the Indians in Uganda is due to the fact that because of their subsidised existence in the Protectorate as a party to colonial exploitation and because of the supreme domination of the British monopolies they could neither markedly improve their position in the society nor were they "proletarianised" in numbers. In all essentials, the Indian community therefore represent the middle class of their society in the mother country, with all the prejudices and aspirations of that class. In 1949—50, the author found that even an ordinary Indian Matriculate in Uganda starts his career as a clerk at a salary of not less than £ 15—20 a month, and a job is available for him in various business and government organisations, or lastly in the ginneries and other concerns owned by the Indians. In India it would have been difficult for him even to secure a job, and, if he got one, he would not get half the amount as salary. Therefore, in Uganda he needs only digest some insults and humiliations from his foreign masters to maintain a standard of living which is certainly higher than what he could expect in India. And, as a reaction, he develops a strong anti-African attitude to justify his superiority over somebody else in the society!

One cannot ignore the fact that there is a good deal of tension today within the Indian community between its socio-economic groups. But, because of their common parasitic existence in Uganda, such tensions are only dissipitated in futile jealousy and mutual bickerings. There is as yet no sign that the people at the bottom of the economic ladder will join the anti-colonial movement of the Africans. As a community, the Indians are awaiting the hour to evacuate this country, and, in the meantime, enriching themselves as best as they can under the servile conditions, without any regard for anybody else.

It should be noted, however, that there may be a change in the situation because of the varying aspirations of the different sections of the community. Those belonging to the Republics of India or Pakistan are more or less determined to leave Uganda when the time comes. They pay their allegiance to the respective States, with which they also maintain regular social, economic and financial relations. But the members of the Khoja community, the followers of His Highness the Aga Khan, who constitute a large section of the Indian community in Uganda today, have taken their leader's advice to remain in East Africa as British subjects. Whether the economic differentiation among these people will in future drive their impoverished members to support the African national or class movement has yet to be seen.

In short, the crux of the matter is that while in the Union of South Africa the simultaneous presence of Indian and African working class could provide the basis for a united resistance movement against the racists and undemocratic forces, no such basis is yet available in Uganda. There is no working class or peasantry other than the African in the Protectorate; on the other hand, the existing parasitic living of the Indian community on the toils of the Africans prevents any healthy contact developing between the two communities.

Hence, on the whole, the situation is that in Uganda the Indian community grew as an off-shoot of a poisonous tree, and will now wither away because the tree does not need further sustinence from the offshoot and the soil has hated it from the start. Therefore, when the present economic role of the Indians in this country is over, nobody will lament their disappearance. The tragedy will however remaion that they were forced from the scene by the connivance of those who are really more to blame — those who so cleverly deflected the anger of the people on this scapegoat.

7. Uganda's Future

Can this move of removing the Indians from Uganda's economic life resolve harmoniously the tension between the British and the Africans? The principle of colonial exploitation will be fully maintained — in fact, it will be accelerated. The people will soon, therefore, be able to realise that the main source of tension is still intact; only a redundant appendage has been cast off to avoid complications, and to beguile them. Hence, it appears very likely that, because of the rapid pauperisation of the people, the tension between the communities of the rulers and the ruled, though out of focus for a short while, will take a more virulent form in future; and, in the absence of a third party on which to lay the blame, the inevitable change in the life of Uganda will soon take place. Uganda will become free.

Moreover, there is the possibility that although an economic basis is lacking in the Protectorate today, the Indians in Uganda, inspired by the glorious campaign launched unitedly by the Africans, the so-called Coloured People, and the Indians in South Africa against Dr. Malan's racial doctrine and the oppressive policy of the White Government of that State, also may ultimately realise that their future in this part of Africa lies not with the colonial perspective of continued enslavement of the peoples of Uganda but with that of the Uganda Africans, namely, the peoples' perspective of freedom, progress and prosperity, which is unfolding itself very rapidly in the post-war years.

There is also the point that in their mother country the people have won the fight against colonial domination in any form whatsoever — direct and indirect, and in the international arena India is now playing an impressive role under the leadership of the Prime Minister of the Indian Republic, Pandit Nehru. As one of the foremost champions for a peaceful co-existence of all nations in the world and for friendship between all peoples on the basis of equality and mutual understanding, Mr. Nehru is now acclaimed as one of the great leaders in the world for freedom and democracy; and because her government follows this progressive path, India has now been raised to a high position in the worldscale. This praiseworthy role of India in the present-day world has stimulated the freedom-urge of the enslaved peoples in Africa and elsewhere, and this no doubt also influences the minds of the Indians in Uganda. It is not improbable therefore that in the future the Indians in Uganda will take up a firm anti-colonial stand in this colony. And that would mean supporting the African movement against the colonial power.

Signs of this change are visible in the neighbouring British colony of Kenya where during the so-called "trial" of the African leader Jomo Kenyatta, the renowned peace-loving lawyer D. N. Pritt was ably assisted by his Indian colleagues from Kenya and India. During this "trial" Mr. Nehru openly declared his concern that the African leader should receive able legal help to defend himself, and so he suggested that Indian lawyers should assist Mr. Pritt. Also, as reported in the London *Times* of August 8, 1955, while opening the Department of African Studies at Delhi University, Mr. Nehru stated:

> "Indians in Africa could expect no help from India if they exploited Africans. He wanted them to live there only so long as they had the good will of the people."

Furthermore, in a recent message to the African students in India, Mr. Nehru declared[99]:

> "I have long felt that the coming years will see emergence of Africa and the African people. The major question is as to how this change will take place. Will it take place peacefully and co-operatively or in conflicts? It is obvious that people, once awake and conscious of their rights, cannot be suppressed for long. The only wise policy, therefore, apart from its humanity, is to recognise the claims of the people of Africa and do justice to them."

Such an attitude of the foremost leader of the Indian people may lead also the Indians in Uganda to take up a positive stand to sympathise with and support the African struggle for freedom. In that case, the inevitable change in Uganda will be even more quick, without the stumbling block of a subsidiary anti-popular interest. And that will mean the dawn of a new life in this part of Africa — a life leading to prosperity and progress for the millions of people now suffocating in misery and degradation, stagnation and decay.

For such a glorious dawn the world public will not have long to wait.

REFERENCES

1 Dutt, R. Palme — "The Crisis of Britain and the British Empire", Lawrence & Wishart, London, 1953, p. 237

2 Quoted in *loc. cit.* 1, p. 238

3 The *Daily Herald*, London, June 26, 1947

4 Huxley, Elspeth — "The Sorcerer's Apprentice: A Journey through East Africa", Chatto and Windus, London, 1948, p. 251

5 Hall, Sir J. — "Some Notes on the Economic Development of Uganda", Foreword to E. B. Worthington's *A Development Plan for Uganda*, Entebbe, Uganda Protectorate, 1949, p. v

6 *ibid.*, p. xii

7 *loc. cit.* 1

8 Quoted in Cheddi Jagan's *Forbidden Freedom*, Lawrence and Wishart, London, 1954, p. 81

9 *loc. cit.* 1, p. 281, etc.

10 *ibid.*, Chapter X, etc.

11 Harris, D. — "The 1948 Revision of the Plan", published in the same volume with E.B. Worthington's *A Development Plan for Uganda*, Entebbe, Uganda Protectorate, 1949, p. vi

12 Worthington, E. B. — "A Development Plan for Uganda", Government of Uganda Protectorate publication, Entebbe, Uganda, 1949, p. 10

13 *loc. cit.* 1, pp. 240—241

14 *loc. cit.* 5, p. v

15 *Proceedings of the Nairobi Scientific and Philosophical Society*, August 12, 1947, Vol. I, Part 1, Nairobi, Kenya Colony, British East Africa

16 (a) Lugard, F. D. — "The Rise of Our East African Empire", Willim Blackwood & Sons, Edinburgh and London, 1893;
(b) Portal, G. — "The British Mission to Uganda in 1893", Edward Arnold, London, 1894;
(c) Johnston, H. H. — „The Uganda Protectorate", Hutchinson, London, 1902;
etc., etc.

17 Roscoe, J. — "The Bakitara or Banyoro", Cambridge University Press, 1923, p. 6

18 *loc. cit.* 5, p. v

19 *loc. cit.* 1, pp. 172—173

20 *loc. cit.* 12, p. 5

21 Quoted in *loc. cit.* 1, p. 245

22 *Proceedings of the Legislative Council of Uganda*, 26 July, 1949, Uganda Government Publication, Uganda
23 Reported in *Uganda Notes*, Kampala, Uganda, December, 1907, p. 119
24 Whitley, N. H. P. — "Report of the Commission of Inquiry into the Disturbances which occurred in Uganda during January 1945", Entebbe, Uganda, 1945, p. 7
25 See the *Appendix* to this book
26 *loc. cit.* 24
27 *Uganda 1953*, Annual Colonial Report, Her Majesty's Stationery Office, London, 1954, p. 8
28 Hailey, Lord — "An African Survey", Oxford University Press, 1938, pp. 1318—1319
29 *An Economic Survey of the Colonial Territories, 1951*, Volume II: "The East African Territories", Her Majesty's Stationery Office, London, 1954, p. 100
30 *loc. cit.* 27, p. 8
31 *ibid.*, p. 59
32 *ibid.*, p. 8
33 *ibid.*, p. 59
34 *ibid.*, p. 59
35 *ibid.*, pp. 7—8
36 *ibid.*, p. 59
37 *ibid.*, p. 58
38 *ibid.*, p. 58
39 *ibid.*, p. 58
40 *ibid.*, p. 58
41 *ibid.*, p. 58
42 *ibid.*, p. 58
43 *ibid.*, p. 59
44 *ibid.*, p. 8
45 Quoted in Cheddi Jagan's *Forbidden Freedom*, Lawrence and Wishart, London, 1954, pp. 81—82
46 Quoted, *ibid.*, p. 82
47 *Uganda 1951*, Annual Colonial Report of the British Government, Her Majesty's Stationery Office, 1952, p. 20
48 Calculated from *loc. cit.* 47, pp. 17—20
49 *loc. cit.* 12, p. 8
50 *loc. cit.* 15
51 *ibid.*
52 *loc. cit.* 29, p. 97
53 Marx, Karl — "The Future Results of British Rule in India", New York Daily Tribune, August 8, 1853
54 *ibid.*
55 Thomas, H. B. and Spencer, A B. — "A History of Uganda Land Surveys and of the Uganda Land Survey Department", Entebbe, Uganda, 1938, p. 70
56 *ibid.*, p. 70
57 See the *Appendix* to this book
58 From the copy of the Scheme given by the President of the African Labour Party in Uganda to the author
59 From the copy of the Plan given by the President of the African Labour Party in Uganda to the author
60 *loc. cit.* 27, p. 46
61 Hailey, Lord — "Native Administration in the British African Territories", His Majesty's Stationery Office, London, 1950, Part I: *East Africa: Uganda, Kenya, Tanganyika*, p. 61
62 *loc. cit.* 4, p. 277
63 *ibid.*, p. 288
64 The *Times* of London, August 22, 1955
65 For an analysis of the *1935 Constitution* and its role in India, see, for instance, R. P. Dutt's *India Today*, People's Publishing House, Bombay, 1949, pp. 464—496, etc.
66 Keith, A. B. — "A Constitutional History of India: 1600—1935", 1936, pp. 473—474; quoted in *loc. cit.* 65, pp. 475—476
67 The *Times* of London, August 13, 1955
68 *ibid.*, August 30, 1955
69 *ibid.*, September 6, 1955
70 *ibid.*, September 6, 1955
71 Eliot, J. R. Mcd. — "Report on an Investigation into Conditions affecting Unskilled Labour, and the Supply thereof, within the Protectorate", Entebbe, Uganda, 1937, *Appendix VI: Tribal Distribution in the Various Industries Visited*
72 Carter, W. M. (Chairman) — "Report of the Commission of Inquiry into the Cotton Industry of Uganda, 1929", Entebbe, Uganda, 1929, pp. 2—3
73 *ibid.*
74 Mair, L. P. — "An African People in the Twentieth Century", Routledge & Sons, London, 1934, p. 202
75 *loc. cit.* 4, p. 264
76 *Cmd. 3234*, His Majesty's Stationery Office, London, 1929
77 *United Nations Year Book: 1947—48*, pp. 774—775
78 *loc. cit.* 29, p. 81
79 See, for instance, Derek Kartun's *Africa, Africa!*, Lawrence and Wishart, London, 1954, pp. 1—2
80 The *Times* of London, August 13, 1955
81 See, for instance, Cheddi Jagan's *Forbidden Freedom*, Lawrence and Wishart, London, 1954
82 The *Times* of London, August 30, 1955
83 For an analysis of the root cause behind the division of the subcontinent of India into the Indian Union and Pakistan, see, for instance, R. P. Dutt's *India Today*

People's Publishing House, Bombay, 1949
84 Coupland, R. — "East Africa and Its Invaders", Oxford at the Clarendon Press, 1938, p. 27
85 *loc. cit.* 16(a)
86 (a) *loc. cit.* 16(a);
(b) *loc. cit.* 16(c)
87 Maini, P. L. — "The Indian Problem in Kenya", M Sc Thesis in the London School of Economics Library, London, 1944, pp 14—16
88 Reported in *Uganda Notes*, Kampala, Uganda, November 1905, pp. 170—171
89 Preston, R. Q. — "Oriental Nairobi", 1938, pp. 48, 75, 77, 81, 91, 105
90 (a) *loc. cit.* 16(a);
(b) *loc. cit.* 16(c);
(c) *loc. cit.* 55, p. 12
91 *loc. cit.* 16(c), Vol. I, pp. 293—295
92 *loc. cit.* 72, pp. 2—3
93 Jackson, Sir Frederick — "Early Days in East Africa", Edward Arnold & Co., London, 1930, p. 327
94 *ibid.*, p. 327
95 *loc. cit.* 24
96 Whitley, N. H. P., (Chairman) — "Report of the Uganda Cotton Industry Commission 1948", Entebbe, Uganda, 1948, p. 17
97 *loc. cit.* 27, pp. 16—17
98 *Proceedings of the Legislative Council of Uganda: 1949*, Uganda Government Publication, Uganda, p. 24
99 Extract from the message sent by Mr Nehru to the annual general conference of the African Students' Association (India), which was held in Bombay from May 9 to May 12, 1956. (cf *India News*, Information Service of India, India House, London, Vol. 8, No. 19, May 12, 1956.)

CONCLUDING REMARKS

In the foregoing pages the author has described the situation in Uganda — past, present and in the near future. The study has endeavoured to bring out the point that neither are the peoples of Uganda living happily under the patronage of the British nor are their sufferings and misery due to some "errors" and "mistakes" commited by the foreign rulers under unavoidable circumstances or unintentionally. Hence a determined effort has to be made to change Uganda's life for the better.

Herein lies another purpose of this study, for, as Lucy Mair had pointed out in her monograph on the Baganda, today, society can take advantage of a scientific analysis of the forces working within it, so that progressive tendencies can be encouraged and reactionary traits rooted out[1]. Such an analysis made in this study shows that the only way really to improve Uganda would be, first of all, to end the colonial exploitation of the native population in what is today the Uganda Protectorate and let the people rule themselves.

How will the different communities react to this demand ? . The Africans because of their direct interest in this matter, as representing the expropriated community, will certainly work to this end. But will the foreigners help them in their aspirations for a progressive life ?

The Asians can, and should, help the Africans by aligning themselves with the national desire of the peoples of Uganda instead of precariously clutching the apron-strings of the colonial rulers. And the possibility of such an alignment is rapidly unfolding itself, for the sentiment frequently expressed by the Asian readers cannot but influence their decision and encourage them to support the peoples' movement all over Africa for freedom and democracy. But for a peaceful solution to the *problem of Uganda* the main responsibility goes to the British, for among the foreigners who count most are they — the rulers of Uganda today. Will they support the natural demand of the colonial peoples ?

The writer has shown in the previous chapters how most of the British administrators and social scientists have so far either been unable to understand the situation or have distorted reality. The role of the administrators and that of those who represent the monopoly interests can be understood, for they form part and parcel of the colonial exploitation. But this, surely, is not true for all British intellectuals who have shown an interest in the situation in Uganda. It is regrettable, therefore, that some of them are playing directly into the hands of the colonial interests. They *justify* the British Government's actions in Uganda and distort or conceal the facts in order to put the blame for all the "troubles" on the Africans or the Asians. A few of them go even beyond this, and advise the foreign rulers how best to exploit the natives under the present circumstances. The oppressed Africans, and the freedom-loving peoples of the world, have nothing to expect from them. History will, in due course, throw them on the rubbish-heap of futility.

However, there are some other intellectuals who, probably being unable to understand reality, are indirectly doing the same thing as the staunch proponents of colonialism. It may be true that these social scientists are subjectively sincere to their science. For instance, explaining the responsibilities of an anthropologist, Malinowski noted[2]:

> "Shall we, therefore, mix politics with science ? In one way, decidedly 'yes', because if knowledge gives foresight and foresight means power, it is a universal stultification of scientific results to insist that they can never be useful by those who have influence. The

importance of culture contact and change as a subject of research has been recognised in most countries where colonial affairs are of practical importance, and anthropology flourishes."

In theory, Malinowski thus showed a truly scientific spirit in the study of culture contact and change. But in practice his attitude and suggestions could only further help the colonial powers which could now get scientific help to "understand" the situation better and thus organise a more suitable form of administration in order to maintain their power and stabilise their interests. It is indeed true that "the importance of culture contact and change as a subject of research has been recognised in most countries where colonial affairs are of practical importance, and anthropology flourishes". Notable among these countries is Britain, where in the post-war period the British Colonial Office has engaged a large number of anthropologists and sociologists to report on the actual state of affairs within various African "tribes". But by examining the practical implications of the attitude and opinion of Malinowski, who still holds a great influence on the British anthropologists, and who was one of the pioneers in making the government and scientists interested in studies of culture contact, one may get an idea of what is likely to be the outcome of these researches in most cases.

It is seen from Malinowski's writings that although subjectively he was willing to disclose the entire truth, he apologised for European vested interests in colonies in order to explain "the dynamics of culture change", as it was revealed to him from the impressions and experience he received during his tour in British East Africa. To quote him[3]:

> "Here we must emphasize again that the administration cannot act entirely as a free agent, that is, without regard for other European vested interests. Whatever the desirability or justice of the latter, they must be taken into account by the anthropologist as determining factors in the total culture contact situation. . . . Therefore, while the presence of European vested interests in the country will obviously affect the practical implementation of the policy of Indirect Rule, there must be safeguards for Native interests and a compromise reached where these clash with European attempts at exploitation."

But, such a compromise, as the foregoing pages have indicated, is impossible to attain. The interests of the Natives indeed depend upon the elimination of "European attempts at exploitation"; but such "attempts" permeate into the entire life of the colony, and, as even Malinowski had to note, "the administration cannot act entirely as a free agent".

Then, in support of the colonial administration, Malinowski advocated creating local tools for the Indirect Rule. He noted[4]:

> "This conflict between European parties is especially relevant when we consider the contradiction inherent in a situation in which the policy of Indirect Rule must support the power and prestige of the chief, while at the same time the influence of the mission undermines it. . . . If the Europeans plan in the void without reference to African realities, there is no real basis for co-operation. On the other hand, if policy is limited entirely by what exists now in African culture, then rigidity and no possibilities for development must be the result."

So much is good as stating the problem, and nobody will disagree with the above lines stating that the African societies must not be preserved as museum-pieces. But how then Malinowski intended to solve the problem? To continue[5]:

> "Thus the basis for the drafting of a program is the fullest understanding by Natives of European aims; and by Europeans of Native difficulties, in so far as the object is to create in Native authority a devoted and dependable ally, controlled, but strong, wealthy, and satisfied."

Nothing could be more explicit.

Furthermore, Malinowski was of the opinion that it is not enough to plant chiefs among the people in order to maintain the facade of their ancient life; it is the duty of an anthropologist to see that these chiefs develop into a strong ruling group subservient to the colonial regime. To quote him[6]:

> "In the economic sphere, we have already discussed the position of the chief at the time of the introduction of Indirect Rule, and the measures which should have been taken to provide a solid financial foundation for his administration.... The real problem of contact which the anthropologist is bound to assess, depends then largely on how to strenghten financially, politically, and legally the present-day chief under present-day conditions."

Such are the recommendations of an eminent scientist like Malinowski for the benefit of the African peoples.

Evidently, even Malinowski, in spite of his great distinction as a scientist and his frequent declaration on the need of a true scientific outlook in anthropological studies, failed to reach an objective conclusion on the African societies, of which the Uganda societies no doubt form a part. As a member of the ruling community in relation to the Africans, he looked more after the interests of the foreign bourgeoisie than those of the exploited Africans, although a truly scientific analysis of the problems of African colonies, such as of Uganda, can align one only on the side of the Africans. However regrettable it is, one is thus compelled to note that while the theoretical principles enunciated by him on the functional approach to social studies mark a step forward in scientific analysis of societies, and while the study of culture-contact in Africa (which Malinowski advocated) could have been, and still can be, of immense value in understanding the *problems* of the African societies, many of his practical recommendations were contrary to his theoretical standpoint. Indeed, one cannot fail to notice that Malinowski's attitude to maintain scientific neutrality with regard to European vested interests in the colonies did not really keep him in a neutral position. It made him an advocate of colonial exploitation. In fact, trying to avoid judging the desirability or the justice of such an attempt on the part of the Europeans, he judged it all the same but in favour of the colonial interests, and not of the subjugated peoples. It may also be recalled that not only did he thus apologise for colonial exploitation, and even suggested some better or more stable methods in the form of improving the status of the native "chiefs", but also praised to the world the goodness of the British Colonial system.

It would therefore be not impertinent to state that these social scientists, at their best, are unconscious of the real character of their studies. This probably results from their approach to the problem, by which they consider what they just see superficially depicts reality. But reality is a composite understanding of the past and present course of social forces and their future trends. As shown in the foregoing pages, the apparent prosperity of the Africans in Uganda does not represent the true situation. This is determined by the forces which had worked in society in the past and are working at present, and by the future interaction of these and other newly evolved forces. A conception of a moving society, therefore, demands study of the *character*, *strength*, and the *direction* of the forces working within it.

Indeed, a society at any stage of its development can be regarded only as maintaining a *dynamic equilibrium*: *equilibrium*, because the magnitude of the forces in different directions have come to a balance; and, *dynamic*, because the character and the strength of the forces will again change the equilibrium and effect changes in society in a particular direction. This concept of dynamic equilibrium of society is inseparable from the fact that however static the life of some people may appear to be and however long a period (relatively speaking) such societies may take to indicate an appreciable change, all types and forms of human society cannot but be in a process of change, and therefore just as it is unrealistic to speak of any society as being in a state of *static equilibrium* so it is wrong to think of it as being in a state of mere *disequilibrium*, for the mere acceptance of the truth that there is nothing as static equilibrium does not explain the *process of change* in society or the *course of development* from one type of society to another and consequently misses the key to the understanding of any living entity. Simplifying with notations, it may be said that it would be quite absurd to consider at any two points of *time* and *space* that $A = A_1$, and insufficient simply to conclude that $A \neq A_1$; the correct way of describing the situation being to consider A and A_1 as representing the relation $A \rightleftharpoons A_1$, where changes from A to A_1 is dependant on the relative magnitude of the interacting forces represented by the sign $\rightleftharpoons$.

That this concept is vital to any *social* study has been illustrated in this book in which it has been shown how the Uganda society was and is representing a dynamic equilibrium at different stages in history because of the interacting forces of African and "western" civilisations. And it is evident therefrom that the *problem of Uganda* today, as it has come out in the forefront because of a particular type of *acculturation* between the Uganda Africans and the British, could not be revealed without this basic concept of dynamic equilibrium of society. Why is it that even some "sincere" scientists belonging to the "western civilisation" are found to overlook this essential fact in *social* changes?

It may not be fortuitous that Malinowski should prefer to use the term "Culture Contact" instead of "Acculturation", because the latter pre-supposes that culture flows mainly from a stronger potential to a weaker one*, and both he and his pupil Mair would find justification in the British introduction of Indirect Rule among the Baganda. According to Malinowski, the process of culture change is not "a mixture of cultural elements (but a) reorganisation on entirely new and specific lines"[8]. With this concept he, of course, rightly rejects the mechanistic concept of transmission of cultural *traits* and *complexes* without reference to the character and composition of a society, but at the same time with his concept of "culture contact" he fails to foresee the *dominance* of one "culture type", or more strictly of one *social formation*, over another. To him, the culture change remains a process of "reorganisation", which is in fact only a part of the whole truth, as the *direction* of the social forces in their movement is not interpreted thereby. Objectively, therefore, using the term "culture contact", instead of "acculturation", Malinowski avoided the concept of a dynamic equilibrium of a society; and, as explained in Chapter II, his pupil Lucy Mair failed to appreciate the real conditions among the Baganda at the time of her inquiry, even though she declared that, "The Functional theory of anthropology in general stresses the importance of studying native life as it is actually lived"[9].

Why is this so? It may be due to the type of education one receives, the ideas one forms in daily life, and the outlook one develops in a country where the ruling class endeavours to keep the *total* reality away from the people. For the existence of a Colonial Power depends upon falsifying the real situation, as it became quite evident the case of British India, for instace. Consequently, a large number of intellectuals of these countries may not be able to go beyond appreciating the apparent character of life today. For Uganda, they can only see that the British administration has raised the African people from a "primitive" to a modern state of living by the introduction of a cash economy and "im-

* The following observation of Phyllis M. Kaberry while editing Malinowski's writings for the posthumous publication of his book "The Dynamics of Culture Change" is of interest in this connection[7].

"Malinowski rarely used the world 'acculturation', and in one of his more recent writings he advocated the adoption of a term coined by Don Fernando Ortiz, namely, 'transculturation', since in this there were 'no implications of one standard dominating all the phases (of culture change), but a transition in which both sides are active, each contributing its quota, each merging into a new reality of civilization'. The phrase generally used by him was 'culture contact and change', . . ." (cf. *loc. cit.* 2, p. vii).

proved" facilities for living. Since they may not interest themselves in seeking the origin or the working of the contending forces in the society, they cannot understand why, in spite of all the "good things" which Britain has done for Uganda, the Africans consider the British rulers as enemies. Some of them who may, in their own way, sincerely desire good for the Africans, feel aggrieved that the latter do not always show gratitude for what the British have done for them.

It is true that the advent of the British in Uganda broke the stagnating economy and thus stirred the native population out of the crisis reflected in the fissiparous developments in the political superstructures of their societies (and especially of the "developed" society like that of the Baganda) without as yet a qualitative change having taken place in their, socio-economic life. From this standpoint, therefore, the following comment from Malinowski is fully justified: namely, "... if policy is limited entirely by what exists now in African culture, then rigidity and no possibilities for development must be the result" [11]. But accepting the above as a general comment also applicable to Uganda in the sense that the African societies should not be preserved as museum-pieces, it may be questioned specifically whether the policy engineered by the colonial rulers led to a progressive development of the Uganda societies or not. And this is a question which has been answered in the negative in this study. How can one then assess the role of Britain in Uganda at the time when the Uganda societies were at a cross-road ? The writer is inclined to think that this can be done only by considering Britain as an unconscious tool of history in Uganda's life. A *tool of history*, because even without British domination the Uganda Africans could have by themselves broken through the stagnating economy, and, moreover, they would then have established a progressive life which the Colonial Power has failed to do as it affects its basic interest. And, *unconscious*, because whatever Britain did for Uganda was done only with the conscious aim of serving her own interest. It has been explained earlier how, for the benefit of the foreign monopolists, Uganda's economy had to be "improved" (its contents changed but the forms retained), better means of transport and communication and towns established, and meagre share of the vast wealth of the country had to be given to the natives of the soil, so that they could survive and serve the foreign masters as the "free" source of labour-power. It is, again, worthy of note that those "progressive" features which were not directly connected with the system of exploitation, are the least developed and are developing at a very slow rate. The Governor of Uganda remarked in 1946 on the inadequate medical facilities and the lack of preventive measures and proper nutrition for the improvement of the health of the African population[11]. It has also been described how meagre are the educational facilities for the Africans, even though these are also not given without ulterior motives, as the visionary Lugard had suggested as early as in 1893.

Evidently, the "improvement" of the Africans was merely a bye-product of colonial exploitation, and was due to the inherent contradictions in the system. To exploit the people and their rich country, their conditions had to be somewhat "improved", and this "improvement" could not but react adversely on the exploiters. As Kartun put it very aptly[12]:

> "It is a familiar irony of history that the movement of the oppressed against their oppressors is always of the oppressor's own making. Africa is no exception."

Everyone, therefore, should realise that there are no grounds for the Africans to be grateful to the British for what the latter have done for them ? Indeed, should they be grateful for the sordid life they have been led to by several decades of colonial rule ? They demand justice.

Now the situation in Uganda has taken an acute form. Colonial vandalism has led Uganda into a blind alley, from which she cannot break out so long as the colonial domination over Uganda continues. History will show whether Britain will endeavour to pull the Uganda society further down or lift it up as an equal member in the world family of nations This she often declares to the international public as her only "mission" in Uganda! When is she going to fulfill this mission?

Furthermore, apart from what the British ruling class is ultimately forced to do in the face of the determined struggle of the Uganda Africans for freedom and democracy, in the light of the present international situation in which the camp of true democracy and universal freedom is securing triumph after triumph the people of Britain cannot fail to understand that it is in their own self interest to see the Uganda Africans free and happy, their land prosperous, their societies progressive. Karl Marx once said: "A people which enslaves another forges its own chains". The truth of this statement could not at any other time be better revealed for Britain than it is today, when their colonial crisis has been transformed into dollar crisis, when their life is unable to rehabilitate itself even a decade after the last war and when the subservience of the British Power to the United States has become a matter which cannot pass unnoticed even by a casual observer of international affairs. Indeed, the situation has come to such a pass that it was written in the *Times Literary Supplement* of the 24th August, 1951:

> "Perhaps America is the only country capable of providing the West with an ideology . . . It is difficult to think that the West could accept any other ideology than an American or American-sponsored one."

And this is so in a country the intellectual standard of which was held so high even by the enslaved peoples that the renowned Indian poet, Rabindra Nath Tagore, wrote that in his early years, enthused by the "English literature which nourished our minds" and impressed by the "evidence of liberal humanity in the character of the English", he "was led to set them on the highest pedestal of my respect"; for "this genorisity in their national character had not yet been vitiated by Imperialist Pride" [13]. But, instead of pursuing her own rich cultural heritage and putting into practice her declarations on freedom and democracy for all people, Britain passed on to the imperialist phase of her capitalist development, with the ultimate result that living on others' toil she is now becoming bankrupt of even her own moral standard.

Moreover, Britain began to lose her prestige in the world because of the contradiction between her declarations and her practice. As Tagore declared in 1941, at the end of his life of eighty years, while in the "first chapter" of his life he "naturally set the English on the throne of my heart", then "came the parting of the ways accompanied with a painful feeling of disillusion when I began increasingly to discover how easily those who accepted the highest truths of civilization disowned them with impunity whenever questions of national self-interest were involved" [14].

This contradiction in Britain's life has to be solved if she wants to regain her prestige in the world. And it is also worthy of note that it was not the ,,national self-interest" but the monopoly interests of the British imperialist bourgeoisie which have brought about the present material and moral crisis of Britain. This is a point which no one can fail to take a note of, while to the British people it is of a vital concern. For, as some recent studies in Britain have shown[15], the Achilles' Heel of Britain's ever-growing problems lies in her imperialist economy and the mode of life that she has built-up by living on the toils of the colonial peoples. Here it is not necessary to elaborate on such analyses, but it is important to stress that while the cream of colonial exploitation is going to bridge the "dollar gap" of Britain, and the remaining profit (of no little worth) is enriching the British monopolists, it is the British people who are suffering under the burden of retaining the colonies. It is they who have to provide in men and money for the imperialists' reign of terror in the colonies in order to smash the national movements. But, in return, the British people are facing an increasingly difficult situation in a parasitic economy, which can no more thrive on colonial spoils. For everywhere the colonial peoples are rising to their feet, and, having been bled dry for a long time, the colonies also can no more furnish great quantities under the ever-accelerated method of recklessly sucking-up their wealth by the imperialist machinery.

From all aspects, colonial exploitation is obsolete today. The British people, therefore, need not support the colonial liberation movements only on humanistic grounds, though there is no reason why one should forget what the Chartist leader, Ernest Jones, said in regard to Britain in May 1851 in his *Notes to the People*:

> "On its colonies the sun never sets, but the blood never dries."

What, however, one cannot fail to recall today is what the Chartist Declaration of 1846 had to say to the British people, for this rings a topical note in regard to what is happening in British colonies today.

> "They, your masters, will take the land — they will fill all the higher situations, civil and military, of the new [today old] colonies — your share will be the slaughter of the combat and the cost of winning and retaining the conquest."

Now, this cost of "retaining the conquest" is devolving upon the shoulders of the common Britons with ever-increasing burden, as the growing crisis of Britain in the post-war years makes it more and more obvious from day to day.

If this state of affairs is allowed to continue unchecked, it can only lead to further worsening of Britain's economy and her position in world society. On the other hand, freedom of the colonies, like Uganda, from Britain's domination would also mean weakening the political and economic power of the monopoly interests at home. And that means a victory for the British people, facilitating their march towards freedom from exploitation. The British people thus cannot ignore the fact that their own interest is intertwined with that of the colonial peoples. It is, therefore, only logical to expect that they will support the movement of their enslaved brethren, like the Uganda Africans, for freedom and democracy, for a future world of progress and prosperity.

And what will the British people get in return, besides serving their own cause for a true democracy and prosperity ? On this point it is worth remembering what Pandit Jawaharlal Nehru wrote when India was a Britis h colony, for his statement in regard to India at that time may well be applicable to Uganda today. To quote him[12]:

> "We have endured for long the painful burden of European domination in Asia. Britain may believe and proclaim that she has done good to India and other Asian countries, but the people of India and other Asian lands think otherwise, and it is after all what we believe that matters now. It is a terribly difficult business to wipe out this past of bitterness and conflict, yet it can be done if there is a complete break from it and the present is made entirely different. Only thus can those psychological conditions be produced which would release mass effort and lead to co-operation in a common endeavour."

As stated then in regard to India, it is not difficult to imagine that if Britain of her own accord solves the *problem of Uganda* by withdrawing her tentacles of colonial rule and exploitation, this will undoubtedly ,,release mass effort and lead to co-operation in a common endeavour", which cannot but be of great benefit to a democratic Britain. For the British people should not forget that a free Uganda maintaining friendly relations with a free and truly democratic Britain can be of great mutual assistance when Uganda embarks upon a mighty programme of nation building. Then there will be a great market for British engineering and other goods which Uganda would need for her properly developing economic and social life. In return, Uganda will willingly provide a friendly Britain with her products which Britain needs. Trade under the banner of friendship and equality will be of great advantage to both the countries in order to develop their free economy, progressive social life, and a prosperous standard of living for all the people.

Foremost of all, we should always bear in mind that a colonial people freed from bondage is a great bulwark for peace and world security. They are the ever-vigilant partisans against war and destruction, for they want to build a new life, not to destroy what is good and progressive in the world. So, the success of colonial liberation movements, in Uganda and elsewhere, contributes to the solid foundation of peace all over the world. And it means, in addition, political weakening of the war mongers, reduction of their war potential, and their eventual decay.

That is the day every human being looks for — a world of peace and security, a world family of nations living in prosperity and happiness in a world society where everybody is equal and free, and society progresses to higher and higher levels. Such a dream does not come true by itself. Everyone has to fight for it, earn it. So the fight of the black people in the heart of Africa becomes intimately connected with that of his white brethren in Britain, and with the rest of mankind of whatever colour and belief.

If such a lesson can be drawn from the foregoing pages, this study has more than fulfilled its purpose.

REFERENCES

1 Mair, L. P. — "An African People in the Twentieth Century", Routledge & Sons, London, 1954, Chapter 1

2 Malinowski, B. — "The Dynamics of Culture Change", Yale University Press, 1945, p. 4

3 *ibid.*, pp. 145—146

4 *ibid.*, pp. 146—147

5 *ibid.*, p. 147

6 *ibid.*, pp. 148—149

7 *ibid.*, p. vii

8 *ibid.*, p. viii

9 Mair, L. P. — "The Place of History in the Study of Culture Contact", published in *Methods of Study of Culture Contact in Africa*, Memorandum XV, International Institute of African Languages and Culture, London, 1938, p. 3

10 *loc. cit.* 2, p. 147

11 Hall, Sir J. — "Some Notes on the Economic Development of Uganda", Foreword to E. B. Worthington's *A Development Plan for Uganda*, Entebbe, Uganda, 1949

12 Kartun, D. — "Africa, Africa!", Lawrence and Wishart, London, 1954, p. 4

13 Tagore, Rabindranath — „Crisis of Civilization", Visva-Bharati, Calcutta, 1950, pp. 3—4

14 *ibid.*, p. 6

15 See, for instance, R. Palme Dutt's *The Crisis of Britain and the British Empire*, Lawrence & Wishart, London, 1953

16 Nehru, Jawaharlal — „What India Wants", India League Publication, London, 1942, p. 8

APPENDIX

On the 1949 "Disturbances" in Uganda

In 1949 some Baganda students at the Makerere College in Uganda were asked by their teacher to write an essay on the 1949 "Disturbances". The essay that they wrote jointly is printed verbatim below. While reading the essay it should be borne in mind that the students were writing this for a teacher who belonged to the ruling community in the Protectorate. It is likely, therefore, that writing a few months after the "disturbances", when everything had not yet returned to normal, they had to be cautions in their account of the incidents during the uprising. Probably for this reason they spoke of "disturbances" and "riots" and emphasised one facet of the "Events" in which the role of some hooligans has been high-lighted, while that of the police, on the one hand, and of the organised Bataka, on the other, have hardly been touched at all. However, whether this supposition is correct or not, the essay gives a graphic account of the historical background to the 1949 "disturbances", and is therefore reproduced below.

The Makerere Students Look at the 1949 Disturbances in Uganda

The causes of the April riots are divisible into two:

I. The remote causes. (a) Katikkiro Wamala's resignation (1945).

Wamala had been a popular Katikkiro, two main points making him outstanding in the eyes of many Baganda:

(I) *Land Bill.* In 1944, the Kabaka, advised by the Governor, introduced to the Lukiko a Bill which would give him the right to acquire land from the landowners. The Kabaka was to acquire land if the Government needed it for public use. Wamala, Chairman of the Lukiko, gave what was called an important speech, by his followers — in which he strongly opposed the Bill. His influence was very strong and the Bill never became Law.

(II) *Bishop Stuart and the Church.* In the same year, the Bishop introduced to the Synod his intention to transfer the Cathedral from the N. A. C. [Native Administrative Council] to the C. M. S. [Church Missionary Society]. He explained why he was intending to do so; and he suggested that a new Cathedral would be built for Baganda at Natete. Wamala was an unofficial member of the Synod, and when he was asked to speak, he started in a violent tone, stating that the Bishop was selfish and that he wanted to rob the Baganda of their Cathedral — the Cathedral "they built with their hands". Wamala's address to the Bishop was not good according to the reports of the Govt. officials; and his speech in the Lukiko on the Land Bill was almost in the same tone; and it was alleged that Wamala was insane. A few days later he was examined by four doctors — three of whom were Europeans, and one a Muganda. The three Europeans agreed on the insanity of Wamala, whereas a Muganda doctor — Mukiibi — opposed them by saying that he was sane. Wamala was forced to resign, and because the Governor suspected Mukiibi to be Wamala's supporter, he ordered for his transfer to Moroto or for his deportation if he did not accept the transfer. The followers of Wamala never believed that he was insane, for they trusted Mukiibi's statement and not the Europeans'. And thus dissatisfaction began. Worse still, was the case passed against Mukiibi — the man who had owned the people's love as a result of his strong opposition against the three European doctors.

(b) *Deportation without trials.* After Wamala's resignation, Nsibirwa was re-appointed Katikkiro (1945). Two months after his appointment, he passed the Land Bill in the Lukiko which had been opposed by Wamala. He had also forced many Chiefs — many of whom were well loved by the people — to resign. On the 5th of September 1945 he was assassinated.

As a result of Nsibirwa's assassination, the Governor ordered the deportation of Wamala, Prince Suna — who was known as the "Prince of the Common people" — a number of Chiefs, and some other leading men, all adding up to seventeen. Many Baganda never believed that these men were justly deported, for they were not tried. But it was rumoured that there were some people who had given evidence secretly against the deportees, and that among those who have the evidence were the newly appointed Chiefs. Did it mean that they gave evidence secretly so that they might be appointed Chiefs ? This, for long, remained a common question among many Baganda, and this was really the beginning of dissatisfaction with the present Baganda Government.

(c) *Death of Wamala and Prince Suna in deportation.* A few months after their deportation, Wamala and Prince Suna died. It was rumoured that they were neglected, and they were given bad food, and there was no information from the Govt. sources about the cause of their deaths. Many people began to suspect some deliberate fatal action of the deceased by the Govt. doctors. And so, future incidents were only to help aggravating old dissatisfaction.

(d) *Katikkiro Kawalya.* Nsibirwa was assassinated before he had signed the Land Act. And so, when Kawalya was appointed Katikkiro (1945), the first thing he did was to sign the Act — much to the opposition of many dissatisfied Baganda. Petitions demanding the repeal of the Act were sent to the Kabaka and the Katikkiro, but no satisfactory answer was given back. In 1945 the Act was the more strengthened by adding one more part stating that the landowner who would violate the Govt.'s action when it wanted to acquire land, was to be penalised with a certain fine of a big sum of shillings and imprisonment for six months. Thus began the dissatisfaction of many landowners.

(e) *The Legco.* The Governor nominated Kawalya to sit on the Legco [Legislative Council 1947] — representing Buganda. Many Baganda liked the idea of representation in the Legco, but not by the Katikkiro — Chairman of the Kabaka's Lukiko. They also did not like the nomination by the Governor, of their representative. They wanted election. Many letters were written to Kawalya, demanding his resignation from the Legco, but he continued to be a member; even to the present day Kawalya is a member of the Legco.

(f) *Paper 210.* Many Baganda had never liked the idea of a close union with the other East African Territories. In fact the late Kabaka Chwa had opposed the union strongly when it was first introduced in 1927. In paper 210, it was laid down that the Commission had the right to acquire land. A Muganda landowner would not like that. The paper was strongly opposed, and there was an intention on the side of some Baganda to boycott the Commission and the Central Assembly. But this could not be done, for Kawalya accepted when was nominated member of the Central Assembly. Many Baganda grew more impatient as a result of Kawalya's conduct. Many suspected that he was probably bribed by the Prot. Govt., otherwise he would not accept nominations to the Legco and then to the Central East African Assembly. It was during about this time when Kawalya was first asked to resign from being Katikkiro, but he never resigned.

(g) *The Rise of Bataka.* The Bataka are originally a hereditary society of the leaders of clans, but to-day it comes to mean any person who has much love for his country. The Bataka had failed to secure a hearing from the local authorities, and so they decided (1947) that they should send their representative to England to present their grievances to the Secy. of State to the Colonies and to some other members of the Parliament. Money was collected throughout the country, and the Bataka could now support their representative in England. Semakula Mulumba was appointed the representative.

The Bataka's petition to the Secretary of State for the Colonies included (I) the abolition of Paper 210, (II) the trial of the deportes, (III) that Uganda Africans should be allowed to sell their produce direct to England, (IV) the repeal of the Land Act of 1945, (V) that Africans should be given more facilities for further education, and (VI) that Uganda (being a Protectorate and not a Colony) should be transferred from the Colonial Office to the Foreign Office.

Mulumba always wrote back that he was gaining some victory. Some Baganda read in newspapers about questions raised in the Parliament as a result of Semakula's work, on the affairs of Uganda. This was an impetus to collect more money for their representative in England. Meetings — at which over 5000 people gathered — were held regularly at Miti's home. During the gatherings, many speeches, some very agitating, were given. Also during these meetings, telegrams from Semakula were usually read, stating what he was doing and what he wished the people here should do. These meetings were regarded by the Govt. officials to be dangerous to the peace of the country, and so the Govt. interfered and prohibited the Bataka from meeting unless they had special permission to do so. But despite the Govt.'s interference the Bataka grew more and more active in their demands, set to work the minds of the Baganda against the present Baganda Chiefs, by circulating throughout the country pamphlets containing Semakula's letters which the Govt. regarded as seditious and therefore dangerous.

(h) *The marriage of Mutesa to Damali.* The Bataka had grown restive as a result of the prevailing political and economic situation in Buganda — as outlined above. Mutesa's intention to marry Damali made the Bataka more restive. Damali belongs to a Nkima clan, and according to custom, the Kabaka should not marry the girl belonging to this clan, for if he did, it would mean that the next Kabaka would also belong to the Nkima clan as the Kabaka and others of the royal family belong to their mother's clans, and the people of this clan are not eligible to kingship. The Bataka were greatly alarmed at this action, and set themselves to stir the minds of the Baganda against Mutesa's marriage to Damali. Despite strong opposition, Mutesa was determined to take Damali for his wife and he did so. The Bataka were defeated, but their defeat meant the delianation of their minds.

(i) *The Church Lands.* In 1941 Bishop Stuart and other Church Trustees made an agreement with the British Govt. giving the rights of mining in the Baganda Church lands to the Govt. This agreement was never published for the public, until 1948, when, by chance, one of the Bataka got hold of a document concerning the agreement. The Bataka asserted that this was a private agreement, and was not liked. The Bishop and the Katikkiro were approached but the answers given could not satisfy the Bataka. Letters were written to the Bishop, to the Governor and to other officials that the agreement should be annulled, but this could not be done, unless both sides, i. e. the Govt. and the Bishop, agreed upon the annulment (but the Govt. would not agree). Ultimately, the Bataka, led by Luwale, demanded the resignation of Bishop Stuart. The Bataka informed Semakula about the matter, and he subsequently wrote to the Bishop, to the Governor, and to some M. P.s in England concerning the Agreement, and other weaknesses of the Bishop and his colleagues.

The letter was circulated among the Bataka and it stirred many Baganda to violent actions. The Bataka decided to drive away the Bishop from their Cathedral, or if that fails, to stop from going to Church in Buganda. The Bataka formed a society of "Bakristayo" (Christians) which was joined by some of the Baganda clergy and their aims were (I) to work at the annulment of the Agreement and (II) to force the Bishop to resign.

(j) *The Lukiko Resolution disowning the Bataka (1948).* Being alarmed at the growing pressure from the Bataka, the three ministers called for a Chief's meeting. This was done and the Lukiko, consisting of 16 Saza chiefs and the three ministers met (May 1948) and passed a resolution that: (I) The Bataka were an illegal society as far as the Buganda Govt. is concerned, and that the Protectorate Govt. could arrest any Bataka who still protested against the Church and the Govt.; (II) Semakula should be arrested in England, for he was the source of all these troubles; (III) The Bataka should not carry on their meetings, and that any people attending these meetings would have offended against the law.

The resolution never worked, for none of these things was carried out. Yet the Bataka attacked the Ministers and the Chiefs on the ground that: (a) 19 people — 3 ministers and 16 chiefs — cannot pass any resolution (according to the 1900 Agreement) unless all the 89 members of the Lukiko were present, and therefore this was a chance for the Bataka to accuse the Ministers and the 16 Chiefs of breaking the 1900 Agreement; (b) any person disowning the Bataka is not a Muganda himself, since the Bataka's society is a hereditary one going back many generations, and therefore the people who passed the resolution were not Baganda, and therefore should resign since Ministers and Chiefs must be Baganda.

The Bataka now set themselves to stir the emotions of Baganda against the 16 Chiefs and the 3 Ministers. They demanded that these should resign. But the Kabaka was still in England, and so the Bataka decided to wait for his return when they would present their case to him against the 16 chiefs and the ministers.

These arguments, outlined above, appear to be too remote to be the cause of the recent riots. But it is well to note that dissatisfaction did not come abruptly.

II. The Immediate Causes:

(I) *The Kabaka's indecisiveness.* The Bataka and their followers waited, with great anxiety, for the Kabaka's return from England, hoping that he would act soon and clear off some of their grievances. So, when he returned in September 1948, the Bataka were quick in presenting their petitions to him —referring to the old troubles which rose during his absence in England. The main points in the petition were:

(a) that the Kabaka should appoint a committee of people to examine the case alleged against 3 ministers and 16 chiefs,

(b) that, if the case goes against the ministers and the chiefs, these should resign from their posts,

(c) that the Buganda Govt. should help the farmers by providing them with some cotton ginneries so that they would be able to gin their own cotton,

(d) that the number of elected representatives in the Lukiko should be increased to 60,

(e) that Chiefs should be appointed by the people,

(f) that Govt. should give reasons why 17 Baganda were deported without trials.

The Kabaka could not give answers to this petition. As for the ginneries, he informed them that it was his Govt.'s intention to provide one ginnery to Baganda, but he did not say when. For the other points, especially about the resignation of the chiefs, it is alleged by the Bataka that he promised them that he would appoint a committee as asked, but other sources say that he said, as he always says, that "I'll consider your petition."

The Bataka now settled down to wait for the Kabaka's decision. For months, they say, he did not answer. The Bataka grew impatient, and sent another petition pressing the same demands. The fate of the petition is unknown, but it is probable that the Kabaka did not answer to it.

The Baganda Cotton Growers. The Union of the Baganda farmers has for long been collecting funds hoping to set up a ginnery of their own. So, in January 1949, they sent Musazi to England to find a market for their cotton. But the Uganda Govt. does not allow individuals to export their own cotton — it is a Govt.'s undertaking. Tbe farmers petitioned the Kabaka to be allowed to gin and export their cotton, but he told them of the difficulties of ginning cotton at present, and the impossibility of exporting it. The farmers still trusted in Musazi that he was doing something beneficial for them in England, and so they cecided not to sell their cotton to the "Indians" until his return. They built a number of otton stores, and collected some of their cotton in these. Others kept their cotton in their houses, waiting fot the day when they would get their own ginnery, no matter when that would happen.

Realizing the loss many farmers were incurring, the Govt. forced them to sell their cotton. But only a few of the farmers sold their cotton; the others still hoped for their ginnery either from their own purse or from the Buganda Govt. But the Govt. went as far as treating those who did not sell their cotton as law-breakers, and an order was passed stating that any person who would not sell his cotton would be offending against the law, and that he would be penalised with a fine or an imprisonment as the case may be.

This action of the Govt. was regarded by the farmers as cruel. The Bataka joined in, and false remours were circulated that the Govt. wanted to take their cotton by force, and that both the Protectorate and the Buganda Govts. aimed at keeping the Baganda poor, and that that is why a ginnery was not provided to them and why they were not allowed to sell their cotton in England.

The February Lukiko Session (1949) and the arrest of Kigongo. For six months since the Bataka petitioned the Kabaka, he did not give them his decision. They wrote to remind him of the petition, but it is doubtful whether he replied. As a consequence, of the Kabaka's indecisiveness, the Bataka published notices stating that the first session of the Lukiko would not take place unless the number of Elected Members was raised to 60, and that the 16 chiefs and the ministers were not members of the Lukiko. This was about a week to the day when the Lukiko would sit. Yet the Kabaka seemed not to be considering the Bataka's petition as he had said he would. The Lukiko Session was to be opened on Feb. 14th (1949), but on 13th of the same month, Kigongo — one of the leaders of the Bataka — sent a telegram to the Kabaka — almost in the form of an order — that the Lukiko would not sit unless the number of Elected Members was 60, and the chiefs in question dismissed. It is doubtful whether the Kabaka took this in serious consideration except that he or any other authority gave orders to arrest Kigongo, for it was considered unconstitutional to send a telegram to the Kabaka on such matters. Kigongo was tried and was imprisoned for two years. This step only helped to aggrevate the matters. For they thought that one of their leaders should not be imprisoned for such a slight case as they considered it to be.

In the morning of Feb. 14th more telegrams were sent to the Kabaka — this time without names — that the Bataka were ready to stop the opening of the Lukiko — with whatever force they could use — unless he considered their petitions. Many Bataka started to gather in big numbers at the Lubiri, and it was at 10 a. m. — just about the hour of opening the

Lukiko, when the Kabaka realised that the Bataka meant what they said, for they had gathered round the Bulange (Lukiko House), their leaders speaking to the Chiefs and others of the crowd who had come to attend the opening of the Lukiko, that the Lukiko would not sit unless the Kabaka considered their petitions. At last the Kabaka sent for the leaders of the Bataka, and postponing the opening of the Lukiko Session one hour later he granted them audience. After a long discussion, he promised that next year (1950) the number of the Elected Members of the Lukiko would be raised to 60. For the dismissal of the Chiefs he said that could not be done in a hurry and so he would consider it seriously. The leaders of the Bataka returned with some satisfaction, and the Lukiko was opened at 11 a. m.; and at the opening of the Lukiko, the Kabaka announced his intention to increase the number of the Elected Members to 60 in 1950. This was a sort of relief to the Bataka.

Thus from February onwards the Bataka were resting with some satisfaction, and hoping that the Kabaka would probably consider the rest of their demands as he had promised them he would. For three months they waited for his decision but he seemed to have forgotten all about what he said. Then they started writing letters to him with copies to the Governor. They said there would be a strike on the 25th of April if the Kabaka did not give his decision. Notices were published informing all the people of that date. At last they decided they would meet at the Lubiri to see the Kabaka and protest against his indecisiveness.

April 26th (1949) — the dispersion of the mob. On 25th April thousands of Bataka gathered at the Lubiri while their representatives had gone to see the Kabaka. They went to present their old petition again, but with one more point added — that Kigongo should be freed from prison. The Kabaka's answer was that while he still considered some of their demands, he could not cope with other — for these were not worth considering. He said he was not to dismiss his chiefs at that moment. The leaders of the Bataka came back to tell the people to return to their homes, for the Kabaka would give his decision later. But their intention was to wait at the palace until the Kabaka dismissed the chiefs in question. So they spent the next night at the Lubiri.

On the 26th April the police took away, by force, the Bataka's food they had brought with them, and at 11 a. m. of the same day, the Bataka were forced to disperse. This act was destined to drive the people into a riot. For the Bataka never believed that any person had the right to drive them away from "their king". Hence the riots.

Events

On Sunday 24th April (1949) the Bataka gathered in thousands at the Kabaka's lake to examine their programme. The programme was that: the next morning (25th) they would go in procession to the Lubiri, and their elected representatives (14) would go and see the Kabaka; that if the Kabaka did not give his decision that day, they would stay at the Lubiri for eight days and that if, after eight days, he did not give his decision, they would be told what to do next.

Lorries were seen from different directions, bringing some Bataka from distant districts, and much food — mostly from Buddu where some had promised to supply food — free of charge — to the gathering Bataka for eight days — was collected at the lake — their headquarters. Thousands of Bataka spent the night at the lake and had their food cooked there.

At 6 in the morning of 25th they went in procession and gathered outside the Lubiri. The Kabaka did not see their representatives until 11 a. m. After about an hour's audience with the Kabaka, the representatives returned. The Kabaka had sent the Katikkiro to read his message to the crowd, but he was not allowed to read it as he was one of the men they were against. So, the representatives announced the message from the Kabaka to the crowd. The message was that the Kabaka would consider some of their demands, and that the people should return to their homes.

But the crowd were not ready to go home — for many of them came from distant districts — until the Kabaka announced his decision about the petition. At about 2 p. m. (25th) — it was estimated that about 8,000 were gathered at the Lubiri. Night came, and the crowd still waited. They spent that night at the Lubiri. The next morning (26th), a Government official gave them an order from the Governor that they should disperse at 10 a. m. or the police forces would use force to disperse them.

10 a. m. came, and the crowd did not disperse. The Assistent Resident came to read them a law which orders the dispersal, by force, of a crowd exceeding 500 people. But they stoned at him as he started reading it. The crowd was going out of control, and so a few bullets were fired at them, hitting one of them in the thigh; and so the crowd split into small groups taking different directions, the riots began now.

These people spread rumours, as they ran away, that six of their number had been shot dead, and so thereby inciting many others to join in the riots. There was not any concerted plan, for every group had taken their way, and acted according to what they thought they should do. The Bataka say that it was not in their plan to commit all those felonies that were committed.

The strike lasted for three days only, but the danger done to property of some people was great. Chiefs' personal houses were burnt down, Indians' shops looted, and many people beaten almost to death.

The rioters used petrol in burning houses. 400 cases of arson were reported. The rioters spread almost all over the parts of Baganda, and in order that they might do quickly the felonies intended to do throughout the country, the rioters used vehicles — motor cars and motorcycles — which they had captured from any people they met — to transport them and their implements; and a person would wonder how they could have done such havoc all over the country in such a short time. The arrival of some garrisons of soldiers from Kenya made it easy to stop the rioters from committing further actions of danger. And by the 30th of April, it was reported that the riots had completely ceased. Many rioters — amounting to over 1000 were arrested by the police and by the soldiers.

Effects

It is too early to speak about the effects of the riots, suffice it to say that much personal property was destroyed, and many people were rendered homeless. It is reported that 500 people still (19th May) stay at Mengo in the Lubiri, for their houses and property were destroyed.

FSC
www.fsc.org
MIX
Papier | Fördert gute Waldnutzung
FSC® C083411

Kolibri 360 GmbH
Ferdinand-Jühlke-Straße 7
99095 Erfurt, Deutschland
produktsicherheit@kolibri360.de